The Princeton Review

Cracking the

AP®

CALCULUS AB EXAM

2015 Edition

David S. Kahn

PrincetonReview.com

PENGUIN RANDOM HOUSE

The Princeton Review
24 Prime Parkway, Suite 201
Natick, MA 01760
E-mail: editorialsupport@review.com

Published in the United States by Random House LLC, New York, and
simultaneously in Canada by Random House of Canada Limited, Toronto.

A Penguin Random House Company.

ISBN: 978-0-8041-2480-5
ISSN: 2334-2633

Editors: Calvin S. Cato and Aaron Riccio
Production Editor: Liz Rutzel
Production Coordinator: Deborah A. Silvestrini

Printed in the United States of America on partially recycled paper.

10 9 8 7 6 5 4 3 2 1

2015 Edition

Editorial
Rob Franek, Senior VP, Publisher
Casey Cornelius, VP Content Development
Mary Beth Garrick, Director of Production
Selena Coppock, Managing Editor
Calvin Cato, Editor
Colleen Day, Editor
Aaron Riccio, Editor
Meave Shelton, Editor
Alyssa Wolff, Editorial Assistant

Random House Publishing Team
Tom Russell, Publisher
Alison Stoltzfus, Publishing Manager
Dawn Ryan, Associate Managing Editor
Ellen Reed, Production Manager
Erika Pepe, Associate Production Manager
Kristin Lindner, Production Supervisor
Andrea Lau, Designer

Acknowledgments

First of all, I would like to thank Arnold Feingold and Peter B. Kahn for once again doing every problem, reading every word, and otherwise lending their invaluable assistance. I also want to thank Calvin Cato and Aaron Riccio for being terrific editors, Gary King for his first-rate reading, analysis, and contributions, and the production team of Liz Rutzel and Deborah Silvestrini. Thanks to Frank, without whose advice I never would have taken this path. Thanks to Jeffrey, Miriam, and Vicki for moral support. Thanks Mom.

Finally, I would like to thank the people who really made all of this effort worthwhile—my students. I hope that I haven't omitted anyone, but if I have, the fault is entirely mine.

Aaron and Sasha, Aaron, Abby B., Abby F., Abby H., Abby L., Abbye, Abigail H., Aidan, Alec G, Alan M., Alec M., Alec R., Alex and Claire, Alex A., Alex B., Alex F., Alex D., Alex G., Alex H., Alex I., Alex S., Alex and Gabe, Alexa, Alexandra, Natalie and Jason, Alexes, Alexis and Brittany, Ali and Jon, Ali and Amy Z., Ali H., Alice, Alice C., Alicia, Alisha, Allie and Lauren, Allison and Andrew, Allison and Matt S., Allison R., Ally, Ally T., Aly and Lauren, Alyssa and Courtney, Amanda, Brittany, and Nick A., Amanda and Pamela B., Amanda C., Amanda H., Amanda M., Amanda R., Amanda S., Amber and Teal, Amparo, Andrea T., Andrea V., Andrew A., Andrew B., Andrew C., Andrew D., Andrew E., Andrew H., Andrew M., Andrew S., Andy and Sarah R., Andy and Allison, Angela, Angela F., Anisha, Ann, Anna C-S., Anna D., Anna and Jon, Anna and Max, Anna L., Anna M., Anna W., Annie, Annie W., Antonio, Anu, April, Ares, Ariadne, Ariane, Ariel, Arielle and Gabrielle, Arthur and Annie, Arya, Asheley and Freddy, Ashley A., Ashley K., Ashley and Sarah, Ashley and Lauren, Avra, Becca A., Becky B., Becky S., Becky H., Ben S., Ben and Andrew Y., Ben D., Ben S., Benjamin D., Benjamin H., Benji, Beth and Sarah, Bethany and Lesley, Betsy and Jon, Blythe, Bianca and Isabella, Bonnie, Bonnie C., Braedan, Brendan, Brett, Brett A., Brette and Josh, Brian, Brian C., Brian N., Brian W., Brian Z., Brigid, Brin, Brittany E., Brooke, Devon, and Megan, Brooke and Lindsay E., Butch, Caitlin, Caitlin F., Caitlin M., Caitlin and Anna S., Camilla and Eloise, Camryn, Caroline A., Caroline H., Caroline S., Caroline and Peter W., Carrie M., Catherine W., Chad, Channing, Charlie, Charlotte, Charlotte B., Charlotte M., Chelsea, Chloe, Chloe K., Chris B., Chris C., Chrissie, Christian, Christine, Christine W., Claire H., Clare, Claudia, Clio, Corey, Corinne, Coryn, Courtney and Keith, Courtney B., Courtney F., Courtney S., Craig, Dan M., Dana J., Dani and Adam, Daniel, Daniel and Jen, Daniella, Daniella C., Danielle and Andreas, Danielle and Nikki D., Danielle G., Danielle H., Danny K., Dara and Stacey, Dara M., Darcy, David B., David R., David S., Deborah and Matthew, Deniz and Destine, Deval, Devin, Devon and Jenna, Dilly, Dong Yi, Dora, Eairinn, Eddy, Elana, Eleanor K., Elexa and Nicky, Elisa, Eliza, Elizabeth, Elizabeth and Mary C., Elizabeth F., Ella, Ellie, Elly B., Emily A., Emily B., Emily and Allison, Emily and Catie A., Emily C., Emily G., Emily H., Emily K., Emily L., Emily and Pete M., Emily R., Emily R-H., Emily S., Emily T., Emma and Sophie, Eric N., Erica F., Erica H., Erica R., Erica S., Eric and Lauren, Erica and Annie, Erika, Erin, Erin I., Ethan, Eugenie, Eva, Evan, Eve H., Eve M., Frank, Gabby, Geoffrey, George and Julian., George M., Gloria, Gracie, Graham and Will, Greg, Greg F., Gussie, Hallie, Hannah C., Hannah J., Hannah R., Hannah and Paige, Harrison, Annabel and Gillian, Harry, Hayley and Tim D., Hazel, Heather D., Heather F., Heather and Gillian, Hernando and Vicki, Hilary and Lindsay, Hilary F., Holli, Holly G., Holly K., Honor, Ian, Ian P., Ingrid, Ira, Isa, Isabel, Isabella and Simone, Isabelle T., Ismini, Isobel, Ivy, Jacob and Kara, Jack, Jack B., Jackie and Vicky, Jackie, Jackie S., Jackson, Jaclyn and Adam, James S., Jamie, Jason P., Jason and Andrew, Jay, Jay K., Jae and Gideon, Jayne and Johnny, Jed D., Jed F., Jeffrey M., Jenna, Jen, Jenn N., Jennifer B., Jennifer W., Jenny K, Jenny and Missy, Jeremy C., Jeremy and Caleb, Jess, Jess P., Jesse, Jessica, Jessica and Eric H., Jessica L., Jessica T., Jessie, Jessie C., Jessie and Perry N., Jill, Jillian, Daniel and Olivia, Jillian S., Jimin, Jimmy C., Jimmy P., Joanna, Joanna C., Joanna G., Joanna M., Joanna and Julia M., Joanna W., Jocelyn, Jody and Kim, Joe B., Johanna, John, John and Dan, Jonah and Zoe, Jonathan G., Jonathan P., Jonathan W., Jordan C., Jordan and Blair F., Jordan G., Jordan P., Jordana, Jordyn, Josh and Jesse, Josh and Noah, Judie and Rob, Julia and Caroline, Julia, Julia G., Julia H., Julia and Charlotte P., Julia T., Julie H., Julie and Dana, Julie P., Juliet, Kara B., Kara O., Kasia and Amy, Kat R., Kate D., Kate F., Kate G., Kate L., Kate P., Kate S., Kate W., Katie C., Katie F., Katherine C., Kathryn, Leslie, and Travis, Katie, Katie M., Katrina, Keith M., Kelly C., Kimberly, Kirsten, Kitty and Alex, Krista, Erika, and Karoline, Kristen, Kristen and Magan F., Laila and Olivia R., Laura F., Laura G., Laura R., Laura S., Laura T., Laura Z., Lauren and Eric, Lauren R., Lauren T., Lauren and Allie, Leah, Lee R., Leigh and Ruthie, Leigh, Lexi S., Lila, Lila M., Lilaj, Lili C., Lily, Lily M-R., LilyHayes, Lisbeth and Charlotte, Lindsay F., Lindsay K., Lindsay L., Lindsay N., Lindsay R, Lindsay and Jessie S., Lindsey and Kari, Lisa, Liz D., Liz H., Liz M., Lizzie A., Lizzi B., Lizzie M., Lizzie W., Lizzy, Lizzy C., Lizzy R., Lizzy T., Louis, Lucas, Luke C., Lucinda, Lucy, Lucy D., Mackenzie, Maddie and JD, Maddie P., Maddy W., Madeline, Magnolia, Mara and Steffie, Marcia, Margaret S., Mariel C., Mariel L., Mariel S., Marielle K., Marielle S., Marietta, Marisa C., Marissa, Marnie and Sam, Mary M., Matt, Matthew B., Matthew K., Matt F., Matt G., Matt and Caroline, Matt V., Matthew G., Max B., Max and Chloe K., Max M., Maxx, Ben, and Sam, Maya and Rohit, Maya N., Meesh, Megan G., Melissa, Melissa and Ashley I., Meredith, Meredith and Gordie B., Meredith and Katie D., Meredith R., Meredith S-K., Michael and Dan, Michael H., Michael R., Michaela, Michal, Mike R., Miles, Milton, Miranda, Moira, Molly L., Molly R., Molly and Annie, Morgan, Morgan C., Morgan, Zoe, and Lila, Morgan K., Nadia and Alexis, NaEun, Nancy, Nanette, Naomi, Natalie and Andrew B., Natasha and Mikaela B., Natasha L., Nathania, Nico, Nick, Nicky S., Nicole B., Nicole J., Nicole N., Nicole P., Nicole S., Nidhi, Nikki G., Nina R., Nina V., Nora, Nushien, Oli and Arni, Oliver, Olivia and Daphne, Olivia P., Omar, Oren, Pam, Paige, Paul A., Peter, Philip and Peter, Phoebe, Pierce, Priscilla, Quanquan, Quinn and Chris, Rachel A, Rachel B., Rachel and Adam, Rachel H., Rachel I., Rachel and Jake, Rachel K., Rachel L., Rachel and Mitchell M., Rachel S., Rachel and Eli, Rachel and Steven, Rachel T., Rachel W., Ramit, Randi and Samantha, Rayna, Rebecca B., Rebecca G., Rebecca and Gaby K., Rebecca R., Rebecca W., Resala, Richard, Tina, and Alice, Richie D., Ricky, Rob L., Romanah, Rose, Ryan, Ryan B., Ryan S., Saahil, Sabrina, Sally, Sam C., Sam L., Sam R., Sam S., Sam W., Samantha M., Samuel C., Samar, Sara H., Sara L., Sara R., Sara S., Sara W., Sarah, Patty, and Kat, Sarah and Beth, Sarah A., Sarah C., Sarah D., Sarah F., Sarah G., Sarah and Michelle K., Sarah L., Sarah M-D., Sarah and Michael, Sarah S., Sascha, Samara, Saya, Selena G., Seung Woo, Shadae, Siegfried, Simon W., Simone, Sinead, Skye F., Skye L., Sofia G., Sofia M., Sonja, Sonja and Talya, Sophia J., Sophia S., Sophie and Tess, Sophie D., Sophie S., Sophie W., Stacey, Stacy, Stephanie, Stephanie L., Stephen C., Sumair, Sunaina, Suzie, Sydney, Sydney S., Tammy and Hayley, Tara and Max, Taylor, Tenley and Galen, Terrence, Tess, Timothy H., Tita, Tom, Torri, Tracy, Tracy K., Tripp H., Tripp W., Tyler, Tyrik, Vana, Vanessa and John, Veda, Vicky B., Victor Z., Victoria, Victoria H., Vinny and Eric, Vivek, Vladimir, Waleed, Will A., Will, Teddy and Ellie, William M., Wyatt and Ryan, Wyna, Xianyuan, Yakir, Yesha, Zach, Zach S., Zachary N., Zoë, Zoe L., and Zoey and Rachel.

Contents

Part I
Using This Book
to Improve Your
AP Score

- Preview: Your Knowledge, Your Expectations
- Your Guide to Using This Book
- How to Begin

PREVIEW: YOUR KNOWLEDGE, YOUR EXPECTATIONS

Your route to a high score on the AP Calculus AB Exam depends a lot on how you plan to use this book. Start thinking about your plan by responding to the following questions.

1. Rate your level of confidence about your knowledge of the content tested by the AP Calculus AB Exam:

   A. Very confident—I know it all
   B. I'm pretty confident, but there are topics for which I could use help
   C. Not confident—I need quite a bit of support
   D. I'm not sure

2. If you have a goal score in mind, circle your goal score for the AP Calculus AB Exam:

   5   4   3   2   1   I'm not sure yet

3. What do you expect to learn from this book? Circle all that apply to you.

   A. A general overview of the test and what to expect
   B. Strategies for how to approach the test
   C. The content tested by this exam
   D. I'm not sure yet

YOUR GUIDE TO USING THIS BOOK

This book is organized to provide as much—or as little—support as you need, so you can use this book in whatever way will be most helpful for improving your score on the AP Calculus AB Exam.

- The remainder of **Part One** will provide guidance on how to use this book and help you determine your strengths and weaknesses.

- **Part Two** of this book will
   o provide information about the structure, scoring, and content of the AP Calculus AB Exam.
   o help you to make a study plan.
   o point you towards additional resources.

- **Part Three** of this book will explore the following strategies:
  - o how to attack multiple-choice questions
  - o how to write a high scoring free-response answer
  - o how to manage your time to maximize the number of points available to you

- **Part Four** of this book covers the content you need for your exam.

- **Part Five** of this book contains practice tests.

You may choose to use some parts of this book over others, or you may work through the entire book. This will depend on your needs and how much time you have. Let's now look at how to make this determination.

HOW TO BEGIN

1. **Take a Test**

   Before you can decide how to use this book, you need to take a practice test. Doing so will give you insight into your strengths and weaknesses, and the test will also help you make an effective study plan. If you're feeling test-phobic, remind yourself that a practice test is a tool for diagnosing yourself—it's not how well you do that matters but how you use information gleaned from your performance to guide your preparation.

   So, before you read further, take the AP Calculus AB Practice Test 1 starting at page 445 of this book. Be sure to do so in one sitting, following the instructions that appear before the test.

2. **Check Your Answers**

   Using the answer key on page 479, count how many multiple-choice questions you got right and how many you missed. Don't worry about the explanations for now, and don't worry about why you missed questions. We'll get to that soon.

3. **Reflect on the Test**

   After you take your first test, respond to the following questions:

   - How much time did you spend on the multiple-choice questions?

   - How much time did you spend on each free-response question?

   - How many multiple-choice questions did you miss?

- Do you feel you had the knowledge to address the subject matter of the free-response questions?

- Do you feel your free responses were well organized and thoughtful?

- Circle the content areas that were most challenging for you and draw a line through the ones in which you felt confident/did well.

   o Functions, Graphs, and Limits

   o Differential Calculus

   o Integral Calculus

   o Applications of Derivatives

   o Applications of Integrals

4. **Read Part Two, and Complete the Self-Evaluation**

Part Two will provide information on how the test is structured and scored. It will also set out areas of content that are tested.

As you read Part Two, reevaluate your answers to the questions above. At the end of Part Two, you will revisit the questions above and refine your answers to them. You will then be able to make a study plan, based on your needs and time available, that will allow you to use this book most effectively.

5. **Engage with Parts Three and Four as Needed**

Notice the word *engage*. You'll get more out of this book if you use it intentionally than if you read it passively, hoping for an improved score through osmosis.

Strategy chapters will help you think about your approach to the question types on this exam. Part Three will open with a reminder to think about how you approach questions now and then close with a reflection section asking you to think about how/whether you will change your approach in the future.

Content chapters are designed to provide a review of the content tested on the AP Calculus AB Exam, including the level of detail you need to know and how the content is tested. You will have the opportunity to assess your mastery of the content of each chapter through test-appropriate questions and a reflection section.

6. **Take Another Test and Assess Your Performance**

Once you feel you have developed the strategies you need and gained the knowledge you lacked, you should take Test 2. You should do so in one sitting, following the instructions at the beginning of the test.

When you are done, check your answers to the multiple-choice sections. See if a teacher will read your long-form calculus responses and provide feedback.

Once you have taken the test, reflect on what areas you still need to work on, and revisit the chapters in this book that address those topics. Through this type of reflection and engagement, you will continue to improve.

7. **Keep Working**

After you have revisited certain chapters in this book, continue the process of testing, reflecting, and engaging with Practice Test 3 on page 621. For extra practice tests, please check out our supplemental title *550 AP Calculus AB & BC Practice Questions*. You want to be increasing your readiness by carefully considering the types of questions you are getting wrong and how you can change your strategic approach to different parts of the test.

As discussed in Part Two, there are other resources available to you, including a wealth of information on AP Central. You can continue to explore areas that can stand to improve and engage in those areas right up to the day of the test.

Part II
About the
AP Calculus
AB Exam

- AB Calculus vs BC Calculus
- The Structure of the Calculus Exams
- Overview of Content Topics
- General Overview of This Book
- How AP Exams Are Used
- Other Resources
- Designing Your Study Plan

Please note that this book will focus only on the topics that will appear on the AP Calculus AB test. If you are looking to prepare for the BC Calculus exam, please purchase *Cracking the AP Calculus BC Exam* in stores now!

AB CALCULUS VS BC CALCULUS

AP Calculus is divided into two types: AB and BC. The former is supposed to be the equivalent of a semester of college calculus; the latter, a year. In truth, AB calculus covers closer to three quarters of a year of college calculus. In fact, the main difference between the two is that BC calculus tests some more theoretical aspects of calculus and it covers a few additional topics. In addition, BC calculus is harder than AB calculus. The AB exam usually tests straightforward problems in each topic. They're not too tricky, and they don't vary very much. The BC exam asks harder questions. But neither exam is tricky in the sense that the SAT is. Nor do they test esoteric aspects of calculus. Rather, both tests tend to focus on testing whether you've learned the basics of differential and integral calculus. The tests are difficult because of the breadth of topics that they cover, not the depth. You will probably find that many of the problems in this book seem easier than the problems you've had in school. This is because your teacher is giving you problems that are harder than those on the AP.

THE STRUCTURE OF THE CALCULUS EXAM

Now, some words about the test itself. The AP exam comes in two parts. First, there is a section of multiple-choice questions covering a variety of calculus topics. The multiple-choice section has two parts. Part A consists of 28 questions; you are not permitted to use a calculator on this section. Part B consists of 17 questions; you are permitted to use a calculator on this part. These two parts comprise a total of 45 questions.

After this, there is a free-response section consisting of six questions, each of which requires you to write out the solutions and the steps by which you solved it. You are permitted to use a calculator for the first two problems but not for the four other problems. Partial credit is given for various steps in the solution of each problem. You'll usually be required to sketch a graph in one of the questions. The College Board does you a big favor here: You may use a graphing calculator. In fact, The College Board recommends it! And they allow you to use programs as well. But here's the truth about calculus: Most of the time, you don't need the calculator anyway. Remember: These are the people who bring you the SAT. Any gift from them should be regarded skeptically!

OVERVIEW OF CONTENT TOPICS

This list is drawn from the topical outline for AP calculus furnished by the College Board. You might find that your teacher covers some additional topics, or omits some, in your course. Some of the topics are very broad, so we cannot guarantee that this book covers these topics exhaustively.

I. Functions, Graphs, and Limits
  A. Analysis of Graphs

  - You should be able to analyze a graph based on "the interplay between geometric and analytic information." The preceding phrase comes directly from the College Board. Don't let it scare you. What the College Board really means is that you should have covered graphing in precalculus, and you should know (a) how to graph and (b) how to read a graph. **This is a precalculus topic and we won't cover it in this book.**

  B. Limits

  - You should be able to calculate limits algebraically or to estimate them from a graph or from a table of data.

  - You do **not** need to find limits using the Delta-Epsilon definition of a limit.

  C. Asymptotes

  - You should understand asymptotes graphically and be able to compare the growth rates of different types of functions (namely polynomial functions, logarithmic functions, and exponential functions). **This is a topic that should have been covered in precalculus, and we won't cover it in this book.**

  - You should understand asymptotes in terms of limits involving infinity.

  D. Continuity

  - You should be able to test the continuity of a function in terms of limits, and you should understand continuous functions graphically.

  - You should understand the intermediate value theorem and the extreme value theorem.

II. Differential Calculus
  A. The Definition of the Derivative

  - You should be able to find a derivative by finding the limit of the difference quotient.

- You should also know the relationship between differentiability and continuity. That is, if a function is differentiable at a point, it's continuous there. But if a function is continuous at a point, it's not necessarily differentiable there.

B. Derivative at a Point

- You should know the Power Rule, the Product Rule, the Quotient Rule, and the Chain Rule.

- You should be able to find the slope of a curve at a point, and the tangent and normal lines to a curve at a point.

- You should also be able to use local linear approximation and differentials to estimate the tangent line to a curve at a point.

- You should be able to find the instantaneous rate of change of a function using the derivative or the limit of the average rate of change of a function.

- You should be able to approximate the rate of change of a function from a graph or from a table of values.

- You should be able to find Higher-Order Derivatives and to use Implicit Differentiation.

C. Derivative of a Function

- You should be able to relate the graph of a function to the graph of its derivative, and vice-versa.

- You should know the relationship between the sign of a derivative and whether the function is increasing or decreasing (positive derivative means increasing; negative means decreasing).

- You should know how to find relative and absolute maxima and minima.

- You should know the Mean Value Theorem for derivatives and Rolle's theorem.

D. Second Derivative

- You should be able to relate the graph of a function to the graph of its derivative and its second derivative, and vice-versa. This is tricky.

- You should know the relationship between concavity and the sign of the second derivative (positive means concave up; negative means concave down).

- You should know how to find points of inflection.

E. Applications of Derivatives

- You should be able to sketch a curve using first and second derivatives and be able to analyze the critical points.

- You should be able to solve Optimization problems (Max/Min problems), and Related Rates problems.

- You should be able to find the derivative of the inverse of a function.

- You should be able to solve Rectilinear Motion problems.

F. Computation of Derivatives

- You should be able to find the derivatives of Trig functions, Logarithmic functions, Exponential functions, and Inverse Trig functions.

III. Integral Calculus

A. Riemann Sums

- You should be able to find the area under a curve using left, right, and midpoint evaluations and the Trapezoid Rule.

- You should know the fundamental theorem of calculus:

$$\int_a^b f(x)dx = F(b) - F(a)$$

B. Applications of Integrals

- You should be able to find the area of a region, the volume of a solid of known cross-section, the volume of a solid of revolution, and the average value of a function.

- You should be able to solve acceleration, velocity, and position problems.

C. Fundamental Theorem of Calculus

- You should know the first and second fundamental theorems of calculus and be able to use them both to find the derivative of an integral and for the analytical and graphical analysis of functions.

D. Techniques of Antidifferentiation

- You should be able to integrate using the power rule and u-Substitution.

E. Applications of Antidifferentiation

- You should be able to find specific antiderivatives using initial conditions.

- You should be able to solve separable differential equations and logistic differential equations.

- You should be able to interpret differential equations via slope fields. Don't be intimidated. These look harder than they are.

GENERAL OVERVIEW OF THIS BOOK

The key to doing well on the exam is to memorize a variety of techniques for solving calculus problems and to recognize when to use them. There's so much to learn in AP calculus that it's difficult to remember everything. Instead, you should be able to derive or figure out how to do certain things based on your mastery of a few essential techniques. In addition, you'll be expected to remember a lot of the math that you did before calculus—particularly trigonometry. You should be able to graph functions, find zeros, derivatives, and integrals with the calculator.

Furthermore, if you can't derive certain formulas, you should memorize them! A lot of students don't bother to memorize the trigonometry special angles and formulas because they can do them on their calculators. This is a big mistake. You'll be expected to be very good with these in calculus, and if you can't recall them easily, you'll be slowed down and the problems will seem much harder. Make sure that you're also comfortable with analytic geometry. If you rely on your calculator to graph for you, you'll get a lot of questions wrong because you won't recognize the curves when you see them.

This advice is going to seem backward compared with what your teachers are telling you. In school you're often yelled at for memorizing things. Teachers tell you to understand the concepts, not just memorize the answers. Well, things are different here. The understanding will come later, after you're comfortable with the mechanics. In the meantime, you should learn techniques and practice them, and, through repetition, you will ingrain them in your memory.

Each chapter is divided into three types of problems: examples, solved problems, and practice problems. The first type is contained in the explanatory portion of the unit. The examples are designed to further your understanding of the subject and to show you how to get the problems right. Each step of the solution to the example is worked out, except for some simple algebraic and arithmetic steps that should come easily to you at this point.

The second type of problems is solved problems. The solutions are worked out in approximately the same detail as the examples. Before you start work on each of these, cover the solution with an index card or something, then check the solution afterward. And you should read through the solution, not just assume that you knew what you were doing because your answer was correct.

The third type is practice problems. Only the answer explanations to these are given. We hope you'll find that each chapter offers enough practice problems for you to be comfortable with the material. The topics that are emphasized on the exam have more problems; those that are de-emphasized have fewer. In other words, if a chapter has only a few practice problems, it's not an important topic on the AP exam and you shouldn't worry too much about it.

HOW AP EXAMS ARE USED

Different colleges use AP Exams in different ways, so it is important that you go to a particular college's website to determine how it uses AP Exams. The three items below represent the main ways in which AP Exam scores can be used.

- **College Credit.** Some colleges will give you college credit if you score well on an AP Exam. These credits count towards your graduation requirements, meaning that you can take fewer courses while in college. Given the cost of college, this could be quite a benefit, indeed.
- **Satisfy Requirements.** Some colleges will allow you to "place out" of certain requirements if you do well on an AP Exam, even if they do not give you actual college credits. For example, you might not need to take an introductory-level course, or perhaps you might not need to take a class in a certain discipline at all.
- **Admissions Plus.** Even if your AP Exam will not result in college credit or even allow you to place out of certain courses, most colleges will respect your decision to push yourself by taking an AP Course or even an AP Exam outside of a course. A high score on an AP Exam shows mastery of more difficult content than is taught in many high school courses, and colleges may take that into account during the admissions process.

OTHER RESOURCES

There are many resources available to help you improve your score on the AP Calculus AB Exam, not the least of which are your **teachers**. If you are taking an AP class, you may be able to get extra attention from your teacher, such as obtaining feedback on your free-response questions. If you are not in an AP course, reach out to a teacher who teaches calculus, and ask if the teacher will review your free-response questions or otherwise help you with content.

Another wonderful resource is **AP Central**, the official site of the AP Exams. The scope of the information at this site is quite broad and includes:

- A course description, which includes details on what content is covered and sample questions
- Sample test questions
- Free-response prompts from previous years

The AP Central home page address is: **http://apcentral.collegeboard.com**.

The AP Calculus AB Exam Course home page address is: **http://apcentral.college board.com/apc/public/courses/teachers_corner/2178.html**.

Additional full-length practice tests and specialized drills for each of the major calculus topics can also be found in *550 AP Calculus AB & BC Practice Questions*. Finally, **The Princeton Review** offers tutoring and small group instruction. Our expert instructors can help you refine your strategic approach and add to your content knowledge. For more information, call 1-800-2REVIEW.

DESIGNING YOUR STUDY PLAN

As part of the Introduction, you identified some areas of potential improvement. Let's now delve further into your performance on Test 1, with the goal of developing a study plan appropriate to your needs and time commitment.

Read the answers and explanations associated with the multiple-choice questions (starting at page 477). After you have done so, respond to the following questions:

- Review the Overview of Content Topics on pages 9–11. Next to each topic, indicate your rank of the topic as follows: "1" means "I need a lot of work on this," "2" means "I need to beef up my knowledge," and "3" means "I know this topic well."

- How many days/weeks/months away is your exam?

- What time of day is your best, most focused study time?

- How much time per day/week/month will you devote to preparing for your exam?

- When will you do this preparation? (Be as specific as possible: Mondays & Wednesdays from 3 to 4 P.M., for example.)

- Based on the answers above, will you focus on strategy (Part Three) or content (Part Four) or both?

- What are your overall goals in using this book?

Part III
Test-Taking Strategies for the AP Calculus AB Exam

PREVIEW ACTIVITY

Review your responses to the first three questions on page 2 of the Introduction, and then respond to the following questions:

- How many multiple-choice questions did you miss even though you knew the answer?

- On how many multiple-choice questions did you guess blindly?

- How many multiple-choice questions did you miss after eliminating some answers and guessing based on the remaining answers?

- Did you find any of the free-response questions easier or harder than the others—and, if so, why?

HOW TO USE THE CHAPTERS IN THIS PART

For the following Strategy chapters, think about what you are doing now before you read the chapters. As you read and engage in the directed practice, be sure to appreciate the ways you can change your approach. At the end of each chapter in Part Three, you will have the opportunity to reflect on how you will change your approach.

Chapter 1
How to Approach
Multiple-Choice
Questions

CRACKING THE MULTIPLE-CHOICE QUESTIONS

Section I of the AP Calculus Exam consists of 45 multiple-choice questions, which you're given 105 minutes to complete. This section is worth 50 percent of your grade.

All the multiple-choice questions will have a similar format: Each will be followed by five answer choices. At times, it may seem that there could be more than one possible correct answer. There is only one! Remember that the committee members who write these questions are calculus teachers. So, when it comes to calculus, they know how students think and what kind of mistakes they make. Answers resulting from common mistakes are often included in the five answer choices to trap you.

Use the Answer Sheet

For the multiple-choice section, you write the answers not in the test booklet but on a separate answer sheet (very similar to the ones we've supplied at the very end of this book). Five oval-shaped bubbles follow the question number, one for each possible answer. *Don't* forget to fill in all your answers on the answer sheet. Don't just mark them in the test booklet. Marks in the test booklet will not be graded. Also, make sure that your filled-in answers correspond to the correct question numbers! Check your answer sheet after every five answers to make sure you haven't skipped any bubbles by mistake.

Should You Guess?

Use Process of Elimination (POE) to rule out answer choices you know are wrong and increase your chances of guessing the right answer. Read all the answer choices carefully. Eliminate the ones that you know are wrong. If you only have one answer choice left, *choose it,* even if you're not completely sure why it's correct. Remember: Questions in the multiple-choice section are graded by a computer, so it doesn't care *how* you arrived at the correct answer.

Even if you can't eliminate answer choices, go ahead and guess. There is no guessing penalty for incorrect answers. You will be assessed only on the total number of correct answers, so be sure to fill in all the bubbles even if you have no idea what the correct answers are. When you get to questions that are too time-consuming, or that you don't know the answer to (and can't eliminate any options), don't just fill in any answer. Use what we call your "letter of the day" (LOTD). Selecting the same answer choice each time you guess will increase your odds of getting a few of those skipped questions right.

Use the Two-Pass System

Remember that you have about two and a quarter minutes per question on this section of the exam. Do not waste time by lingering too long over any single question. If you're having trouble, move on to the next question. After you finish all the questions, you can come back to the ones you skipped.

The best strategy is to go through the multiple-choice section twice. The first time, do all the questions that you can answer fairly quickly—the ones in which you feel confident about the correct answer. On this first pass, skip the questions that seem to require more thinking or the ones you need to read two or three times before you understand them. Circle the questions that you've skipped in the question booklet so that you can find them easily in the second pass. You must *be very careful* with the answer sheet by making sure the filled-in answers correspond correctly to the questions.

Once you have gone through all the questions, go back to the ones that you skipped in the first pass. But don't linger too long on any one question even in the second pass. Spending too much time wrestling over a hard question can cause two things to happen: One, you may run out of time and miss out on answering easier questions in the later part of the exam. Two, your anxiety might start building up, and this could prevent you from thinking clearly, which would make answering other questions even more difficult. If you simply don't know the answer, or can't eliminate any of them, just use your LOTD and move on.

REFLECT
Respond to the following questions:

- How long will you spend on multiple-choice questions?

- How will you change your approach to multiple-choice questions?

- What is your multiple-choice guessing strategy?

Chapter 2
How to Approach
Free-Response
Questions

CRACKING FREE-RESPONSE QUESTIONS

Section II is worth 50 percent of your grade on the AP Calculus Exam. This section is composed of two parts. Part A contains two free-response questions (you may use a calculator on this part); Part B contains four free-response questions where there are no calculators allowed. You're given a total of 90 minutes for this section.

Clearly Explain and Justify Your Answers

Remember that your answers to the free-response questions are graded by *readers* and not by computers. Communication is a very important part of AP Calculus. Compose your answers in precise sentences. Just getting the correct numerical answer is not enough. You should be able to *explain* your reasoning behind the technique that you selected and *communicate* your answer in the context of the problem. Even if the question does not explicitly say so, always explain and *justify* every step of your answer, including the final answer. Do not expect the graders to read between the lines. Explain everything as though somebody with no knowledge of calculus is going to read it. Be sure to present your solution in a systematic manner using solid logic and appropriate language. And remember: Although you won't earn points for neatness, the graders can't give you a grade if they can't read and understand your solution!

Use Only the Space You Need

Do not try to fill up the space provided for each question. The space given is usually more than enough. The people who design the tests realize that some students write in big letters or make mistakes and need extra space for corrections. So if you have a complete solution, don't worry about the extra space. Writing more will not earn you extra credit. In fact, many students tend to go overboard and shoot themselves in the foot by making a mistake after they've already written the right answer.

Read the Whole Question!

Some questions might have several subparts. Try to answer them all, and don't give up on the question if one part is giving you trouble. For example, if the answer to part (b) depends on the answer to part (a), but you think you got the answer to part (a) wrong, you should still go ahead and do part (b) using your answer to part (a) as required. Chances are that the grader will not mark you wrong twice, unless it is obvious from your answer that you should have discovered your mistake.

REFLECT

Respond to the following questions:

- How much time will you spend on each free-response question?

- How will you change your approach to the free-response questions?

- Will you seek further help, outside of this book (such as a teacher, tutor, or AP Central), on how to approach the calculus exam?

Part IV
Content Review
for the AP
Calculus AB
Exam

HOW TO USE THE CHAPTERS IN THIS PART

For the following content chapters, you may need to come back to them more than once. Your goal is to obtain mastery of the content you are missing, and a single read of a chapter may not be sufficient. At the end of each chapter, you will have an opportunity to reflect on whether you truly have mastered the content of that chapter.

Chapter 3
Limits

WHAT IS A LIMIT?

In order to understand calculus, you need to know what a "limit" is. A limit is the value a function (which usually is written "$f(x)$" on the AP exam) approaches as the variable within that function (usually "x") gets nearer and nearer to a particular value. In other words, when x is very close to a certain number, what is $f(x)$ very close to?

Let's look at an example of a limit: What is the limit of the function $f(x) = x^2$ as x approaches 2? In limit notation, the expression "the limit of $f(x)$ as x approaches 2" is written like this: $\lim_{x \to 2} f(x)$. In order to evaluate the limit, let's check out some values of $\lim_{x \to 2} f(x)$ as x increases and gets closer to 2 (without ever exactly getting there).

> When $x = 1.9$, $f(x) = 3.61$.
> When $x = 1.99$, $f(x) = 3.9601$.
> When $x = 1.999$, $f(x) = 3.996001$.
> When $x = 1.9999$, $f(x) = 3.99960001$.

As x increases and approaches 2, $f(x)$ gets closer and closer to 4. This is called the **left-hand limit** and is written: $\lim_{x \to 2^-} f(x)$. Notice the little minus sign!

What about when x is bigger than 2?

> When $x = 2.1$, $f(x) = 4.41$.
> When $x = 2.01$, $f(x) = 4.0401$.
> When $x = 2.001$, $f(x) = 4.004001$.
> When $x = 2.0001$, $f(x) = 4.00040001$.

As x decreases and approaches 2, $f(x)$ still approaches 4. This is called the **right-hand limit** and is written like this: $\lim_{x \to 2^+} f(x)$. Notice the little plus sign!

We got the same answer when evaluating both the left- and right-hand limits, because when x is 2, $f(x)$ is 4. You should always check both sides of the independent variable because, as you'll see shortly, sometimes you don't get the same answer. Therefore, we write that $\lim_{x \to 2} x^2 = 4$.

We didn't really need to look at all of these decimal values to know what was going to happen when x got really close to 2. But it's important to go through the exercise because, typically, the answers get a lot more complicated. Let's do a few examples.

Example 1: Find $\lim\limits_{x \to 5} x^2$.

The approach is simple: Plug in 5 for x, and you get 25.

Example 2: Find $\lim\limits_{x \to 3} x^3$.

Here the answer is 27.

There are some simple algebraic rules of limits that you should know. These are

$$\lim_{x \to a} kf(x) = k \lim_{x \to a} f(x)$$

Example: $\lim\limits_{x \to 5} 3x^2 = 3 \lim\limits_{x \to 5} x^2 = 75$

$$\text{If } \lim_{x \to a} f(x) = L_1 \text{ and } \lim_{x \to a} g(x) = L_2 \text{, then}$$
$$\lim_{x \to a} \left[f(x) + g(x) \right] = L_1 + L_2$$

Example: $\lim\limits_{x \to 5} \left[x^2 + x^3 \right] = \lim\limits_{x \to 5} x^2 + \lim\limits_{x \to 5} x^3 = 150$

$$\text{If } \lim_{x \to a} f(x) = L_1 \text{ and } \lim_{x \to a} g(x) = L_2 \text{, then}$$
$$\lim_{x \to a} \left[f(x) \cdot g(x) \right] = L_1 \cdot L_2$$

Example: $\lim\limits_{x \to 5} \left[\left(x^2 + 1 \right) \sqrt{x - 1} \right] = \lim\limits_{x \to 5} \left(x^2 + 1 \right) \lim\limits_{x \to 5} \sqrt{x - 1} = 52$

Example 3: Find $\lim_{x \to 0}\left(x^2 + 5x\right)$.

Plug in 0, and you get 0.

So far, so good. All you do to find the limit of a simple polynomial is plug in the number that the variable is approaching and see what the answer is. Naturally, the process can get messier—especially if x approaches zero.

Example 4: Find $\lim_{x \to 0} \dfrac{1}{x^2}$.

If you plug in some very small values for x, you'll see that this function approaches ∞.

And it doesn't matter whether x is positive or negative, you still get ∞. Look at the

graph of $y = \dfrac{1}{x^2}$:

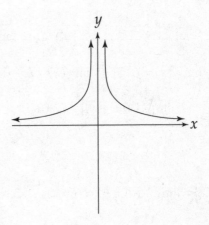

On either side of $x = 0$ (the y-axis), the curve approaches ∞.

Example 5: Find $\lim_{x \to 0} \dfrac{1}{x}$.

Here you have a problem. If you plug in some very small positive values for x (0.1, 0.01, 0.001, and so on), you approach ∞. In other words, $\lim_{x \to 0^+} \dfrac{1}{x} = \infty$. But, if you

plug in some very small negative values for x (−0.1, −0.01, −0.001, and so on) you

approach $-\infty$. That is, $\lim_{x \to 0^-} \dfrac{1}{x} = -\infty$. Because the right-hand limit is not equal to

the left-hand limit, the limit does not exist.

Look at the graph of $\dfrac{1}{x}$.

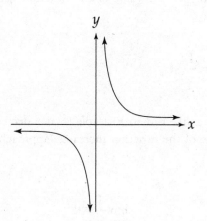

You can see that on the left side of $x = 0$, the curve approaches $-\infty$, and on the right side of $x = 0$, the curve approaches ∞. There are some very important points that we need to emphasize from the last two examples.

Why do we state the limit in Example 4 but not for Example 5? Because when we have $\dfrac{k}{x^2}$, the function is always positive no matter what the sign of x is and thus the function has the same limit from the left and the right. But when we have $\dfrac{k}{x}$, the function's sign depends on the sign of x, and you get a different limit from each side.

Let's look at a few examples in which the independent variable approaches infinity.

(1) If the left-hand limit of a function is not equal to the right-hand limit of the function, then the limit does not exist.

(2) A limit equal to infinity is not the same as a limit that does not exist, but sometimes you will see the expression "no limit," which serves both purposes. If $\lim\limits_{x \to a} f(x) = \infty$, the limit, technically, does not exist.

(3) If k is a positive constant, then $\lim\limits_{x \to 0^+} \dfrac{k}{x} = \infty$, $\lim\limits_{x \to 0^-} \dfrac{k}{x} = -\infty$, and $\lim\limits_{x \to 0} \dfrac{k}{x}$ does not exist.

(4) If k is a positive constant, then

$$\lim_{x \to 0^+} \frac{k}{x^2} = \infty, \ \lim_{x \to 0^-} \frac{k}{x^2} = \infty, \text{ and}$$

$$\lim_{x \to 0} \frac{k}{x^2} = \infty.$$

Example 6: Find $\lim\limits_{x \to \infty} \dfrac{1}{x}$.

As x gets bigger and bigger, the value of the function gets smaller and smaller. Therefore, $\lim\limits_{x \to \infty} \dfrac{1}{x} = 0$.

Example 7: Find $\lim\limits_{x \to -\infty} \dfrac{1}{x}$.

It's the same situation as the one in Example 6; as x decreases (approaches negative infinity), the value of the function increases (approaches zero). We write the following:

$$\lim_{x \to -\infty} \frac{1}{x} = 0$$

We don't have the same problem here that we did when x approached zero because "positive zero" is the same thing as "negative zero," whereas positive infinity is different from negative infinity.

Here's another rule.

> If k and n are constants, $|x| > 1$, and $n > 0$, then $\lim\limits_{x \to \infty} \dfrac{k}{x^n} = 0$,
> and $\lim\limits_{x \to -\infty} \dfrac{k}{x^n} = 0$.

Example 8: Find $\lim\limits_{x \to \infty} \dfrac{3x+5}{7x-2}$.

When you have variables in both the top and the bottom, you can't just plug ∞ into the expression. You'll get $\dfrac{\infty}{\infty}$. We solve this by using the following technique:

When an expression consists of a polynomial divided by another polynomial, divide each term of the numerator and the denominator by the highest power of x that appears in the expression.

The highest power of x in this case is x^1, so we divide every term in the expression (both top and bottom) by x, like so

$$\lim_{x \to \infty} \frac{3x+5}{7x-2} = \lim_{x \to \infty} \frac{\dfrac{3x}{x} + \dfrac{5}{x}}{\dfrac{7x}{x} - \dfrac{2}{x}} = \lim_{x \to \infty} \frac{3 + \dfrac{5}{x}}{7 - \dfrac{2}{x}}$$

Now when we take the limit, the two terms containing x approach zero. We're left with $\dfrac{3}{7}$.

Example 9: Find $\displaystyle\lim_{x\to\infty}\dfrac{8x^2-4x+1}{16x^2+7x-2}$

Divide each term by x^2. You get

$$\lim_{x\to\infty}\frac{8-\dfrac{4}{x}+\dfrac{1}{x^2}}{16+\dfrac{7}{x}-\dfrac{2}{x^2}}=\frac{8}{16}=\frac{1}{2}$$

Example 10: Find $\displaystyle\lim_{x\to\infty}\dfrac{-3x^{10}-70x^5+x^3}{33x^{10}+200x^8-1000x^4}$.

Divide each term by x^{10}.

$$\lim_{x\to\infty}\frac{-3x^{10}-70x^5+x^3}{33x^{10}+200x^8-1000x^4}=\lim_{x\to\infty}\frac{-3-\dfrac{70}{x^5}+\dfrac{1}{x^7}}{33+\dfrac{200}{x^2}-\dfrac{1000}{x^6}}=-\frac{3}{33}=-\frac{1}{11}$$

The other powers don't matter because they're all going to disappear. Now we have three new rules for evaluating the limit of a rational expression as x approaches infinity.

(1) If the highest power of x in a rational expression is in the numerator, then the limit as x approaches infinity is infinity.

Remember to focus your attention on the highest power of x.

Example: $\displaystyle\lim_{x\to\infty}\dfrac{5x^7-3x}{16x^6-3x^2}=\infty$

(2) If the highest power of x in a rational expression is in the denominator, then the limit as x approaches infinity is zero.

Example: $\lim\limits_{x \to \infty} \dfrac{5x^6 - 3x}{16x^7 - 3x^2} = 0$

> (3) If the highest power of x in a rational expression is the same in both the numerator and denominator, then the limit as x approaches infinity is the coefficient of the highest term in the numerator divided by the coefficient of the highest term in the denominator.

Example: $\lim\limits_{x \to \infty} \dfrac{5x^7 - 3x}{16x^7 - 3x^2} = \dfrac{5}{16}$

LIMITS OF TRIGONOMETRIC FUNCTIONS

At some point during the exam, you'll have to find the limit of certain trig expressions, usually as x approaches either zero or infinity. There are four standard limits that you should memorize—with those, you can evaluate all of the trigonometric limits that appear on the test. As you'll see throughout this book, calculus requires that you remember all of your trig from previous years.

> Rule No. 1: $\lim\limits_{x \to 0} \dfrac{\sin x}{x} = 1$ (x is in radians, *not* degrees)

This may seem strange, but if you look at the graphs of $f(x) = \sin x$ and $f(x) = x$, they have approximately the same slope near the origin (as x gets closer to zero). Because x and the sine of x are about the same as x approaches zero, their quotient will be very close to one. Furthermore, because $\lim\limits_{x \to 0} \cos x = 1$ (review cosine values if you don't get this!), we know that $\lim\limits_{x \to 0} \tan x = \lim\limits_{x \to 0} \dfrac{\sin x}{\cos x} = 0$.

Remember that the $\lim\limits_{x \to 0} \sin x = 0$.

Now we will find a second rule. Let's evaluate the limit $\lim_{x \to 0} \dfrac{\cos x - 1}{x}$. First, multiply the top and bottom by $\cos x + 1$. We get $\lim_{x \to 0} \left(\dfrac{\cos x - 1}{x} \right) \left(\dfrac{\cos x + 1}{\cos x + 1} \right)$. Now simplify the limit to: $\lim_{x \to 0} \dfrac{\cos^2 x - 1}{x(\cos x + 1)}$. Next, we can use the trigonometric identity $\sin^2 x = 1 - \cos^2 x$ and rewrite the limit as: $\lim_{x \to 0} \dfrac{-\sin^2 x}{x(\cos x + 1)}$. Now, break this into two limits: $\lim_{x \to 0} \dfrac{-\sin x}{x} \dfrac{\sin x}{(\cos x + 1)}$. The first limit is -1 (see Rule No. 1) and the second is 0 (why?), so the limit is 0.

Rule No. 2: $\lim_{x \to 0} \dfrac{\cos x - 1}{x} = 0$

Example 11: Find $\lim_{x \to 0} \dfrac{\sin 3x}{x}$.

Use a simple trick: Multiply the top and bottom of the expression by 3. This gives us: $\lim_{x \to 0} \dfrac{3 \sin 3x}{3x}$. Next, substitute a letter for $3x$; for example, a. Now, we get the following:

$$\lim_{a \to 0} \frac{3 \sin a}{a} = 3 \lim_{a \to 0} \frac{\sin a}{a} = 3(1) = 3$$

Example 12: Find $\lim_{x \to 0} \dfrac{\sin 5x}{\sin 4x}$.

Now we get a bit more sophisticated. First, divide both the numerator and the denominator by x, like so

$$\lim_{x \to 0} \frac{\dfrac{\sin 5x}{x}}{\dfrac{\sin 4x}{x}}$$

Next, multiply the top and bottom of the numerator by 5, and the top and bottom of the denominator by 4, which gives us

$$\lim_{x\to 0}\frac{\dfrac{5\sin 5x}{5x}}{\dfrac{4\sin 4x}{4x}}$$

From the work we did in Example 11, we can see that this limit is $\dfrac{5}{4}$.

Guess what! You have two more rules!

Rule No. 3: $\lim\limits_{x\to 0}\dfrac{\sin ax}{x}=a$

Rule No. 4: $\lim\limits_{x\to 0}\dfrac{\sin ax}{\sin bx}=\dfrac{a}{b}$

Example 13: Find $\lim\limits_{x\to 0}\dfrac{x^2}{1-\cos^2 x}$.

Using trigonometric identities, you can replace $(1-\cos^2 x)$ with $\sin^2 x$.

$$\lim_{x\to 0}\frac{x^2}{1-\cos^2 x}=\lim_{x\to 0}\frac{x^2}{\sin^2 x}=\lim_{x\to 0}\left(\frac{x}{\sin x}\cdot\frac{x}{\sin x}\right)=1\cdot 1=1$$

Here are other examples for you to try, with answers right beneath them. Give 'em a try, and check your work.

PROBLEM 1. Find $\lim\limits_{x\to 3}\dfrac{x-3}{x+2}$.

Answer: If you plug in 3 for x, you get $\lim\limits_{x\to 3}\dfrac{3-3}{3+2}=\dfrac{0}{5}=0$.

PROBLEM 2. Find $\lim\limits_{x\to 3}\dfrac{x+2}{x-3}$.

If plugging in the value of x results in the denominator equaling zero and you cannot factor the quotient anymore, then check the left and right hand limits to find the limit of the expression.

Answer: The left-hand limit is: $\lim\limits_{x\to 3^-}\dfrac{x+2}{x-3}=-\infty$

The right-hand limit is: $\lim\limits_{x\to 3^+} \dfrac{x+2}{x-3} = \infty$

These two limits are not the same. Therefore, the limit does not exist.

PROBLEM 3. Find $\lim\limits_{x\to 3} \dfrac{x+2}{(x-3)^2}$.

Answer: The left-hand limit is: $\lim\limits_{x\to 3^-} \dfrac{x+2}{(x-3)^2} = \infty$

The right-hand limit is: $\lim\limits_{x\to 3^+} \dfrac{x+2}{(x-3)^2} = \infty$

These two limits are the same, so the limit is ∞.

PROBLEM 4. Find $\lim\limits_{x\to -4} \dfrac{x^2+6x+8}{x+4}$.

Answer: If you plug -4 into the top and bottom, you get $\dfrac{0}{0}$. You have to factor the top into $(x+2)(x+4)$ to get this: $\lim\limits_{x\to -4} \dfrac{(x+2)(x+4)}{(x+4)}$

Now it's time to cancel like terms: $\lim\limits_{x\to -4} \dfrac{(x+2)(x+4)}{(x+4)} = \lim\limits_{x\to -4}(x+2) = -2$

PROBLEM 5. Find $\lim\limits_{x\to \infty} \dfrac{15x^2-11x}{22x^2+4x}$.

Answer: Divide each term by x^2.

$$\lim\limits_{x\to \infty} \dfrac{15x^2-11x}{22x^2+4x} = \lim\limits_{x\to \infty} \dfrac{15-\dfrac{11}{x}}{22+\dfrac{4}{x}} = \dfrac{15}{22}$$

PROBLEM 6. Find $\lim\limits_{x\to 0} \dfrac{4x}{\tan x}$.

Answer: Replace $\tan x$ with $\dfrac{\sin x}{\cos x}$, which changes the expression into

$$\lim\limits_{x\to 0} \dfrac{4x}{\tan x} = \lim\limits_{x\to 0} \dfrac{4x}{\dfrac{\sin x}{\cos x}} = \lim\limits_{x\to 0} \dfrac{4x\cos x}{\sin x}$$

Note: Pay careful attention to this next solved problem. It will be very important when you work on problems in Chapter 6.

Because $\lim_{x \to 0} \dfrac{\sin x}{x} = 1$, the $\lim_{x \to 0} \dfrac{x}{\sin x} = 1$ as well. Thus, because $\lim_{x \to 0} \dfrac{x}{\sin x} = 1$ and $\lim_{x \to 0} \cos x = 1$, the answer is 4.

Problem 7. Find $\lim_{h \to 0} \dfrac{(5+h)^2 - 25}{h}$.

Answer: First, expand and simplify the numerator.

$$\lim_{h \to 0} \dfrac{(5+h)^2 - 25}{h} = \lim_{h \to 0} \dfrac{25 + 10h + h^2 - 25}{h} = \lim_{h \to 0} \dfrac{10h + h^2}{h}.$$

Next, factor h out of the numerator and the denominator.

$$\lim_{h \to 0} \dfrac{10h + h^2}{h} = \lim_{h \to 0} \dfrac{h(10 + h)}{h} = \lim_{h \to 0}(10 + h).$$

Taking the limit you get: $\lim_{h \to 0}(10 + h) = 10$.

PRACTICE PROBLEM SET 1

Try these 30 problems to test your skill with limits. The answers are in Chapter 19.

1. $\lim_{x \to 8}(x^2 - 5x - 11) =$ $64 - 40 - 11 = 13$

2. $\lim_{x \to 5}\left(\dfrac{x+3}{x^2 - 15}\right) =$ $\dfrac{8}{10} = \dfrac{4}{5}$

3. $\lim_{x \to 0}\pi^2 =$ π^2

4. $\lim_{x \to 3}\left(\dfrac{x^2 - 2x - 3}{x - 3}\right) =$ $\dfrac{(x-3)(x+1)}{(x-3)} = 4$

5. $\lim_{x \to \infty}\left(\dfrac{10x^2 + 25x + 1}{x^4 - 8}\right) =$

6. $\lim_{x \to \infty}\left(\dfrac{x^4 - 8}{10x^2 + 25x + 1}\right) =$

7. $\lim\limits_{x\to\infty}\left(\dfrac{x^4-8}{10x^4+25x+1}\right)=$

8. $\lim\limits_{x\to\infty}\left(\dfrac{\sqrt{5x^4+2x}}{x^2}\right)=$

9. $\lim\limits_{x\to6^+}\left(\dfrac{x+2}{x^2-4x-12}\right)=$

10. $\lim\limits_{x\to6^-}\left(\dfrac{x+2}{x^2-4x-12}\right)=$

11. $\lim\limits_{x\to6}\left(\dfrac{x+2}{x^2-4x-12}\right)=$

12. $\lim\limits_{x\to0^+}\left(\dfrac{x}{|x|}\right)=$

13. $\lim\limits_{x\to0^-}\left(\dfrac{x}{|x|}\right)=$

14. $\lim\limits_{x\to7^+}\left(\dfrac{x}{x^2-49}\right)=$ $\lim\limits_{x\to7^+}\left(\dfrac{x}{(x-7)(x+7)}\right)=$ DNE (∞)

15. $\lim\limits_{x\to7}\left(\dfrac{x}{x^2-49}\right)=$

16. $\lim\limits_{x\to7}\dfrac{x}{(x-7)^2}=$

17. Let $f(x)=\begin{cases}x^2-5,\ x\le3\\x+2,\ x>3\end{cases}$ $\lim\limits_{x\to3^-}=4$ $\lim\limits_{x\to3^+}f(x)=5$

    $\lim\limits_{x\to3}f(x)=DNE$

    Find: (a) $\lim\limits_{x\to3^-}f(x)$; (b) $\lim\limits_{x\to3^+}f(x)$; and (c) $\lim\limits_{x\to3}f(x)$

18. Let $f(x) = \begin{cases} x^2 - 5, & x \le 3 \\ x + 1, & x > 3 \end{cases}$

Find: (a) $\lim\limits_{x \to 3^-} f(x)$; (b) $\lim\limits_{x \to 3^+} f(x)$; and (c) $\lim\limits_{x \to 3} f(x)$

19. Find $\lim\limits_{x \to \frac{\pi}{4}} 3\cos x$.

20. Find $\lim\limits_{x \to 0} 3\dfrac{x}{\cos x}$.

21. Find $\lim\limits_{x \to 0} 3\dfrac{x}{\sin x}$.

22. Find $\lim\limits_{x \to 0} \dfrac{\sin 3x}{\sin 8x}$.

23. Find $\lim\limits_{x \to 0} \dfrac{\tan 7x}{\sin 5x}$.

24. Find $\lim\limits_{x \to \infty} \sin x$.

25. Find $\lim\limits_{x \to \infty} \sin\dfrac{1}{x}$.

26. Find $\lim\limits_{x \to 0} \dfrac{x^2 \sin x}{1 - \cos^2 x}$.

27. Find $\lim\limits_{x \to 0} \dfrac{\sin^2 7x}{\sin^2 11x}$.

28. Find $\lim\limits_{h \to 0} \dfrac{(3+h)^2 - 9}{h}$.

29. Find $\lim\limits_{h \to 0} \dfrac{\sin(x+h) - \sin x}{h}$.

30. Find $\lim\limits_{h \to 0} \dfrac{\dfrac{1}{x+h} - \dfrac{1}{x}}{h}$.

Chapter 4
Continuity

Every AP exam has a few questions on continuity, so it's important to understand the basic idea of what it means for a function to be continuous. The concept is very simple: If the graph of the function doesn't have any breaks or holes in it within a certain interval, the function is continuous over that interval.

Simple polynomials are continuous everywhere; it's the other ones—trigonometric, rational, piecewise—that might have continuity problems. Most of the test questions concern these last types of functions. In order to learn how to test whether a function is continuous, you'll need some more mathematical terminology.

THE DEFINITION OF CONTINUITY

In order for a function $f(x)$ to be continuous at a point $x = c$, it must fulfill *all three* of the following conditions:

Condition 1: $f(c)$ exists.

Condition 2: $\lim\limits_{x \to c} f(x)$ exists.

Condition 3: $\lim\limits_{x \to c} f(x) = f(c)$

Let's look at a simple example of a continuous function.

Example 1: Is the function $f(x) = \begin{cases} x+1, & x < 2 \\ 2x-1, & x \geq 2 \end{cases}$ continuous at the point $x = 2$?

Condition 1: Does $f(2)$ exist?

Yes. It's equal to $2(2) - 1 = 3$.

Condition 2: Does $\lim\limits_{x \to 2} f(x)$ exist?

You need to look at the limit from both sides of 2. The left-hand limit is: $\lim\limits_{x \to 2^-} f(x) = 2 + 1 = 3$. The right-hand limit is: $\lim\limits_{x \to 2^+} f(x) = 2(2) - 1 = 3$.

Because the two limits are the same, the limit exists.

Condition 3: Does $\lim_{x \to 2} f(x) = f(2)$?

The two equal each other, so yes; the function is continuous at $x = 2$.

A simple and important way to check whether a function is continuous is to sketch the function. If you can't sketch the function without lifting your pencil from the paper at some point, then the function is not continuous.

Now let's look at some examples of functions that are not continuous.

Example 2: Is the function $f(x) = \begin{cases} x+1, & x < 2 \\ 2x - 1, & x > 2 \end{cases}$ continuous at $x = 2$?

Condition 1: Does $f(2)$ exist?

Nope. The function of x is defined if x is greater than or less than 2, but not if x is equal to 2. Therefore, the function is not continuous at $x = 2$. Notice that we don't have to bother with the other two conditions. Once you find a problem, the function is automatically not continuous, and you can stop.

Example 3: Is the function $f(x) = \begin{cases} x+1, & x < 2 \\ 2x+1, & x \geq 2 \end{cases}$ continuous at $x = 2$?

Condition 1: Does $f(x)$ exist?

Yes. It is equal to $2(2) + 1 = 5$.

Condition 2: Does $\lim_{x \to 2} f(x)$ exist?

The left-hand limit is: $\lim_{x \to 2^-} f(x) = 2 + 1 = 3$.

The right-hand limit is: $\lim_{x \to 2^+} f(x) = 2(2) + 1 = 5$.

The two limits don't match, so the limit doesn't exist and the function is not continuous at $x = 2$.

Example 4: Is the function $f(x) = \begin{cases} x+1, & x < 2 \\ x^2, & x = 2 \\ 2x-1, & x > 2 \end{cases}$ continuous at $x = 2$?

Condition 1: Does $f(2)$ exist?

Yes. It's equal to $2^2 = 4$.

Condition 2: Does $\lim\limits_{x \to 2} f(x)$ exist?

The left-hand limit is: $\lim\limits_{x \to 2^-} f(x) = 2 + 1 = 3$.

The right-hand limit is: $\lim\limits_{x \to 2^+} f(x) = 2(2) - 1 = 3$.

Because the two limits are the same, the limit exists.

Condition 3: Does $\lim\limits_{x \to 2} f(x) = f(2)$?

The $\lim\limits_{x \to 2} f(x) = 3$, but $f(2) = 4$. Because these aren't equal, the answer is "no" and the function is not continuous at $x = 2$.

TYPES OF DISCONTINUITIES

There are four types of discontinuities you have to know: jump, point, essential, and removable.

> A **jump** discontinuity occurs when the curve "breaks" at a particular place and starts somewhere else. In other words, $\lim\limits_{x \to a^-} f(x) \neq \lim\limits_{x \to a^+} f(x)$.

An example of jump discontinuity looks like this.

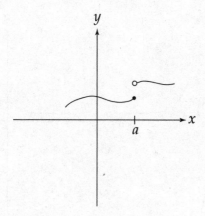

A **point** discontinuity occurs when the curve has a "hole" in it from a missing point because the function has a value at that point that is "off the curve." In other words, $\lim\limits_{x \to a} f(x) \neq f(a)$.

Here's what a point discontinuity looks like.

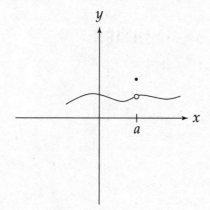

An **essential** discontinuity occurs when the curve has a vertical asymptote.

This is an example of an essential discontinuity.

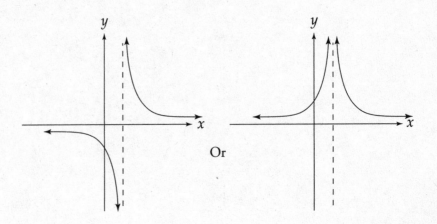

Or

A **removable** discontinuity occurs when you have a rational expression with common factors in the numerator and denominator. Because these factors can be canceled, the discontinuity is "removable."

Here's an example of a removable discontinuity.

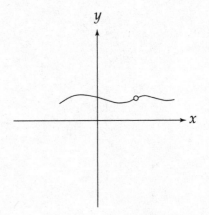

This curve looks very similar to a point discontinuity, but notice that with a removable discontinuity, $f(x)$ is not defined at the point, whereas with a point discontinuity, $f(x)$ is defined there.

Now that you know what these four types of discontinuities look like, let's see what types of functions are not everywhere continuous.

Example 5: Consider the following function:

$$f(x) = \begin{cases} x+3, \ x \le 2 \\ x^2, \ x > 2 \end{cases}$$

The left-hand limit is 5 as x approaches 2, and the right-hand limit is 4 as x approaches 2. Because the curve has different values on each side of 2, the curve is discontinuous at $x = 2$. We say that the curve "jumps" at $x = 2$ from the left-hand curve to the right-hand curve because the left and right-hand limits differ. It looks like the following:

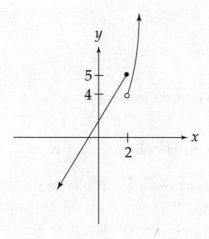

This is an example of a jump discontinuity.

Example 6: Consider the following function:

$$f(x) = \begin{cases} x^2, \ x \ne 2 \\ 5, \ x = 2 \end{cases}$$

Because $\lim_{x\to 2} f(x) \neq f(2)$; the function is discontinuous at $x = 2$. The curve is continuous everywhere except at the point $x = 2$. It looks like the following:

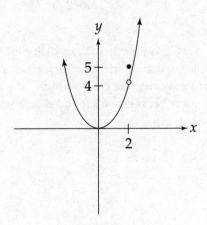

This is an example of a point discontinuity.

Example 7: Consider the following function: $f(x) = \dfrac{5}{x-2}$

The function is discontinuous because it's possible for the denominator to equal zero (at $x = 2$). This means that $f(2)$ doesn't exist, and the function has an asymptote at $x = 2$. In addition, $\lim_{x\to 2^-} f(x) = -\infty$ and $\lim_{x\to 2^+} f(x) = \infty$.

The graph looks like the following:

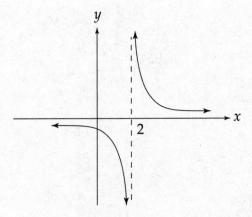

This is an example of an essential discontinuity.

Example 8: Consider the following function:

$$f(x) = \frac{x^2 - 8x + 15}{x^2 - 6x + 5}$$

If you factor the top and bottom, you can see where the discontinuities are.

$$f(x) = \frac{x^2 - 8x + 15}{x^2 - 6x + 5} = \frac{(x-3)(x-5)}{(x-1)(x-5)}$$

The function has a zero in the denominator when $x = 1$ or $x = 5$, so the function is discontinuous at those two points. But you can cancel the term $(x - 5)$ from both the numerator and the denominator, leaving you with

$$f(x) = \frac{x-3}{x-1}$$

Now the reduced function *is* continuous at $x = 5$. Thus, the original function has a removable discontinuity at $x = 5$. Furthermore, if you now plug $x = 5$ into the reduced function, you get

$$f(5) = \frac{2}{4} = \frac{1}{2}$$

The discontinuity is at $x = 5$, and there's a hole at $\left(5, \frac{1}{2}\right)$. In other words, if the original function were continuous at $x = 5$, it would have the value $\frac{1}{2}$. Notice that this is the same as: $\lim\limits_{x \to 5} f(x)$.

These are the types of discontinuities that you can expect to encounter on the AP examination. Here are some sample problems and their solutions. Cover the answers as you work, then check your results.

PROBLEM 1. Is the function $f(x) = \begin{cases} 2x^3 - 1, & x < 2 \\ 6x - 3, & x \geq 2 \end{cases}$ continuous at $x = 2$?

Answer: Test the conditions necessary for continuity.

Condition 1: $f(2) = 9$, so we're okay so far.

Condition 2: The $\lim\limits_{x \to 2^-} f(x) = 15$ and the $\lim\limits_{x \to 2^+} f(x) = 9$. These two limits don't agree, so the $\lim\limits_{x \to 2} f(x)$ doesn't exist and the function is not continuous at $x = 2$.

PROBLEM 2. Is the function $f(x) = \begin{cases} x^2 + 3x + 5, & x < 1 \\ 6x + 3, & x \geq 1 \end{cases}$ continuous at $x = 1$?

Answer: Condition 1: $f(1) = 9$.

Condition 2: The $\lim\limits_{x \to 1^-} f(x) = 9$ and the $\lim\limits_{x \to 1^+} f(x) = 9$.

Therefore, the $\lim\limits_{x \to 1} f(x)$ exists and is equal to 9.

Condition 3: $\lim\limits_{x \to 1} f(x) = f(1) = 9$.

The function satisfies all three conditions, so it is continuous at $x = 1$.

PROBLEM 3. For what value of a is the function $f(x) = \begin{cases} ax + 5, & x < 4 \\ x^2 - x, & x \geq 4 \end{cases}$ continuous at $x = 4$?

Answer: Because $f(4) = 12$, the function passes the first condition.

For Condition 2 to be satisfied, the $\lim\limits_{x \to 4^-} f(x) = 4a + 5$ must equal the

$\lim\limits_{x \to 4^+} f(x) = 12$. So set $4a + 5 = 12$. If $a = \dfrac{7}{4}$, the limit will exist at $x = 4$ and the

other two conditions will also be fulfilled. Therefore, the value $a = \dfrac{7}{4}$ makes the

function continuous at $x = 4$.

PROBLEM 4. Where does the function $f(x) = \dfrac{2x^2 - 7x - 15}{x^2 - x - 20}$ have: (a) an essential discontinuity; and (b) a removable discontinuity?

Answer: If you factor the top and bottom of this fraction, you get

$$f(x) = \frac{2x^2 - 7x - 15}{x^2 - x - 20} = \frac{(2x + 3)(x - 5)}{(x + 4)(x - 5)}$$

Thus, the function has an essential discontinuity at $x = -4$. If we then cancel the term $(x - 5)$, and substitute $x = 5$ into the reduced expression, we get $f(5) = \dfrac{13}{9}$. Therefore, the function has a removable discontinuity at $\left(5, \dfrac{13}{9}\right)$.

Note: Don't confuse coordinate parentheses with interval notation. In interval notation, square brackets include endpoints and parentheses do not. For example, the interval $2 \le x \le 4$ is written [2, 4] and the interval $2 < x < 4$ is written (2, 4).

PRACTICE PROBLEM SET 2

Now try these problems. The answers are in Chapter 19.

1. Is the function $f(x) = \begin{cases} x+7, & x < 2 \\ 9, & x = 2 \\ 3x+3, & x > 2 \end{cases}$ continuous at $x = 2$?

2. Is the function $f(x) = \begin{cases} 4x^2 - 2x, & x < 3 \\ 10x - 1, & x = 3 \\ 30, & x > 3 \end{cases}$ continuous at $x = 3$?

3. Is the function $f(x) = \begin{cases} 5x+7, & x < 3 \\ 7x+1, & x > 3 \end{cases}$ continuous at $x = 3$?

4. Is the function $f(x) = \sec x$ continuous everywhere?

5. Is the function $f(x) = \sec x$ continuous on the interval $\left[-\dfrac{\pi}{2}, \dfrac{\pi}{2}\right]$?

6. Is the function $f(x) = \sec x$ continuous on the interval $\left(-\dfrac{\pi}{2}, \dfrac{\pi}{2}\right)$?

7. For what value(s) of k is the function $f(x) = \begin{cases} 3x^2 - 11x - 4, & x \le 4 \\ kx^2 - 2x - 1, & x > 4 \end{cases}$ continuous at $x = 4$?

8. For what value(s) of k is the function $f(x) = \begin{cases} -6x - 12, & x < -3 \\ k^2 - 5k, & x = -3 \\ 6, & x > -3 \end{cases}$ continuous at $x = -3$?

9. At what point is the removable discontinuity for the function
$f(x) = \dfrac{x^2 + 5x - 24}{x^2 - x - 6}$?

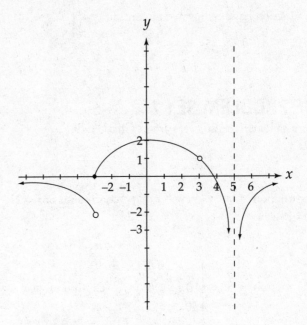

10. Given the graph of $f(x)$ above, find

(a) $\lim\limits_{x \to -\infty} f(x)$

(b) $\lim\limits_{x \to \infty} f(x)$

(c) $\lim\limits_{x \to 3^-} f(x)$

(d) $\lim\limits_{x \to 3^+} f(x)$

(e) $f(3)$

(f) Any discontinuities.

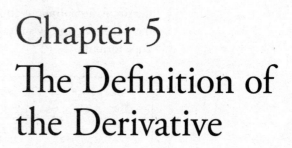

Chapter 5
The Definition of
the Derivative

The main tool that you'll use in differential calculus is called the **derivative**. All of the problems that you'll encounter in differential calculus make use of the derivative, so your goal should be to become an expert at finding, or "taking," derivatives by the end of Chapter 6. However, before you learn a simple way to take a derivative, your teacher will probably make you learn how derivatives are calculated by teaching you something called the "Definition of the Derivative."

DERIVING THE FORMULA

The best way to understand the definition of the derivative is to start by looking at the simplest continuous function: a line. As you should recall, you can determine the slope of a line by taking two points on that line and plugging them into the slope formula.

$$m = \frac{y_2 - y_1}{x_2 - x_1} \qquad\qquad m \text{ stands for slope.}$$

Notice that you can use the coordinates in reverse order and still get the same result. It doesn't matter in which order you do the subtraction as long as you're consistent.

For example, suppose a line goes through the points (3, 7) and (8, 22). First, you subtract the y-coordinates $(22 - 7) = 15$. Next, subtract the corresponding x-coordinates $(8 - 3) = 5$. Finally, divide the first number by the second: $\frac{15}{5} = 3$. The result is the slope of the line: $m = 3$.

Let's look at the graph of that line. The slope measures the steepness of the line, which looks like the following:

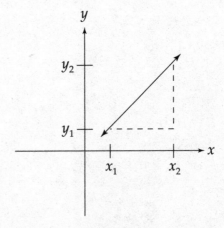

You probably remember your teachers referring to the slope as the "rise" over the "run." The rise is the difference between the y-coordinates, and the run is the difference between the x-coordinates. The slope is the ratio of the two.

Now for a few changes in notation. Instead of calling the x-coordinates x_1 and x_2, we're going to call them x_1 and $x_1 + h$, where h is the difference between the two x-coordinates. Second, instead of using y_1 and y_2, we use $f(x_1)$ and $f(x_1 + h)$. So now the graph looks like the following:

Sometimes, instead of h, some books use Δx.

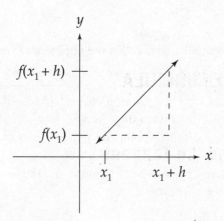

The picture is exactly the same—only the notation has changed.

The Slope of a Curve

Suppose that instead of finding the slope of a line, we wanted to find the slope of a curve. Here, the slope formula no longer works because the distance from one point to the other is along a curve, not a straight line. But we could find the approximate slope if we took the slope of the line between the two points. This is called the **secant line**.

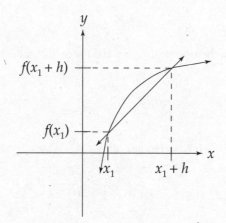

The formula for the slope of the secant line is

$$\frac{f(x_1 + h) - f(x_1)}{h}$$

Remember this formula! This is called the **Difference Quotient**.

The Secant and the Tangent

As you can see, the farther apart the two points are, the less the slope of the line corresponds to the slope of the curve.

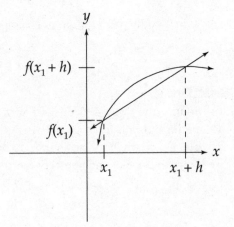

Conversely, the closer the two points are, the more accurate the approximation is.

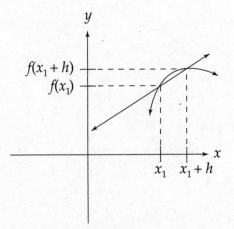

In fact, there is one line, called the **tangent line**, that touches the curve at exactly one point. The slope of the tangent line is equal to the slope of the curve at exactly this point. The object of using the above formula, therefore, is to shrink h down to an infinitesimally small amount. If we could do that, then the difference between $(x_1 + h)$ and x_1 would be a point.

Keep in mind that there are infinitely many tangents for any curve because there are infinitely many points on the curve.

Graphically, it looks like the following:

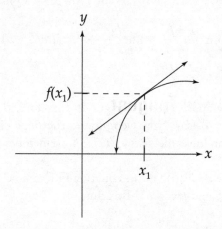

How do we perform this shrinking act? By using the limits we discussed in Chapter 3. We set up a limit during which h approaches zero, like the following:

$$\lim_{h \to 0} \frac{f(x_1 + h) - f(x_1)}{h}$$

This is the **definition of the derivative**, and we call it $f'(x)$.

Notice that the equation is just a slightly modified version of the difference quotient, with different notation. The only difference is that we're finding the slope between two points that are infinitesimally close to each other.

Example 1: Find the slope of the curve $f(x) = x^2$ at the point $(2, 4)$.

This means that $x_1 = 2$ and $f(2) = 2^2 = 4$. If we can figure out $f(x_1 + h)$, then we can find the slope. Well, how did we find the value of $f(x)$? We plugged x_1 into the equation $f(x) = x^2$. To find $f(x_1 + h)$ we plug $x_1 + h$ into the equation, which now looks like this

$$f(x_1 + h) = (2 + h)^2 = 4 + 4h + h^2$$

Now plug this into the slope formula.

$$\lim_{h \to 0} \frac{f(x_1 + h) - f(x_1)}{h} = \lim_{h \to 0} \frac{4 + 4h + h^2 - 4}{h} = \lim_{h \to 0} \frac{4h + h^2}{h}$$

Next simplify by factoring h out of the top.

$$\lim_{h \to 0} \frac{4h + h^2}{h} = \lim_{h \to 0} \frac{h(4 + h)}{h} = \lim_{h \to 0} (4 + h)$$

Taking the limit as h approaches 0, we get 4. Therefore, the slope of the curve $y = x^2$ at the point (2, 4) is 4. Now we've found the slope of a curve at a certain point, and the notation looks like this: $f'(2) = 4$. Remember this notation!

Example 2: Find the derivative of the equation in Example 1 at the point (5, 25). This means that $x_1 = 5$ and $f(x) = 25$. This time,

$$(x_1 + h)^2 = (5 + h)^2 = 25 + 10h + h^2$$

Now plug this into the formula for the derivative.

$$\lim_{h \to 0} \frac{f(x_1 + h) - f(x_1)}{h} = \lim_{h \to 0} \frac{25 + 10h + h^2 - 25}{h} = \lim_{h \to 0} \frac{10h + h^2}{h}$$

Once again, simplify by factoring h out of the top.

$$\lim_{h \to 0} \frac{10h + h^2}{h} = \lim_{h \to 0} \frac{h(10 + h)}{h} = \lim_{h \to 0} (10 + h)$$

Taking the limit as h goes to 0, you get 10. Therefore, the slope of the curve $y = x^2$ at the point (5, 25) is 10, or: $f'(5) = 10$.

Using this pattern, let's forget about the arithmetic for a second and derive a formula.

Example 3: Find the slope of the equation $f(x) = x^2$ at the point $\left(x_1, \, x_1^2\right)$.

Follow the steps in the last two problems, but instead of using a number, use x_1. This means that $f(x_1) = x_1^2$ and $(x_1 + h)^2 = x_1^2 + 2x_1h + h^2$. Then the derivative is

$$\lim_{h \to 0} \frac{x_1^2 + 2x_1h + h^2 - x_1^2}{h} = \lim_{h \to 0} \frac{2x_1h + h^2}{h}$$

Factor h out of the top.

$$\lim_{h \to 0} \frac{h(2x_1 + h)}{h} = \lim_{h \to 0}(2x_1 + h)$$

Now take the limit as h goes to 0: you get $2x_1$. Therefore, $f'(x_1) = 2x_1$.

This example gives us a general formula for the derivative of this curve. Now we can pick any point, plug it into the formula, and determine the slope at that point. For example, the derivative at the point $x = 7$ is 14. At the point $x = \frac{7}{3}$, the derivative is $\frac{14}{3}$.

DIFFERENTIABILITY

One of the important requirements for the differentiability of a function is that the function be continuous. But, even if a function is continuous at a point, the function is not necessarily differentiable there. Check out the graph below.

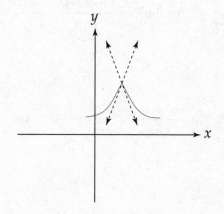

If a function has a "sharp corner," you can draw more than one tangent line at that point, and because the slopes of these tangent lines are not equal, the function is not differentiable there.

Another possible problem occurs when the tangent line is vertical (which can also occur at a cusp) because a vertical line has an infinite slope. For example, if the derivative of a function is $\dfrac{1}{x+1}$, it doesn't have a derivative at $x = -1$.

Try these problems on your own, then check your work against the answers immediately beneath each problem.

PROBLEM 1. Find the derivative of $f(x) = 3x^2$ at (4, 48).

Answer: $f(4 + h) = 3(4 + h)^2 = 48 + 24h + 3h^2$. Use the definition of the derivative.

$$f'(4) = \lim_{h \to 0} \frac{48 + 24h + 3h^2 - 48}{h}$$

Simplify.

$$\lim_{h \to 0} \frac{24h + 3h^2}{h} = \lim_{h \to 0}(24 + 3h) = 24$$

The slope of the curve at the point (4, 48) is 24.

PROBLEM 2. Find the derivative of $f(x) = 3x^2$.

Answer: $f(x + h) = 3(x + h)^2 = 3x^2 + 6xh + 3h^2$. Use the definition of the derivative.

$$f'(x) = \lim_{h \to 0} \frac{3x^2 + 6xh + 3h^2 - 3x^2}{h}$$

Simplify.

$$\lim_{h \to 0} \frac{6xh + 3h^2}{h} = \lim_{h \to 0}(6x + 3h) = 6x$$

The derivative is $6x$.

PROBLEM 3. Find the derivative of $f(x) = x^3$.

Answer: $f(x + h) = (x + h)^3 = x^3 + 3x^2h + 3xh^2 + h^3$. First, use the definition of the derivative.

$$f'(x) = \lim_{h \to 0} \frac{x^3 + 3x^2h + 3xh^2 + h^3 - x^3}{h}$$

And simplify.

$$\lim_{h \to 0} \frac{3x^2h + 3xh^2 + h^3}{h} = \lim_{h \to 0} \left(3x^2 + 3xh + h^2\right) = 3x^2$$

The derivative is $3x^2$.

This next one will test your algebraic skills. Don't say we didn't warn you!

PROBLEM 4. Find the derivative of $f(x) = \sqrt{x}$.

Answer: $f(x + h) = \sqrt{x + h}$.

Use the definition of the derivative.

$$f'(x) = \lim_{h \to 0} \frac{\sqrt{x + h} - \sqrt{x}}{h}$$

Notice that this one doesn't cancel as conveniently as the other problems did. In order to simplify this expression, we have to multiply both the top and the bottom of the expression by $\sqrt{x + h} + \sqrt{x}$ (the conjugate of the numerator).

$$f'(x) = \lim_{h \to 0} \frac{\sqrt{x + h} - \sqrt{x}}{h}\left(\frac{\sqrt{x + h} + \sqrt{x}}{\sqrt{x + h} + \sqrt{x}}\right) = \lim_{h \to 0} \frac{x + h - x}{h\left(\sqrt{x + h} + \sqrt{x}\right)} = \lim_{h \to 0} \frac{h}{h\left(\sqrt{x + h} + \sqrt{x}\right)}$$

Simplify.

$$\lim_{h \to 0} \frac{1}{\left(\sqrt{x + h} + \sqrt{x}\right)} = \frac{1}{2\sqrt{x}}$$

The derivative is $\dfrac{1}{2\sqrt{x}}$.

PRACTICE PROBLEM SET 3

Now find the derivative of the following expressions. The answers are in Chapter 19.

1. $f(x) = 5x$ at $x = 3$

2. $f(x) = 4x$ at $x = -8$

3. $f(x) = 2x^2$ at $x = 5$

4. $f(x) = 5x^2$ at $x = -1$

5. $f(x) = 8x^2$

6. $f(x) = -10x^2$

7. $f(x) = 20x^2$ at $x = a$

8. $f(x) = 2x^3$ at $x = -3$

9. $f(x) = -3x^3$

10. $f(x) = x^4$

11. $f(x) = x^5$

12. $f(x) = 2\sqrt{x}$ at $x = 9$

13. $f(x) = 5\sqrt{2x}$ at $x = 8$

14. $f(x) = \sin x$ at $x = \dfrac{\pi}{3}$

15. $f(x) = \cos x$

16. $f(x) = x^2 + x$

17. $f(x) = x^3 + 3x + 2$

18. $f(x) = \dfrac{1}{x}$

19. $f(x) = ax^2 + bx + c \qquad : 2a + b$

20. $f(x) = \dfrac{1}{x^2} \qquad : -\dfrac{2}{x^3}$

Chapter 6
Basic
Differentiation

In calculus, you'll be asked to do two things: differentiate and integrate. In this section, you're going to learn differentiation. Integration will come later, in the second half of this book. Before we go about the business of learning how to take derivatives, however, here's a brief note about notation. Read this!

NOTATION

There are several different notations for derivatives in calculus. We'll use two different types interchangeably throughout this book, so get used to them now.

We'll refer to functions three different ways: $f(x)$, u or v, and y. For example, we might write: $f(x) = x^3$, $g(x) = x^4$, $h(x) = x^5$. We'll also use notation like: $u = \sin x$ and $v = \cos x$. Or we might use: $y = \sqrt{x}$. Usually, we pick the notation that causes the least confusion.

The derivatives of the functions will use notation that depends on the function, as shown in the following table:

Function	First Derivative	Second Derivative
$f(x)$	$f'(x)$	$f''(x)$
$g(x)$	$g'(x)$	$g''(x)$
y	y' or $\dfrac{dy}{dx}$	y'' or $\dfrac{d^2y}{dx^2}$

In addition, if we refer to a derivative of a function in general (for example, $ax^2 + bx + c$), we might enclose the expression in parentheses and use either of the following notations:

$$\left(ax^2 + bx + c\right)', \text{ or } \frac{d}{dx}\left(ax^2 + bx + c\right)$$

Sometimes math books refer to a derivative using either D_x or f_x. We're not going to use either of them.

THE POWER RULE

In the last chapter, you learned how to find a derivative using the definition of the derivative, a process that is very time-consuming and sometimes involves a lot of complex algebra. Fortunately, there's a shortcut to taking derivatives, so you'll never have to use the definition again—except when it's a question on an exam!

The basic technique for taking a derivative is called the **Power Rule**.

Rule No. 1: If $y = x^n$, then $\dfrac{dy}{dx} = nx^{n-1}$

That's it. Wasn't that simple? Of course, this and all of the following rules can be derived easily from the definition of the derivative. Look at these next few examples of the Power Rule in action.

Example 1: If $y = x^5$, then $\dfrac{dy}{dx} = 5x^4$.

Example 2: If $y = x^{20}$, then $\dfrac{dy}{dx} = 20x^{19}$.

Example 3: If $f(x) = x^{-5}$, then $f'(x) = -5x^{-6}$.

Example 4: If $u = x^{\frac{1}{2}}$, then $\dfrac{du}{dx} = \dfrac{1}{2}x^{-\frac{1}{2}}$.

Example 5: If $y = x^1$, then $\dfrac{dy}{dx} = 1x^0 = 1$. (Because x^0 is 1!)

Example 6: If $y = x^0$, then $\dfrac{dy}{dx} = 0$.

Notice that when the power of the function is negative, the power of the derivative is more negative.

When the power is a fraction, you should be careful to get the subtraction right (you'll see the powers $\dfrac{1}{2}, \dfrac{1}{3}, \dfrac{3}{2}, -\dfrac{1}{2}$, and $-\dfrac{1}{3}$ often, so be comfortable with subtracting 1 from them).

When the power is 1, the derivative is just a constant. When the power is 0, the derivative is 0.

This leads to the next three rules.

Rule No. 2: If $y = x$, then $\dfrac{dy}{dx} = 1$

Rule No. 3: If $y = kx$, then $\dfrac{dy}{dx} = k$ (where k is a constant)

Rule No. 4: If $y = k$, then $\dfrac{dy}{dx} = 0$ (where k is a constant)

Note: For future reference, a, b, c, n, and k always stand for constants.

Example 7: If $y = 8x^4$, then $\dfrac{dy}{dx} = 32x^3$.

Example 8: If $y = 5x^{100}$, then $y' = 500x^{99}$.

Example 9: If $y = -3x^{-5}$, then $\dfrac{dy}{dx} = 15x^{-6}$.

Example 10: If $f(x) = 7x^{\frac{1}{2}}$, then $f'(x) = \dfrac{7}{2}x^{-\frac{1}{2}}$.

Example 11: If $y = x\sqrt{15}$, then $\dfrac{dy}{dx} = \sqrt{15}$.

Example 12: If $y = 12$, then $\dfrac{dy}{dx} = 0$.

If you have any questions about any of these 12 examples (especially the last two), review the rules. Now for one last rule.

THE ADDITION RULE

If $y = ax^n + bx^m$, where a and b are constants, then

$$\frac{dy}{dx} = a\left(nx^{n-1}\right) + b\left(mx^{m-1}\right)$$

This handy rule works for subtraction, too.

Example 13: If $y = 3x^4 + 8x^{10}$, then $\dfrac{dy}{dx} = 12x^3 + 80x^9$.

Example 14: If $y = 7x^{-4} + 5x^{-\frac{1}{2}}$, then $\dfrac{dy}{dx} = -28x^{-5} - \dfrac{5}{2}x^{-\frac{3}{2}}$.

Example 15: If $y = 5x^4(2 - x^3)$, then $\dfrac{dy}{dx} = 40x^3 - 35x^6$.

Example 16: If $y = (3x^2 + 5)(x - 1)$, then

$$y = 3x^3 - 3x^2 + 5x - 5 \text{ and } \frac{dy}{dx} = 9x^2 - 6x + 5.$$

Example 17: If $y = ax^3 + bx^2 + cx + d$, then $\dfrac{dy}{dx} = 3ax^2 + 2bx + c$.

After you've worked through all 17 of these examples, you should be able to take the derivative of any polynomial with ease.

As you may have noticed from the examples above, in calculus, you are often asked to convert from fractions and radicals to negative powers and fractional powers. In addition, don't freak out if your answer doesn't match any of the answer choices. Because answers to problems are often presented in simplified form, your answer may not be simplified enough.

There are two basic expressions that you'll often be asked to differentiate. You can make your life easier by memorizing the following derivatives:

If $y = \dfrac{k}{x}$, then $\dfrac{dy}{dx} = -\dfrac{k}{x^2}$

If $y = k\sqrt{x}$, then $\dfrac{dy}{dx} = \dfrac{k}{2\sqrt{x}}$

HIGHER ORDER DERIVATIVES

This may sound like a big deal, but it isn't. This term refers only to taking the derivative of a function more than once. You don't have to stop at the first derivative of a function; you can keep taking derivatives. The derivative of a first derivative is called the second derivative. The derivative of the second derivative is called the third derivative, and so on.

Generally, you'll have to take only first and second derivatives.

Notice how we simplified the derivatives in the latter example? You should be able to do this mentally.

Function	First Derivative	Second Derivative
x^6	$6x^5$	$30x^4$
$8\sqrt{x}$	$\dfrac{4}{\sqrt{x}}$	$-2x^{-\frac{3}{2}}$

Here are some sample problems involving the rules we discussed above. As you work, cover the answers with an index card, and then check your work after you're done. By the time you finish them, you should know the rules by heart.

PROBLEM 1. If $y = 50x^5 + \dfrac{3}{x} - 7x^{-\frac{5}{3}}$, then $\dfrac{dy}{dx} =$

Answer: $\dfrac{dy}{dx} = 50\left(5x^4\right) + \left(-\dfrac{3}{x^2}\right) - 7\left(-\dfrac{5}{3}\right)x^{-\frac{8}{3}} = 250x^4 - \dfrac{3}{x^2} + \dfrac{35}{3}x^{-\frac{8}{3}}$

PROBLEM 2. If $y = 9x^4 + 6x^2 - 7x + 11$, then $\dfrac{dy}{dx} =$

Answer: $\dfrac{dy}{dx} = 9\left(4x^3\right) + 6\left(2x\right) - 7\left(1\right) + 0 = 36x^3 + 12x - 7$

PROBLEM 3. If $f(x) = 6x^{\frac{3}{2}} - 12\sqrt{x} - \dfrac{8}{\sqrt{x}} + 24x^{-\frac{3}{2}}$, then $f'(x) =$

Answer:

$f'(x) = 6\left(\dfrac{3}{2}x^{\frac{1}{2}}\right) - \left(\dfrac{12}{2\sqrt{x}}\right) - 8\left(-\dfrac{1}{2}x^{-\frac{3}{2}}\right) + 24\left(-\dfrac{3}{2}x^{-\frac{5}{2}}\right) = 9\sqrt{x} - \dfrac{6}{\sqrt{x}} + 4x^{-\frac{3}{2}} - 36x^{-\frac{5}{2}}$

How'd you do? Did you notice the changes in notation? How about the fractional powers, radical signs, and x's in denominators? You should be able to switch back and forth between notations, between fractional powers and radical signs, and between negative powers in a numerator and positive powers in a denominator.

PRACTICE PROBLEM SET 4

Find the derivative of each expression and simplify. The answers are in Chapter 19.

1. $(4x^2 + 1)^2$

2. $(x^5 + 3x)^2$

3. $11x^7$

4. $8x^{10}$

5. $18x^3 + 12x + 11$

6. $\dfrac{1}{2}(x^{12} + 17)$

7. $-\dfrac{1}{3}(x^9 + 2x^3 - 9)$

8. π^5

9. $\dfrac{1}{a}\left(\dfrac{1}{b}x^2 - \dfrac{2}{a}x - \dfrac{d}{x}\right)$

10. $-8x^{-8} + 12\sqrt{x}$

11. $6x^{-7} - 4\sqrt{x}$

12. $x^{-5} + \dfrac{1}{x^8}$

13. $\sqrt{x} + \dfrac{1}{x^3}$

14. $(6x^2 + 3)(12x - 4)$

15. $(3 - x - 2x^3)(6 + x^4)$

16. $e^{10} + \pi^3 - 7$

17. $\left(\dfrac{1}{x} + \dfrac{1}{x^2}\right)\left(\dfrac{4}{x^3} - \dfrac{6}{x^4}\right)$

18. $\sqrt{x} + \dfrac{1}{\sqrt{3}}$

19. $(x^2 + 8x - 4)(2x^{-2} + x^{-4})$

20. 0

21. $(x + 1)^3$

22. $\sqrt{x} + \sqrt[3]{x} + \sqrt[3]{x^2}$ $= \dfrac{1}{2\sqrt{x}} + \dfrac{1}{3\sqrt[3]{x^2}} + \dfrac{2}{3\sqrt[3]{x}}$

23. $x(2x + 7)(x - 2)$

24. $\sqrt{x}\left(\sqrt[3]{x} + \sqrt[5]{x}\right)$

25. $ax^5 + bx^4 + cx^3 + dx^2 + ex + f$

THE PRODUCT RULE

Now that you know how to find derivatives of simple polynomials, it's time to get more complicated. What if you had to find the derivative of this?

$$f(x) = (x^3 + 5x^2 - 4x + 1)(x^5 - 7x^4 + x)$$

You could multiply out the expression and take the derivative of each term, like

$$f(x) = x^8 - 2x^7 - 39x^6 + 29x^5 - 6x^4 + 5x^3 - 4x^2 + x$$

And the derivative is

$$f'(x) = 8x^7 - 14x^6 - 234x^5 + 145x^4 - 24x^3 + 15x^2 - 8x + 1$$

Needless to say, this process is messy. Naturally, there's an easier way. When a function involves two terms multiplied by each other, we use the **Product Rule**.

$$\text{The Product Rule: If } f(x) = uv, \text{ then } f'(x) = u\frac{dv}{dx} + v\frac{du}{dx}$$

To find the derivative of two things multiplied by each other, you multiply the first function by the derivative of the second, and add that to the second function multiplied by the derivative of the first.

Let's use the Product Rule to find the derivative of our example.

$$f'(x) = (x^3 + 5x^2 - 4x + 1)(5x^4 - 28x^3 + 1) + (x^5 - 7x^4 + x)(3x^2 + 10x - 4)$$

If we were to simplify this, we'd get the same answer as before. But here's the best part: We're not going to simplify it. One of the great things about the AP exam is that when it's difficult to simplify an expression, you almost never have to. Nonetheless, you'll often need to simplify expressions when you're taking second derivatives, or when you use the derivative in some other equation. Practice simplifying whenever possible.

Example 1: $f(x) = (9x^2 + 4x)(x^3 - 5x^2)$

$$f'(x) = (9x^2 + 4x)(3x^2 - 10x) + (x^3 - 5x^2)(18x + 4)$$

Example 2: $y = \left(\sqrt{x} + 4\sqrt[3]{x}\right)\left(x^5 - 11x^8\right)$

$$y' = \left(\sqrt{x} + 4\sqrt[3]{x}\right)\left(5x^4 - 88x^7\right) + \left(x^5 - 11x^8\right)\left(\frac{1}{2\sqrt{x}} + \frac{4}{3\sqrt[3]{x^2}}\right)$$

Example 3: $y = \left(\frac{1}{x} + \frac{1}{x^2} - \frac{1}{x^3}\right)\left(\frac{1}{x} - \frac{1}{x^3} + \frac{1}{x^5}\right)$

$$y' = \left(\frac{1}{x} + \frac{1}{x^2} - \frac{1}{x^3}\right)\left(-\frac{1}{x^2} + \frac{3}{x^4} - \frac{5}{x^6}\right) + \left(\frac{1}{x} - \frac{1}{x^3} + \frac{1}{x^5}\right)\left(-\frac{1}{x^2} - \frac{2}{x^3} + \frac{3}{x^4}\right)$$

With the product rule, the order of these two operations doesn't matter. It does matter with other rules, though, so it helps to use the same order each time.

THE QUOTIENT RULE

What happens when you have to take the derivative of a function that is the quotient of two other functions? You guessed it: Use the **Quotient Rule**.

The Quotient Rule: If $f(x) = \dfrac{u}{v}$, then $f'(x) = \dfrac{v\dfrac{du}{dx} - u\dfrac{dv}{dx}}{v^2}$

In this rule, as opposed to the Product Rule, the order in which you take the derivatives is very important, because you're subtracting instead of adding. It's always the bottom function times the derivative of the top minus the top function times the derivative of the bottom. Then divide the whole thing by the bottom function squared. A good way to remember this is to say the following:

$$\frac{"LoDeHi - HiDeLo"}{(Lo)^2}$$

You could also write

$$f(x) = \frac{u}{v} \text{ as } f(x) = u\frac{1}{v}$$

For the Quotient Rule, remember that order matters!

Then you could derive the Quotient Rule using the Product Rule.

$$f'(x) = u\left(-\frac{1}{v^2}\frac{dv}{dx}\right) + \frac{du}{dx}\frac{1}{v} = \frac{v\dfrac{du}{dx} - u\dfrac{dv}{dx}}{v^2}$$

Here are some more examples.

Example 4: $f(x) = \dfrac{\left(x^5 - 3x^4\right)}{\left(x^2 + 7x\right)}$

$$f'(x) = \frac{\left(x^2 + 7x\right)\left(5x^4 - 12x^3\right) - \left(x^5 - 3x^4\right)\left(2x + 7\right)}{\left(x^2 + 7x\right)^2}$$

Example 5: $y = \dfrac{\left(x^{-3} - x^{-8}\right)}{\left(x^{-2} + x^{-6}\right)}$

$$\frac{dy}{dx} = \frac{\left(x^{-2} + x^{-6}\right)\left(-3x^{-4} + 8x^{-9}\right) - \left(x^{-3} - x^{-8}\right)\left(-2x^{-3} - 6x^{-7}\right)}{\left(x^{-2} + x^{-6}\right)^2}$$

We're not going to simplify these, although the Quotient Rule often produces expressions that simplify more readily than those involving the Product Rule. Sometimes it's helpful to simplify, but avoid it otherwise. When you have to find a second derivative, however, you do have to simplify the quotient. If this is the case, the AP exam usually will give you a simple expression to deal with, such as in the example below.

Example 6: $y = \dfrac{3x+5}{5x-3}$

$$\frac{dy}{dx} = \frac{(5x-3)(3) - (3x+5)(5)}{(5x-3)^2} = \frac{(15x-9) - (15x+25)}{(5x-3)^2} = \frac{-34}{(5x-3)^2}$$

In order to take the derivative of $\dfrac{dy}{dx}$, you have to use the Chain Rule.

THE CHAIN RULE

The most important rule in this chapter (and sometimes the most difficult one) is called the **Chain Rule**. It's used when you're given composite functions—that is, a function inside of another function. You'll always see one of these on the AP exam, so it's important to know the Chain Rule cold.

A composite function is usually written as: $f(g(x))$.

For example: If $f(x) = \dfrac{1}{x}$ and $g(x) = \sqrt{3x}$, then $f\left(g(x)\right) = \dfrac{1}{\sqrt{3x}}$

We could also find: $g\left(f(x)\right) = \sqrt{\dfrac{3}{x}}$

When finding the derivative of a composite function, we take the derivative of the "outside" function, with the inside function g considered as the variable, leaving the "inside" function alone. Then, we multiply this by the derivative of the "inside" function, with respect to its variable x.

Here is another way to write the Chain Rule.

$$\text{The Chain Rule: } \text{If } y = f\big(g(x)\big), \text{ then } y' = \left(\frac{df\big(g(x)\big)}{dg}\right)\left(\frac{dg}{dx}\right)$$

This rule is tricky, so here are several examples. The last couple incorporate the Product Rule and the Quotient Rule.

Example 7: If $y = (5x^3 + 3x)^5$, then $\dfrac{dy}{dx} = 5(5x^3 + 3x)^4(15x^2 + 3)$

We just dealt with the derivative of something to the fifth power, like this:

$$y = (g)^5, \text{ so } \frac{dy}{dg} = 5(g)^4, \text{ where } g = 5x^3 + 3x$$

Then we multiplied by the derivative of g: $(15x^2 + 3)$.

Always do it this way. The process has several successive steps, like peeling away the layers of an onion until you reach the center.

Example 8: If $y = \sqrt{x^3 - 4x}$, then $\dfrac{dy}{dx} = \dfrac{1}{2}\left(x^3 - 4x\right)^{-\frac{1}{2}}\left(3x^2 - 4\right)$

Again, we took the derivative of the outside function, leaving the inside alone. Then we multiplied by the derivative of the inside.

Example 9: If $y = \sqrt{\left(x^5 - 8x^3\right)\left(x^2 + 6x\right)}$, then

$$\frac{dy}{dx} = \frac{1}{2}\left[\left(x^5 - 8x^3\right)\left(x^2 + 6x\right)\right]^{-\frac{1}{2}}\left[\left(x^5 - 8x^3\right)(2x + 6) + \left(x^2 + 6x\right)\left(5x^4 - 24x^2\right)\right]$$

Messy, isn't it? That's because we used the Chain Rule and the Product Rule. Now for one with the Chain Rule and the Quotient Rule.

Example 10: If $y = \left(\dfrac{2x+8}{x^2-10x} \right)^5$, then

$$\frac{dy}{dx} = 5\left[\frac{2x+8}{x^2-10x} \right]^4 \left[\frac{\left(x^2-10x\right)(2)-(2x+8)(2x-10)}{\left(x^2-10x\right)^2} \right]$$

Example 11: If $y = \sqrt{5x^3+x}$, then $\dfrac{dy}{dx} = \dfrac{1}{2}\left(5x^3+x\right)^{-\frac{1}{2}}\left(15x^2+1\right)$

Now we use the Product Rule and the Chain Rule to find the second derivative.

$$\frac{d^2y}{dx^2} = \frac{1}{2}\left(5x^3+x\right)^{-\frac{1}{2}}(30x) + \left(15x^2+1\right)\left[-\frac{1}{4}\left(5x^3+x\right)^{-\frac{3}{2}}\left(15x^2+1\right) \right]$$

You can also simplify this further, if necessary.

There's another representation of the Chain Rule that you need to learn.

> If $y = y(v)$ and $v = v(x)$, then $\dfrac{dy}{dx} = \dfrac{dy}{dv}\dfrac{dv}{dx}$

Example 12: $y = 8v^2 - 6v$ and $v = 5x^3 - 11x$, then

$$\frac{dy}{dx} = (16v - 6)(15x^2 - 11)$$

Then substitute for v.

$$\frac{dy}{dx} = \left(16\left(5x^3-11x\right)-6\right)\left(15x^2-11\right) = \left(80x^3-176x-6\right)\left(15x^2-11\right)$$

Here are some solved problems. Cover the answers first, then check your work.

As you can see, these grow quite complex, so we simplify these only as a last resort. If you must simplify, the AP exam will have only a very simple Chain Rule problem.

PROBLEM 1. Find $\dfrac{dy}{dx}$ if $y = \left(5x^4 + 3x^7\right)\left(x^{10} - 8x\right)$.

Answer: $\dfrac{dy}{dx} = \left(5x^4 + 3x^7\right)\left(10x^9 - 8\right) + \left(x^{10} - 8x\right)\left(20x^3 + 21x^6\right)$

PROBLEM 2. Find $\dfrac{dy}{dx}$ if $y = \left(x^3 + 3x^2 + 3x + 1\right)\left(x^2 + 2x + 1\right)$.

Answer: $\dfrac{dy}{dx} = \left(x^3 + 3x^2 + 3x + 1\right)\left(2x + 2\right) + \left(x^2 + 2x + 1\right)\left(3x^2 + 6x + 3\right)$

PROBLEM 3. Find $\dfrac{dy}{dx}$ if $y = \left(\sqrt{x} + \dfrac{1}{x}\right)\left(\sqrt[3]{x^2} - \dfrac{1}{x^3}\right)$.

Answer: $\dfrac{dy}{dx} = \left(\sqrt{x} + \dfrac{1}{x}\right)\left(\dfrac{2}{3}x^{-\frac{1}{3}} + \dfrac{3}{x^4}\right) + \left(\sqrt[3]{x^2} - \dfrac{1}{x^3}\right)\left(\dfrac{1}{2\sqrt{x}} - \dfrac{1}{x^2}\right)$

PROBLEM 4. Find $\dfrac{dy}{dx}$ if $y = \left(x^3 + 1\right)\left(x^2 + 5x - \dfrac{1}{x^5}\right)$.

Answer: $\dfrac{dy}{dx} = \left(x^3 + 1\right)\left(2x + 5 + \dfrac{5}{x^6}\right) + \left(x^2 + 5x - \dfrac{1}{x^5}\right)\left(3x^2\right)$

PROBLEM 5. Find $\dfrac{dy}{dx}$ if $y = \dfrac{2x - 4}{x^2 - 6}$.

Answer: $\dfrac{dy}{dx} = \dfrac{\left(x^2 - 6\right)(2) - (2x - 4)(2x)}{\left(x^2 - 6\right)^2} = \dfrac{-2x^2 + 8x - 12}{\left(x^2 - 6\right)^2}$

PROBLEM 6. Find $\dfrac{dy}{dx}$ if $y = \dfrac{x^2 + 1}{x^2 + x + 4}$.

Answer: $\dfrac{dy}{dx} = \dfrac{\left(x^2 + x + 4\right)(2x) - \left(x^2 + 1\right)(2x + 1)}{\left(x^2 + x + 4\right)^2} = \dfrac{x^2 + 6x - 1}{\left(x^2 + x + 4\right)^2}$

Problem 7. Find $\dfrac{dy}{dx}$ if $y = \dfrac{x+5}{x-5}$.

Answer: $\dfrac{dy}{dx} = \dfrac{(x-5)(1)-(x+5)(1)}{(x-5)^2} = \dfrac{-10}{(x-5)^2}$

Problem 8. Find $\dfrac{dy}{dx}$ if $y = (x^4 + x)^2$.

Answer: $\dfrac{dy}{dx} = 2\left(x^4 + x\right)\left(4x^3 + 1\right)$

Problem 9. Find $\dfrac{dy}{dx}$ if $y = \left(\dfrac{x+3}{x-3}\right)^3$.

Answer: $\dfrac{dy}{dx} = 3\left(\dfrac{x+3}{x-3}\right)^2\left(\dfrac{(x-3)(1)-(x+3)(1)}{(x-3)^2}\right) = -18\left(\dfrac{(x+3)^2}{(x-3)^4}\right)$

Problem 10. Find $\dfrac{dy}{dx}$ at $x = 1$ if $y = \left[\left(x^3 + x\right)\left(x^4 - x^2\right)\right]^2$.

Answer: $\dfrac{dy}{dx} = 2\left[\left(x^3 + x\right)\left(x^4 - x^2\right)\right]\left[\left(x^3 + x\right)\left(4x^3 - 2x\right)+\left(x^4 - x^2\right)\left(3x^2 + 1\right)\right]$

Once again, plug in right away. Never simplify until after you've substituted.

At $x = 1$, $\dfrac{dy}{dx} = 0$.

PRACTICE PROBLEM SET 5

Simplify when possible. The answers are in Chapter 19.

1. Find $f'(x)$ if $f(x) = \left(\dfrac{4x^3 - 3x^2}{5x^7 + 1} \right)$.

2. Find $f'(x)$ if $f(x) = \left(x^2 - 4x + 3\right)(x + 1)$.

3. Find $f'(x)$ if $f(x) = (x + 1)^{10}$.

4. Find $f'(x)$ if $f(x) = 8\sqrt{\left(x^4 - 4x^2\right)}$.

5. Find $f'(x)$ if $f(x) = \left(\dfrac{x}{x^2 + 1} \right)^3$.

6. Find $f'(x)$ if $f(x) = \sqrt[4]{\left(\dfrac{2x - 5}{5x + 2} \right)}$.

7. Find $f'(x)$ if $f(x) = \dfrac{4x^8 - \sqrt{x}}{8x^4}$.

8. Find $f'(x)$ if $f(x) = \left(x + \dfrac{1}{x} \right)\left(x^2 - \dfrac{1}{x^2} \right)$.

9. Find $f'(x)$ if $f(x) = \left(\dfrac{x}{x + 1} \right)^4$.

10. Find $f'(x)$ if $f(x) = \left(x^2 + x\right)^{100}$.

11. Find $f'(x)$ if $f(x) = \sqrt{\dfrac{x^2 + 1}{x^2 - 1}}$.

12. Find $f'(x)$ at $x = 2$ if $f(x) = \dfrac{(x + 4)(x - 8)}{(x + 6)(x - 6)}$.

13. Find $f'(x)$ at $x = 1$ if $f(x) = \dfrac{x^6 + 4x^3 + 6}{\left(x^4 - 2\right)^2}$.

14. Find $f'(x)$ at $x = 1$ if $f(x) = \left[\dfrac{x - \sqrt{x}}{x + \sqrt{x}}\right]^2$.

15. Find $f'(x)$ if $f(x) = \dfrac{x^2 - 3}{(x - 3)}$.

16. Find $f'(x)$ at $x = 1$ if $f(x) = \left(x^4 - x^2\right)\left(2x^3 + x\right)$.

17. Find $f'(x)$ at $x = 2$ if $f(x) = \dfrac{x^2 + 2x}{x^4 - x^3}$.

18. Find $f'(x)$ if $f(x) = \sqrt{x^4 + x^2}$.

19. Find $f'(x)$ at $x = 1$ if $f(x) = \dfrac{x}{\left(1 + x^2\right)^2}$.

20. Find $\dfrac{dy}{dx}$ if $y = u^2 - 1$ and $u = \dfrac{1}{x - 1}$.

21. Find $\dfrac{dy}{dx}$ at $x = 1$ if $y = \dfrac{t^2 + 2}{t^2 - 2}$ and $t = x^3$.

22. Find $\dfrac{dy}{dt}$ if $y = \left(x^6 - 6x^5\right)\left(5x^2 + x\right)$ and $x = \sqrt{t}$.

23. Find $\dfrac{du}{dv}$ at $v = 2$ if $u = \sqrt{x^3 + x^2}$ and $x = \dfrac{1}{v}$.

24. Find $\dfrac{dy}{dx}$ at $x = 1$ if $y = \dfrac{1 + u}{1 + u^2}$ and $u = x^2 - 1$.

25. Find $\dfrac{du}{dv}$ if $u = y^3$, $y = \dfrac{x}{x + 8}$, and $x = v^2$.

DERIVATIVES OF TRIG FUNCTIONS

There are a lot of trigonometry problems in calculus and on the AP Calculus Exam. If you're not sure of your trig, you should definitely go to the Appendix and review the unit on Prerequisite Math. You'll need to remember your trig formulas, the values of the special angles, and the trig ratios, among other stuff.

In addition, angles are *always* referred to in radians. You can forget all about using degrees.

You should know the derivatives of all six trig functions. The good news is that the derivatives are pretty easy, and all you have to do is memorize them. Because the AP exam might ask you about this, though, let's use the definition of the derivative to figure out the derivative of sin x.

If $f(x) = \sin x$, then $f(x + h) = \sin(x + h)$.

Substitute this into the definition of the derivative.

$$\lim_{h \to 0} \frac{f(x+h) - f(x)}{h} = \lim_{h \to 0} \frac{\sin(x+h) - \sin x}{h}$$

Remember that $\sin(x + h) = \sin x \cos h + \cos x \sin h$. Now simplify it.

$$\lim_{h \to 0} \frac{\sin x \cos h + \cos x \sin h - \sin x}{h}$$

Next, rewrite this as

$$\lim_{h \to 0} \frac{\sin x (\cos h - 1) + \cos x \sin h}{h} = \lim_{h \to 0} \frac{\sin x (\cos h - 1)}{h} + \lim_{h \to 0} \frac{\cos x \sin h}{h}$$

Next, use some of the trigonometric limits that you memorized back in Chapter 3. Specifically,

$$\lim_{h \to 0} \frac{(\cos h - 1)}{h} = 0 \text{ and } \lim_{h \to 0} \frac{\sin h}{h} = 1$$

This gives you

$$\lim_{h \to 0} \frac{\sin x (\cos h - 1)}{h} + \lim_{h \to 0} \frac{\cos x \sin h}{h} = \sin x (0) + \cos x (1) = \cos x$$

$$\frac{d}{dx} \sin x = \cos x$$

Example 1: Find the derivative of $\sin \left(\dfrac{\pi}{2} - x \right)$.

$$\frac{d}{dx} \sin \left(\frac{\pi}{2} - x \right) = \cos \left(\frac{\pi}{2} - x \right)(-1) = -\cos \left(\frac{\pi}{2} - x \right)$$

Use some of the rules of trigonometry you remember from last year. Because

$$\sin \left(\frac{\pi}{2} - x \right) = \cos x \text{ and } \cos \left(\frac{\pi}{2} - x \right) = \sin x,$$

you can substitute into the above expression and get

$$\frac{d}{dx} \cos x = -\sin x$$

Now, let's derive the derivatives of the other four trigonometric functions.

Example 2: Find the derivative of $\dfrac{\sin x}{\cos x}$.

Use the Quotient Rule.

$$\frac{d}{dx} \frac{\sin x}{\cos x} = \frac{(\cos x)(\cos x) - (\sin x)(-\sin x)}{(\cos x)^2} = \frac{\cos^2 x + \sin^2 x}{\cos^2 x} = \frac{1}{\cos^2 x} = \sec^2 x$$

Because $\dfrac{\sin x}{\cos x} = \tan x$, you should get

$$\frac{d}{dx}\tan x = \sec^2 x$$

Example 3: Find the derivative of $\dfrac{\cos x}{\sin x}$.

Use the Quotient Rule.

$$\frac{d}{dx}\frac{\cos x}{\sin x} = \frac{(\sin x)(-\sin x)-(\cos x)(\cos x)}{(\sin x)^2} = \frac{-\left(\cos^2 x + \sin^2 x\right)}{\sin^2 x} = -\frac{1}{\sin^2 x} = -\csc^2 x$$

Because $\dfrac{\cos x}{\sin x} = \cot x$, you get: $\dfrac{d}{dx}\cot x = -\csc^2 x$.

Example 4: Find the derivative of $\dfrac{1}{\cos x}$.

Use the Reciprocal Rule.

$$\frac{d}{dx}\frac{1}{\cos x} = \frac{-1}{(\cos x)^2}\left(-\sin x\right) = \frac{\sin x}{\cos^2 x} = \frac{1}{\cos x}\frac{\sin x}{\cos x} = \sec x \tan x$$

Because $\dfrac{1}{\cos x} = \sec x$, you get

$$\frac{d}{dx}\sec x = \sec x \tan x$$

Example 5: Find the derivative of $\dfrac{1}{\sin x}$.

You get the idea by now.

$$\frac{d}{dx}\frac{1}{\sin x} = \frac{-1}{(\sin x)^2}(\cos x) = \frac{-\cos x}{\sin^2 x} = \frac{-1}{\sin x}\frac{\cos x}{\sin x} = -\csc x \cot x$$

Because $\dfrac{1}{\sin x} = \csc x$, you get

$$\frac{d}{dx}\csc x = -\csc x \cot x$$

There you go. We have now found the derivatives of all six of the trigonometric functions. (A chart of them appears at the end of the book.) Now memorize them. You'll thank us later.

Let's do some more examples.

Example 6: Find the derivative of $\sin(5x)$.

$$\frac{d}{dx}\sin(5x) = \cos(5x)(5) = 5\cos(5x)$$

Example 7: Find the derivative of $\sec(x^2)$.

$$\frac{d}{dx}\sec(x^2) = \sec(x^2)\tan(x^2)(2x)$$

Example 8: Find the derivative of $\csc(x^3 - 5x)$.

$$\frac{d}{dx}\csc(x^3 - 5x) = -\csc(x^3 - 5x)\cot(x^3 - 5x)(3x^2 - 5).$$

These derivatives are almost like formulas. You just follow the pattern and use the Chain Rule when appropriate.

Here are some solved problems. Do each problem, covering the answer first, then checking your answer.

PROBLEM 1. Find $f'(x)$ if $f(x) = \sin(2x^3)$.

Answer: Follow the rule: $f'(x) = \cos(2x^3)6x^2$

PROBLEM 2. Find $f'(x)$ if $f(x) = \cos\left(\sqrt{3x}\right)$.

Answer: $f'(x) = -\sin\left(\sqrt{3x}\right)\left[\dfrac{1}{2}(3x)^{-\frac{1}{2}}(3)\right] = \dfrac{-3\sin\left(\sqrt{3x}\right)}{2\sqrt{3x}}$

PROBLEM 3. Find $f'(x)$ if $f(x) = \tan\left(\dfrac{x}{x+1}\right)$.

Answer: $f'(x) = \sec^2\left(\dfrac{x}{x+1}\right)\left[\dfrac{(x+1)-x}{(x+1)^2}\right] = \left(\dfrac{1}{(x+1)^2}\right)\sec^2\left(\dfrac{x}{x+1}\right)$

PROBLEM 4. Find $f'(x)$ if $f(x) = \csc(x^3 + x + 1)$.

Answer: Follow the rule: $f'(x) = -\csc(x^3 + x + 1)\cot(x^3 + x + 1)(3x^2 + 1)$

GRAPHICAL DERIVATIVES

Sometimes, we are given the graph of a function and we are asked to graph the derivative. We do this by analyzing the sign of derivative at various places on the graph and then sketching a graph of the derivative from that information. Let's start with something simple. Suppose we have the graph of $y = f(x)$ below and we are asked to sketch the graph of its derivative.

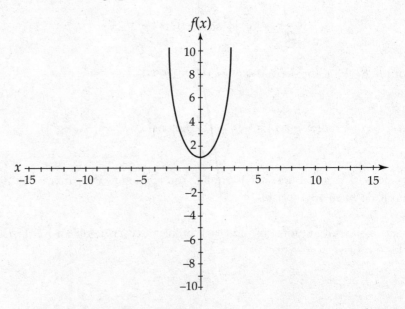

First, note that the derivative is zero at the point (0, 1) because the tangent line is horizontal there. Next, the derivative is negative for all $x < 0$ because the tangent lines to the curve have negative slopes everywhere on the interval $(-\infty, 0)$. Finally, the derivative is positive for all $x > 0$ because the tangent lines to the curve have positive slopes everywhere on the interval $(0, \infty)$. Now we can make a graph of the derivative. It will go through the origin (because the derivative is 0 at $x = 0$), it will be negative on the interval $(-\infty, 0)$ and it will be positive on the interval $(0, \infty)$. The graph looks something like the following:

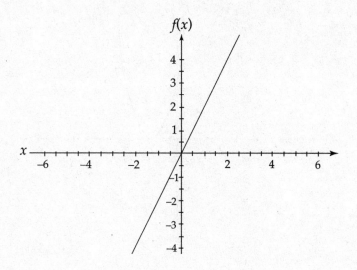

Note that it's not important for your graph to be exact. All we are doing here is sketching the derivative. The important parts of the graph are where the derivative is positive, negative, and zero.

Now let's try something a little harder. Suppose we have the following graph and we are asked to sketch the graph of the derivative:

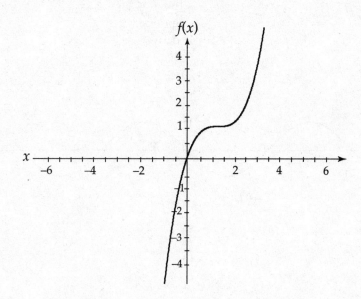

Notice that the tangent line looks as if it's horizontal at $x = 1$. This means that the graph of the derivative is zero there. Next, the curve is increasing for all other values of x, so the graph of the derivative will be positive. As we go from left to right on the graph, notice that the slope starts out very steep, so the derivatives are large positive numbers. As we approach $x = 1$, the curve starts to flatten out, so the derivatives will approach zero but will still be positive. Then, the slope is zero at $x = 1$. Then the curve gets steep again. If we sketch the derivative, we get something like the following:

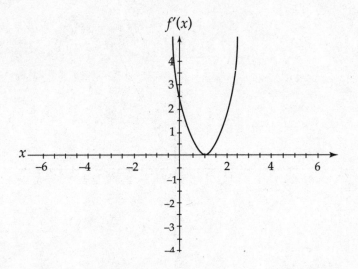

Here is an important one to understand. Suppose we have the graph of $y = \sin x$.

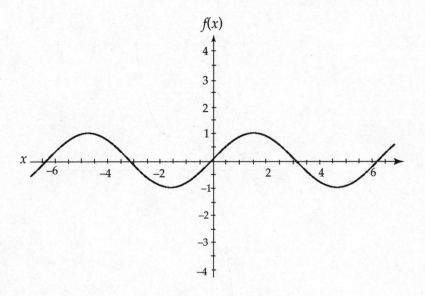

Notice that the slope of the tangent line will be horizontal at all of the maxima and minima of the graph. Because the slope of a horizontal line is zero, this means that the derivative will be zero at those values $(\pm\frac{\pi}{2},\frac{3\pi}{2},\ldots)$. Next, notice that the slope of the curve is about 1 as the curve goes through the origin. This should make sense if you recall that $\lim\limits_{x\to0}\dfrac{\sin x}{x}=1$. The slope of the curve is about -1 as the curve goes through $x=\pi$. And so on. If we now sketch the derivative, it looks something like the following:

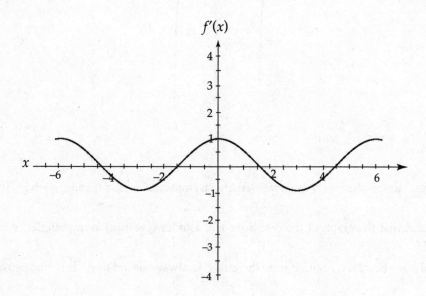

Notice that this is the graph of $y=\cos x$. This should be obvious because the derivative of $\sin x$ is $\cos x$.

Now let's do a hard one. Suppose we have the following graph:

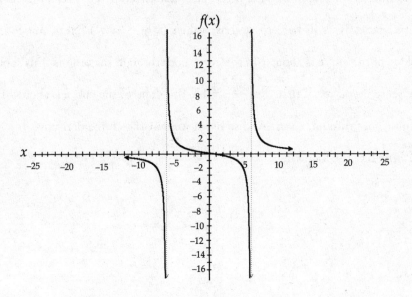

First, notice that we have two vertical asymptotes at $x = 6$ and $x = -6$. This means that the graph of the derivative will also have vertical asymptotes at $x = 6$ and $x = -6$. Next, notice that the curve is always decreasing. This means that the graph of the derivative will always be negative. Moving from left to right, the graph starts out close to flat, so the derivative will be close to zero. Then, the graph gets very steep and points downward, so the graph of the derivative will be negative and getting more negative. Then, we have the asymptote $x = -6$. Next, the graph begins very steep and negative and starts to flatten out as we approach the origin. At the origin, the slope of the graph is approximately $-\frac{1}{2}$. This means that the graph of the derivative will increase until it reaches $\left(0, -\frac{1}{2}\right)$. Then, the graph starts to get steep again as we approach the other asymptote $x = 6$. Thus, the graph will get more negative again. Finally, to the right of the asymptote $x = 6$ the graph starts out steep and negative and flattens out, approaching zero. This means that the graph of the derivative will start out very negative and will approach zero. If we now sketch the derivative, it looks something like the following:

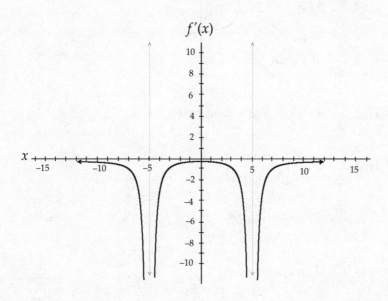

PRACTICE PROBLEM SET 6

Now try these problems. The answers are in Chapter 19.

1. Find $\dfrac{dy}{dx}$ if $y = \sin^2 x$. $= 2\sin x \cos x$

2. Find $\dfrac{dy}{dx}$ if $y = \cos x^2$. $= -\sin x^2 (2x)$

3. Find $\dfrac{dy}{dx}$ if $y = (\tan x)(\sec x)$. $= \sec^2 x \sec x + \tan x \sec x \tan x$

4. Find $\dfrac{dy}{dx}$ if $y = \cot 4x$. $= -\csc 4x \cot 4x \cdot 4$

5. Find $\dfrac{dy}{dx}$ if $y = \sqrt{\sin 3x}$. $= \dfrac{1}{2\sqrt{\sin 3x}} \cdot \cos 3x \cdot 3$

6. Find $\dfrac{dy}{dx}$ if $y = \dfrac{1 + \sin x}{1 - \sin x}$.

7. Find $\dfrac{dy}{dx}$ if $y = \csc^2 x^2$.

8. Find $\dfrac{dy}{dx}$ if $y = 2 \sin 3x \cos 4x$.

9. Find $\dfrac{d^4 y}{dx^4}$ if $y = \sin 2x$.

10. Find $\dfrac{dy}{dx}$ if $y = \sin t - \cos t$ and $t = 1 + \cos^2 x$.

11. Find $\dfrac{dy}{dx}$ if $y = \left(\dfrac{\tan x}{1 - \tan x}\right)^2$.

12. Find $\dfrac{dr}{d\theta}$ if $r = \sec\theta \tan 2\theta$.

13. Find $\dfrac{dr}{d\theta}$ if $r = \cos(1 + \sin\theta)$.

14. Find $\dfrac{dr}{d\theta}$ if $r = \dfrac{\sec\theta}{1 + \tan\theta}$.

15. Find $\dfrac{dy}{dx}$ if $y = \left(1 + \cot\left(\dfrac{2}{x}\right)\right)^{-2}$.

16. Find $\dfrac{dy}{dx}$ if $y = \sin\left(\cos\left(\sqrt{x}\right)\right)$.

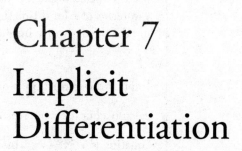

Chapter 7
Implicit
Differentiation

HOW TO DO IT

By now, it should be easy for you to take the derivative of an equation such as $y = 3x^5 - 7x$. If you're given an equation such as $y^2 = 3x^5 - 7x$, you can still figure out the derivative by taking the square root of both sides, which gives you y in terms of x. This is known as finding the derivative **explicitly**. It's messy, but possible.

If you have to find the derivative of $y^2 + y = 3x^5 - 7x$, you don't have an easy way to get y in terms of x, so you can't differentiate this equation using any of the techniques you've learned so far. That's because each of those previous techniques needs to be used on an equation in which y is in terms of x. When you can't isolate y in terms of x (or if isolating y makes taking the derivative a nightmare), it's time to take the derivative **implicitly**.

Implicit differentiation is one of the simpler techniques you need to learn to do in calculus, but for some reason it gives many students trouble. Suppose you have the equation $y^2 = 3x^5 - 7x$. This means that the value of y is a function of the value of x. When we take the derivative, $\dfrac{dy}{dx}$, we're looking at the rate at which y changes as x changes. Thus, given $y = x^2 + x$, when we write

$$\frac{dy}{dx} = 2x + 1$$

we're saying that "the rate" at which y changes, with respect to how x changes, is $2x + 1$.

Now, suppose you want to find $\dfrac{dx}{dy}$. As you might imagine

$$\frac{dx}{dy} = \frac{1}{\dfrac{dy}{dx}}$$

So here, $\dfrac{dx}{dy} = \dfrac{1}{2x+1}$. But notice that this derivative is in terms of x, not y, and you need to find the derivative with respect to y. This derivative is an **implicit** one. When you can't isolate the variables of an equation, you often end up with a derivative that is in terms of both variables.

Another way to think of this is that there is a hidden term in the derivative, $\dfrac{dx}{dx}$, and when we take the derivative, what we really get is

$$\frac{dy}{dx} = 2x\left(\frac{dx}{dx}\right) + 1\left(\frac{dx}{dx}\right)$$

A fraction that has the same term in its numerator and denominator is equal to 1, so we write

$$\frac{dy}{dx} = 2x(1) + 1(1) = 2x + 1$$

Every time we take a derivative of a term with x in it, we multiply by the term $\dfrac{dx}{dx}$, but because this is 1, we ignore it. Suppose however, that we wanted to find out how y changes with respect to t (for time). Then we would have

$$\frac{dy}{dt} = 2x\left(\frac{dx}{dt}\right) + 1\left(\frac{dx}{dt}\right)$$

If we wanted to find out how y changes with respect to r, we would have

$$\frac{dy}{dr} = 2x\left(\frac{dx}{dr}\right) + 1\left(\frac{dx}{dr}\right)$$

and if we wanted to find out how y changes with respect to y, we would have

$$\frac{dy}{dy} = 2x\left(\frac{dx}{dy}\right) + 1\left(\frac{dx}{dy}\right) \text{ or } 1 = 2x\frac{dx}{dy} + \frac{dx}{dy}$$

This is how we really do differentiation. Remember the following:

$$\frac{dx}{dy} = \frac{1}{\dfrac{dy}{dx}}$$

When you have an equation of x in terms of y, and you want to find the derivative with respect to y, simply differentiate. But if the equation is of y in terms of x, find $\dfrac{dy}{dx}$ and take its reciprocal to find $\dfrac{dx}{dy}$. Go back to our original example.

$$y^2 + y = 3x^5 - 7x$$

You should use implicit differentiation any time you can't write a function explicitly in terms of the variable that we want to take the derivative with respect to.

To take the derivative according to the information in the last paragraph, you get

$$2y\left(\frac{dy}{dx}\right) + 1\left(\frac{dy}{dx}\right) = 15x^4\left(\frac{dx}{dx}\right) - 7\left(\frac{dx}{dx}\right)$$

Notice how each variable is multiplied by its appropriate $\frac{d}{dx}$. Now, remembering that $\frac{dx}{dx} = 1$, rewrite the expression this way: $2y\left(\frac{dy}{dx}\right) + 1\left(\frac{dy}{dx}\right) = 15x^4 - 7$.

Next, factor $\frac{dy}{dx}$ out of the left-hand side: $\frac{dy}{dx}(2y+1) = 15x^4 - 7$.

Isolating $\frac{dy}{dx}$ gives you $\frac{dy}{dx} = \frac{15x^4 - 7}{(2y+1)}$.

This is the derivative you're looking for. Notice how the derivative is defined in terms of y **and** x. Up until now, $\frac{dy}{dx}$ has been strictly in terms of x. This is why the differentiation is "implicit."

Confused? Let's do a few examples and you will get the hang of it.

Example 1: Find $\frac{dy}{dx}$ if $y^3 - 4y^2 = x^5 + 3x^4$.

Using implicit differentiation, you get

$$3y^2\left(\frac{dy}{dx}\right) - 8y\left(\frac{dy}{dx}\right) = 5x^4\left(\frac{dx}{dx}\right) + 12x^3\left(\frac{dx}{dx}\right)$$

Remember that $\frac{dx}{dx} = 1$: $\frac{dy}{dx}\left(3y^2 - 8y\right) = 5x^4 + 12x^3$.

After you factor out $\frac{dy}{dx}$, divide both sides by $3y^2 - 8y$.

$$\frac{dy}{dx} = \frac{5x^4 + 12x^3}{\left(3y^2 - 8y\right)}$$

Note: Now that you understand that the derivative of an x term with respect to x will always be multiplied by $\frac{dx}{dx}$, and that $\frac{dx}{dx} = 1$, we won't write $\frac{dx}{dx}$ anymore. You should understand that the term is implied.

Example 2: Find $\dfrac{dy}{dx}$ if $\sin y^2 - \cos x^2 = \cos y^2 + \sin x^2$.

Use implicit differentiation.

$$\cos y^2 \left(2y\dfrac{dy}{dx} \right) + \sin x^2 \left(2x \right) = -\sin y^2 \left(2y\dfrac{dy}{dx} \right) + \cos x^2 \left(2x \right)$$

Then simplify.

$$2y\cos y^2 \left(\dfrac{dy}{dx} \right) + 2x\sin x^2 = -2y\sin y^2 \left(\dfrac{dy}{dx} \right) + 2x\cos x^2$$

Next, put all of the terms containing $\dfrac{dy}{dx}$ on the left and all of the other terms on the right.

$$2y\cos y^2 \left(\dfrac{dy}{dx} \right) + 2y\sin y^2 \left(\dfrac{dy}{dx} \right) = -2x\sin x^2 + 2x\cos x^2$$

Next, factor out $\dfrac{dy}{dx}$.

$$\dfrac{dy}{dx} \left(2y\cos y^2 + 2y\sin y^2 \right) = -2x\sin x^2 + 2x\cos x^2$$

And isolate $\dfrac{dy}{dx}$.

$$\dfrac{dy}{dx} = \dfrac{-2x\sin x^2 + 2x\cos x^2}{\left(2y\cos y^2 + 2y\sin y^2 \right)}$$

This can be simplified further to the following:

$$\dfrac{dy}{dx} = \dfrac{-x\left(\sin x^2 - \cos x^2 \right)}{y\left(\cos y^2 + \sin y^2 \right)}$$

Example 3: Find $\dfrac{dy}{dx}$ if $3x^2 + 5xy^2 - 4y^3 = 8$.

Implicit differentiation should result in the following:

$$6x + \left[5x\left(2y\dfrac{dy}{dx} \right) + (5)y^2 \right] - 12y^2 \left(\dfrac{dy}{dx} \right) = 0$$

Did you notice the use of the Product Rule to find the derivative of $5xy^2$? The AP exam loves to make you do this. All of the same differentiation rules that you've learned up until now still apply. We're just adding another technique.

You can simplify this to

$$6x + 10xy\frac{dy}{dx} + 5y^2 - 12y^2\frac{dy}{dx} = 0$$

Next, put all of the terms containing $\frac{dy}{dx}$ on the left and all of the other terms on the right.

$$10xy\frac{dy}{dx} - 12y^2\frac{dy}{dx} = -6x - 5y^2$$

Next, factor out $\frac{dy}{dx}$.

$$\left(10xy - 12y^2\right)\frac{dy}{dx} = -6x - 5y^2$$

Then, isolate $\frac{dy}{dx}$.

$$\frac{dy}{dx} = \frac{-6x - 5y^2}{\left(10xy - 12y^2\right)}$$

Example 4: Find the derivative of $3x^2 - 4y^2 + y = 9$ at (2, 1).

You need to use implicit differentiation to find $\frac{dy}{dx}$.

$$6x - 8y\left(\frac{dy}{dx}\right) + \left(\frac{dy}{dx}\right) = 0$$

Now, instead of rearranging to isolate $\frac{dy}{dx}$, plug in (2, 1) immediately and solve for the derivative.

$$6(2) - 8(1)\left(\frac{dy}{dx}\right) + \left(\frac{dy}{dx}\right) = 0$$

Be smart about your problem solving. Just because you can simplify something doesn't mean that you should. In a case like this, plugging into this form of the derivative is more effective.

Simplify: $12 - 7\left(\frac{dy}{dx}\right) = 0$, so $\frac{dy}{dx} = \frac{12}{7}$

Getting the hang of implicit differentiation yet? We hope so, because these next examples are slightly harder.

Example 5: Find the derivative of $\dfrac{2x-5y^2}{4y^3-x^2} = -x$ at (1, 1).

First, cross-multiply.

$$2x-5y^2 = -x\left(4y^3-x^2\right)$$

Distribute.

$$2x - 5y^2 = -4xy^3 + x^3$$

Take the derivative.

$$2-10y\frac{dy}{dx} = -4x\left(3y^2\frac{dy}{dx}\right)-4y^3+3x^2$$

Do not simplify now. Rather, plug in (1, 1) right away. This will save you from the algebra.

$$2-10(1)\frac{dy}{dx} = -4(1)\left(3(1)^2\frac{dy}{dx}\right)-4(1)^3+3(1)^2$$

Now solve for $\dfrac{dy}{dx}$.

$$2-10\frac{dy}{dx} = -12\frac{dy}{dx}-1$$

$$2\frac{dy}{dx} = -3$$

$$\frac{dy}{dx} = \frac{-3}{2}$$

SECOND DERIVATIVES

Sometimes, you'll be asked to find a second derivative implicitly.

Example 6: Find $\dfrac{d^2y}{dx^2}$ if $y^2 + 2y = 4x^2 + 2x$.

Differentiating implicitly, you get

$$2y\frac{dy}{dx} + 2\frac{dy}{dx} = 8x + 2$$

Remember: When it is required to take a second derivative, the first derivative should be simplified first.

Next, simplify and solve for $\dfrac{dy}{dx}$.

$$\frac{dy}{dx} = \frac{4x+1}{y+1}$$

Now, it's time to take the derivative again.

$$\frac{d^2y}{dx^2} = \frac{4(y+1) - (4x+1)\left(\dfrac{dy}{dx}\right)}{(y+1)^2}$$

Finally, substitute for $\dfrac{dy}{dx}$.

$$\frac{4(y+1) - (4x+1)\left(\dfrac{4x+1}{y+1}\right)}{(y+1)^2} = \frac{4(y+1)^2 - (4x+1)^2}{(y+1)^3}$$

Try these solved problems without looking at the answers. Then check your work.

PROBLEM 1. Find $\dfrac{dy}{dx}$ if $x^2 + y^2 = 6xy$.

Answer: Differentiate with respect to x.

$$2x + 2y\frac{dy}{dx} = 6x\frac{dy}{dx} + 6y$$

Group all of the $\frac{dy}{dx}$ terms on the left and the other terms on the right.

$$2y\frac{dy}{dx} - 6x\frac{dy}{dx} = 6y - 2x$$

Now factor out $\frac{dy}{dx}$.

$$\frac{dy}{dx}(2y - 6x) = 6y - 2x$$

Therefore, the first derivative is the following:

$$\frac{dy}{dx} = \frac{6y - 2x}{2y - 6x} = \frac{3y - x}{y - 3x}$$

PROBLEM 2. Find $\frac{dy}{dx}$ if $x - \cos y = xy$

Answer: Differentiate with respect to x.

$$1 + \sin y\frac{dy}{dx} = x\frac{dy}{dx} + y$$

Grouping the terms, you get

$$\sin y\frac{dy}{dx} - x\frac{dy}{dx} = y - 1$$

Now factor out $\frac{dy}{dx}$.

$$\frac{dy}{dx}(\sin y - x) = y - 1$$

The derivative is

$$\frac{dy}{dx} = \frac{y - 1}{\sin y - x}$$

PROBLEM 3. Find the derivative of each variable with respect to t of $x^2 + y^2 = z^2$.

Answer: $2x\dfrac{dx}{dt} + 2y\dfrac{dy}{dt} = 2z\dfrac{dz}{dt}$

PROBLEM 4. Find the derivative of each variable with respect to t of $V = \dfrac{1}{3}\pi r^2 h$.

Answer: $\dfrac{dV}{dt} = \dfrac{1}{3}\pi\left(r^2\dfrac{dh}{dt} + 2r\dfrac{dr}{dt}h\right)$

PROBLEM 5. Find $\dfrac{d^2y}{dx^2}$ if $y^2 = x^2 - 2x$.

Answer: First, take the derivative with respect to x.

$$2y\frac{dy}{dx} = 2x - 2$$

Then, solve for $\dfrac{dy}{dx}$.

$$\frac{dy}{dx} = \frac{2x-2}{2y} = \frac{x-1}{y}$$

The second derivative with respect to x becomes

$$\frac{d^2y}{dx^2} = \frac{y(1) - (x-1)\dfrac{dy}{dx}}{y^2}$$

Now substitute for $\dfrac{dy}{dx}$ and simplify.

$$\frac{d^2y}{dx^2} = \frac{y - (x-1)\left(\dfrac{x-1}{y}\right)}{y^2} = \frac{y^2 - (x-1)^2}{y^3}$$

PRACTICE PROBLEM SET 7

Use implicit differentiation to find the following derivatives. The answers are in Chapter 19.

1. Find $\dfrac{dy}{dx}$ if $x^3 - y^3 = y$.

2. Find $\dfrac{dy}{dx}$ if $x^2 - 16xy + y^2 = 1$.

3. Find $\dfrac{dy}{dx}$ at $(2, 1)$ if $\dfrac{x+y}{x-y} = 3$.

4. Find $\dfrac{dy}{dx}$ if $\cos y - \sin x = \sin y - \cos x$.

5. Find $\dfrac{dy}{dx}$ if $16x^2 - 16xy + y^2 = 1$ at $(1, 1)$.

6. Find $\dfrac{dy}{dx}$ if $x^{\frac{1}{2}} + y^{\frac{1}{2}} = 2y^2$ at $(1, 1)$.

7. Find $\dfrac{dy}{dx}$ if $x\sin y + y\sin x = \dfrac{\pi}{2\sqrt{2}}$ at $\left(\dfrac{\pi}{4}, \dfrac{\pi}{4}\right)$.

8. Find $\dfrac{d^2 y}{dx^2}$ if $x^2 + 4y^2 = 1$.

9. Find $\dfrac{d^2 y}{dx^2}$ if $\sin x + 1 = \cos y$.

10. Find $\dfrac{d^2 y}{dx^2}$ if $x^2 - 4x = 2y - 2$.

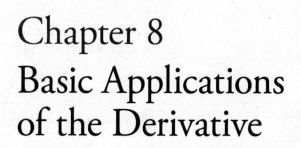

Chapter 8
Basic Applications
of the Derivative

EQUATIONS OF TANGENT LINES AND NORMAL LINES

Finding the equation of a line tangent to a certain curve at a certain point is a standard calculus problem. This is because, among other things, the derivative is the slope of a tangent line to a curve at a particular point. Thus, we can find the equation of the tangent line to a curve if we have the equation of the curve and the point at which we want to find the tangent line. Then all we have to do is take the derivative of the equation, plug in the x-coordinate of the point to find the slope, then use the point and the slope to find the equation of the line. Let's take this one step at a time.

Suppose we have a point (x_1, y_1) and a slope m. Then the equation of the line through that point with that slope is

$$(y - y_1) = m(x - x_1)$$

You should remember this formula from algebra. If not, memorize it!

Next, suppose that we have an equation $y = f(x)$, where (x_1, y_1) satisfies that equation. Then $f'(x_1) = m$, and we can plug all of our values into the equation for a line and get the equation of the tangent line. This is much easier to explain with a simple example.

Example 1: Find the equation of the tangent line to the curve $y = 5x^2$ at the point (3, 45).

First of all, notice that the point satisfies the equation: when $x = 3$, $y = 45$. Now, take the derivative of the equation.

$$\frac{dy}{dx} = 10x$$

By the way, the notation for plugging in a point is $\big|_{x=}$. Learn to recognize it!

Now, if you plug in $x = 3$, you'll get the slope of the curve at that point.

$$\frac{dy}{dx}\bigg|_{x=3} = 10(3) = 30$$

Thus, we have the slope and the point, and the equation is

$$(y - 45) = 30(x - 3)$$

It's customary to simplify the equation if it's not too onerous.

$$y = 30x - 45$$

Example 2: Find the equation of the tangent line to $y = x^3 + x^2$ at $(3, 36)$.

The derivative looks like the following:

$$\frac{dy}{dx} = 3x^2 + 2x$$

So, the slope is

$$\left.\frac{dy}{dx}\right|_{x=3} = 3(3)^2 + 2(3) = 33$$

The equation looks like the following:

$$(y - 36) = 33(x - 3), \text{ or } y = 33x - 63$$

Naturally, there are a couple of things that can be done to make the problems harder. First of all, you can be given only the x-coordinate. Second, the equation can be more difficult to differentiate.

In order to find the y-coordinate, all you have to do is plug the x-value into the equation for the curve and solve for y. Remember this: You'll see it again!

Example 3: Find the equation of the tangent line to $y = \dfrac{2x+5}{x^2-3}$ at $x = 1$.

First, find the y-coordinate.

$$y\,(1) = \frac{2(1)+5}{1^2-3} = -\frac{7}{2}$$

Second, take the derivative.

$$\frac{dy}{dx} = \frac{(x^2-3)(2)-(2x+5)(2x)}{(x^2-3)^2}$$

You're probably dreading having to simplify this derivative. Don't waste your time! Plug in $x = 1$ right away.

$$\left.\frac{dy}{dx}\right|_{x=1} = \frac{\left(1^2-3\right)(2)-\left(2(1)+5\right)\left(2(1)\right)}{\left(1^2-3\right)^2} = \frac{-4-14}{4} = -\frac{9}{2}$$

Now, we have a slope and a point, so the equation is

$$y + \frac{7}{2} = -\frac{9}{2}(x-1), \text{ or } 2y = -9x + 2$$

Remember: the slope of a perpendicular line is just the negative reciprocal of the slope of the tangent line.

Sometimes, instead of finding the equation of a tangent line, you will be asked to find the equation of a normal line. A **normal** line is simply the line perpendicular to the tangent line at the same point. You follow the same steps as with the tangent line, but you use the slope that will give you a perpendicular line.

Example 4: Find the equation of the line normal to $y = x^5 - x^4 + 1$ at $x = 2$.

First, find the y-coordinate.

$$y(2) = 2^5 - 2^4 + 1 = 17$$

Second, take the derivative.

$$\frac{dy}{dx} = 5x^4 - 4x^3$$

Third, find the slope at $x = 2$.

$$\left.\frac{dy}{dx}\right|_{x=2} = 5(2)^4 - 4(2)^3 = 48$$

Fourth, take the negative reciprocal of 48, which is $-\frac{1}{48}$.

Finally, the equation becomes

$$y - 17 = -\frac{1}{48}(x-2)$$

Try these solved problems. Do each problem, covering the answer first, then checking your answer.

PROBLEM 1. Find the equation of the tangent line to the graph of $y = 4 - 3x - x^2$ at the point $(2, -6)$.

Answer: First, take the derivative of the equation.

$$\frac{dy}{dx} = -3 - 2x$$

Now, plug in $x = 2$ to get the slope of the tangent line.

$$\frac{dy}{dx} = -3 - 2(2) = -7$$

Third, plug the slope and the point into the equation for the line.

$$y - (-6) = -7\,(x - 2)$$

This simplifies to $y = -7x + 8$.

PROBLEM 2. Find the equation of the normal line to the graph of $y = 6 - x - x^2$ at $x = -1$.

Answer: Plug $x = -1$ into the original equation to get the y-coordinate.

$$y = 6 + 1 - 1 = 6$$

Once again, take that derivative.

$$\frac{dy}{dx} = -1 - 2x$$

Now plug in $x = -1$ to get the slope of the tangent.

$$\frac{dy}{dx} = -1 - 2(-1) = 1$$

Use the negative reciprocal of the slope in the second step to get the slope of the normal line.

$$m = -1$$

Finally, plug the slope and the point into the equation for the line.

$$y - 6 = -1\,(x + 1)$$

This simplifies to $y = -x + 5$.

PROBLEM 3. Find the equations of the tangent and normal lines to the graph of $y = \dfrac{10x}{x^2+1}$ at the point (2, 4).

Answer: This problem will put your algebra to the test. You have to use the Quotient Rule to take the derivative of this mess.

$$\frac{dy}{dx} = \frac{\left(x^2+1\right)(10) - (10x)(2x)}{\left(x^2+1\right)^2}$$

Second, plug in $x = 2$ to get the slope of the tangent.

$$\frac{dy}{dx} = \frac{(5)(10)-(20)(4)}{5^2} = -\frac{30}{25} = -\frac{6}{5}$$

Now, plug the slope and the point into the equation for the tangent line.

$$y - 4 = -\frac{6}{5}(x-2)$$

That simplifies to $6x + 5y = 32$. The equation of the normal line must then be

$$y - 4 = \frac{5}{6}(x-2)$$

That, in turn, simplifies to $-5x + 6y = 14$.

PROBLEM 4. The curve $y = ax^2 + bx + c$ passes through the point (2, 4) and is tangent to the line $y = x + 1$ at (0, 1). Find a, b, and c.

Answer: The curve passes through (2, 4), so if you plug in $x = 2$, you'll get $y = 4$. Therefore,

$$4 = 4a + 2b + c$$

Second, the curve also passes through the point (0, 1), so $c = 1$.

Because the curve is tangent to the line $y = x + 1$ at (0, 1), they must both have the same slope at that point. The slope of the line is 1. The slope of the curve is the first derivative.

$$\frac{dy}{dx} = 2ax + b$$

$$\left.\frac{dy}{dx}\right|_{x=0} = 2a(0) + b = b$$

At (0, 1), $\frac{dy}{dx} = b$. Therefore, $b = 1$.

Now that you know b and c, plug them back into the equation from the first step and solve for a.

$$4 = 4a + 2 + 1, \text{ and } a = \frac{1}{4}$$

PROBLEM 5. Find the points on the curve $y = 2x^3 - 3x^2 - 12x + 20$ where the tangent is parallel to the x-axis.

Answer: The x-axis is a horizontal line, so it has slope zero. Therefore, you want to know where the derivative of this curve is zero. Take the derivative.

$$\frac{dy}{dx} = 6x^2 - 6x - 12$$

Set it equal to zero and solve for x. Get accustomed to doing this: It's one of the most common questions in differential calculus.

$$\frac{dy}{dx} = 6x^2 - 6x - 12 = 0$$

$$6(x^2 - x - 2) = 0$$

$$6(x - 2)(x + 1) = 0$$

$$x = 2 \text{ or } x = -1$$

Third, find the y-coordinates of these two points.

$$y = 2(8) - 3(4) - 12(2) + 20 = 0$$

$$y = 2(-1) - 3(1) - 12(-1) + 20 = 27$$

Therefore, the points are (2, 0) and (−1, 27).

PRACTICE PROBLEM SET 8

Now try these problems. The answers are in Chapter 19.

1. Find the equation of the tangent to the graph of $y = 3x^2 - x$ at $x = 1$.

2. Find the equation of the tangent to the graph of $y = x^3 - 3x$ at $x = 3$.

3. Find the equation of the normal to the graph of $y = \sqrt{8x}$ at $x = 2$.

4. Find the equation of the tangent to the graph of $y = \dfrac{1}{\sqrt{x^2 + 7}}$ at $x = 3$.

5. Find the equation of the normal to the graph of $y = \dfrac{x+3}{x-3}$ at $x = 4$.

6. Find the equation of the tangent to the graph of $y = 4 - 3x - x^2$ at $(0, 4)$.

7. Find the equation of the tangent to the graph of $y = 2x^3 - 3x^2 - 12x + 20$ at $x = 2$.

8. Find the equation of the tangent to the graph of $y = \dfrac{x^2 + 4}{x - 6}$ at $x = 5$.

9. Find the equation of the tangent to the graph of $y = \sqrt{x^3 - 15}$ at $(4, 7)$.

10. Find the equation of the tangent to the graph of $y = (x^2 + 4x + 4)^2$ at $x = -2$.

11. Find the values of x where the tangent to the graph of $y = 2x^3 - 8x$ has a slope equal to the slope of $y = x$.

12. Find the equation of the normal to the graph of $y = \dfrac{3x+5}{x-1}$ at $x = 3$.

13.   Find the values of x where the normal to the graph of $(x-9)^2$ is parallel to the y-axis.

14.   Find the coordinates where the tangent to the graph of $y = 8 - 3x - x^2$ is parallel to the x-axis.

15.   Find the values of a, b, and c where the curves $y = x^2 + ax + b$ and $y = cx + x^2$ have a common tangent line at $(-1, 0)$.

$$y' = 2x + a$$
$$y' = c + 2x$$

THE MEAN VALUE THEOREM FOR DERIVATIVES

Remember that in order for The Mean Value Theorem for Derivatives to work, the curve must be continuous on the interval <u>and</u> at the endpoints.

If $y = f(x)$ is continuous on the interval $[a, b]$, and is differentiable everywhere on the interval (a, b), then there is at least one number c between a and b such that

$$f'(c) = \frac{f(b) - f(a)}{b - a}$$

In other words, there's some point in the interval where the slope of the tangent line equals the slope of the secant line that connects the endpoints of the interval. (The function has to be continuous at the endpoints of the interval, but it doesn't have to be differentiable at the endpoints. Is this important? Maybe to mathematicians, but probably not to you!) You can see this graphically in the following figure:

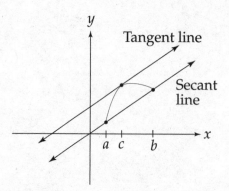

Example 1: Suppose you have the function $f(x) = x^2$, and you're looking at the interval $[1, 3]$. The Mean Value Theorem for derivatives (this is often abbreviated MVTD) states that there is some number c such that

$$f'(c) = \frac{3^2 - 1^2}{3 - 1} = 4$$

Note: This is a great way to self-check your work. Always look at whether your answer makes sense.

Because $f'(x) = 2x$, plug in c for x and solve: $2c = 4$ so $c = 2$. Notice that 2 is in the interval. This is what the MVTD predicted! If you don't get a value for c within the interval, something went wrong; either the function is not continuous and differentiable in the required interval, or you made a mistake.

Example 2: Consider the function $f(x) = x^3 - 12x$ on the interval [–2, 2]. The MVTD states that there is a c such that

$$f'(c) = \frac{\left(2^3 - 24\right) - \left((-2)^3 + 24\right)}{2 - (-2)} = -8$$

Then $f'(c) = 3c^2 - 12 = -8$ and $c = \pm\dfrac{2}{\sqrt{3}}$ (which is approximately ±1.155).

Notice that here there are two values of c that satisfy the MVTD. That's allowed. In fact, there can be infinitely many values, depending on the function.

Example 3: Consider the function $f(x) = \dfrac{1}{x}$ on the interval [–2, 2].

Follow the MVTD.

$$f'(c) = \frac{\dfrac{1}{2} - \left(-\dfrac{1}{2}\right)}{2 - (-2)} = \frac{1}{4}$$

Then

$$f'(c) = \frac{-1}{c^2} = \frac{1}{4}$$

There is no value of c that will satisfy this equation! We expected this. Why? Because $f(x)$ is not continuous at $x = 0$, which is in the interval. Suppose the interval had been [1, 3], eliminating the discontinuity. The result would have been

$$f'(c) = \frac{\dfrac{1}{3} - (1)}{3 - 1} = -\frac{1}{3} \text{ and } f'(c) = \frac{-1}{c^2} = -\frac{1}{3}; \ c = \pm\sqrt{3}$$

$c = -\sqrt{3}$ is not in the interval, but $c = \sqrt{3}$ is. The answer is $c = \sqrt{3}$.

Example 4: Consider the function $f(x) = x^2 - x - 12$ on the interval [–3, 4].

Follow the MVTD.

$$f'(c) = \frac{0 - 0}{7} = 0 \text{ and } f'(c) = 2c - 1 = 0 \text{ , so } c = \frac{1}{2}$$

In this last example, you discovered where the derivative of the equation equaled zero. This is going to be the single most common problem you'll encounter in differential calculus. So now, we've got an important tip for you.

> When you don't know what to do, take the derivative of the equation and set it equal to zero!!!

Remember this advice for the rest of AP calculus.

ROLLE'S THEOREM

Now let's learn Rolle's theorem, which is a special case of the MVTD.

> If $y = f(x)$ is continuous on the interval $[a, b]$, and is differentiable everywhere on the interval (a, b), and if $f(a) = f(b) = 0$, then there is at least one number c between a and b such that $f'(c) = 0$.

Graphically, this means that a continuous, differentiable curve has a horizontal tangent between any two points where it crosses the x-axis.

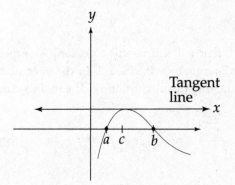

Example 4 was an example of Rolle's theorem, but let's do another.

Example 5: Consider the function $f(x) = \dfrac{x^2}{2} - 6x$ on the interval $[0, 12]$.

First, show that

$$f(0) = \frac{0}{2} - 6(0) = 0 \text{ and } f(12) = \frac{144}{2} - 6(12) = 0$$

Then find

$$f'(x) = x - 6, \text{ so } f'(c) = c - 6$$

If you set this equal to zero (remember what we told you!), you get $c = 6$. This value of c falls in the interval, so the theorem holds for this example.

As with the MVTD, you'll run into problems with the theorem when the function is not continuous and differentiable over the interval. This is where you need to look out for a trap set by ETS. Otherwise, just follow what we did here and you won't have any trouble with either Rolle's theorem or the MVTD. Try these example problems, and cover the responses until you check your work.

PROBLEM 1. Find the values of c that satisfy the MVTD for $f(x) = x^2 + 2x - 1$ on the interval $[0, 1]$.

Answer: First, find $f(0)$ and $f(1)$.

$$f(0) = 0^2 + 2(0) - 1 = -1 \text{ and } f(1) = 1^2 + 2(1) - 1 = 2$$

Then,

$$\frac{2 - (-1)}{1 - 0} = \frac{3}{1} = 3 = f'(c)$$

Next, find $f'(x)$.

$$f'(x) = 2x + 2$$

Thus, $f'(c) = 2c + 2 = 3$, and $c = \dfrac{1}{2}$.

PROBLEM 2. Find the values of c that satisfy the MVTD for $f(x) = x^3 + 1$ on the interval $[1, 2]$.

Answer: Find $f(1) = 1^3 + 1 = 2$ and $f(2) = 2^3 + 1 = 9$. Then,

$$\frac{9 - 2}{2 - 1} = 7 = f'(c)$$

Next, $f'(x) = 3x^2$, so $f'(c) = 3c^2 = 7$ and $c = \pm\sqrt{\dfrac{7}{3}}$.

Notice that there are two answers for c, but only one of them is in the interval. The answer is $c = \sqrt{\dfrac{7}{3}}$.

PROBLEM 3. Find the values of c that satisfy the MVTD for $f(x) = x + \dfrac{1}{x}$ on the interval $[-4, 4]$.

Answer: First, because the function is not continuous on the interval, there may not be a solution for c. Let's show that this is true. Find $f(-4) = -4 - \dfrac{1}{4} = -\dfrac{17}{4}$ and $f(4) = 4 + \dfrac{1}{4} = \dfrac{17}{4}$. Then,

$$\frac{\dfrac{17}{4} - \left(-\dfrac{17}{4}\right)}{4 - (-4)} = \frac{17}{16} = f'(c)$$

Next, $f'(x) = 1 - \dfrac{1}{x^2}$. Therefore, $f'(c) = 1 - \dfrac{1}{c^2} = \dfrac{17}{16}$.

There's no solution to this equation.

PROBLEM 4. Find the values of c that satisfy Rolle's theorem for $f(x) = x^4 - x$ on the interval $[0, 1]$.

Answer: Show that $f(0) = 0^4 - 0 = 0$ and that $f(1) = 1^4 - 1 = 0$.

Next, find $f'(x) = 4x^3 - 1$. By setting $f'(c) = 4c^3 - 1 = 0$ and solving, you'll see that $c = \sqrt[3]{\dfrac{1}{4}}$, which is in the interval.

PRACTICE PROBLEM SET 9

Now try these problems. The answers are in Chapter 19.

1. Find the values of c that satisfy the MVTD for $f(x) = 3x^2 + 5x - 2$ on the interval $[-1, 1]$.

2. Find the values of c that satisfy the MVTD for $f(x) = x^3 + 24x - 16$ on the interval $[0, 4]$.

3. Find the values of c that satisfy the MVTD for $f(x) = x^3 + 12x^2 + 7x$ on the interval $[-4, 4]$.

4. Find the values of c that satisfy the MVTD for $f(x) = \dfrac{6}{x} - 3$ on the interval $[1, 2]$.

5. Find the values of c that satisfy the MVTD for $f(x) = \dfrac{6}{x} - 3$ on the interval $[-1, 2]$.

6. Find the values of c that satisfy Rolle's theorem for $f(x) = x^2 - 8x + 12$ on the interval $[2, 6]$.

7. Find the values of c that satisfy Rolle's theorem for $f(x) = x^3 - x$ on the interval $[-1, 1]$.

8. Find the values of c that satisfy Rolle's theorem for $f(x) = x(1 - x)$ on the interval $[0, 1]$.

9. Find the values of c that satisfy Rolle's theorem for $f(x) = 1 - \dfrac{1}{x^2}$ on the interval $[-1, 1]$.

10. Find the values of c that satisfy Rolle's theorem for $f(x) = x^{\frac{2}{3}} - x^{\frac{1}{3}}$ on the interval $[0, 1]$.

Chapter 9
Maxima and
Minima

Here's another chapter of material involving more ways to apply the derivative to several other types of problems. This stuff focuses mainly on using the derivative to aid in graphing a function, etc.

APPLIED MAXIMA AND MINIMA PROBLEMS

One of the most common applications of the derivative is to find a maximum or minimum value of a function. These values can be called extreme values, optimal values, or critical points. Each of these problems involves the same, very simple principle.

> A maximum or a minimum of a function occurs at a point where the derivative of a function is zero, or where the derivative fails to exist.

At a point where the first derivative equals zero, the curve has a horizontal tangent line, at which point it could be reaching either a "peak" (maximum) or a "valley" (minimum).

There are a few exceptions to every rule. This rule is no different.

If the derivative of a function is zero at a certain point, it is usually a maximum or minimum—but not always.

There are two different kinds of maxima and minima: relative and absolute. A **relative** or **local** maximum or minimum means that the curve has a horizontal tangent line at that point, but it is not the highest or lowest value that the function attains. In the figure to the right, the two indicated points are relative maxima/minima.

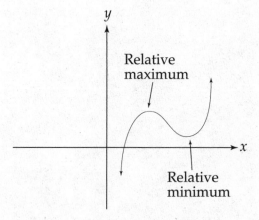

An **absolute** maximum or minimum occurs either at an artificial point or an end point. In the figure below, the two indicated points are absolute maxima/minima. A relative maximum can also be an absolute maximum.

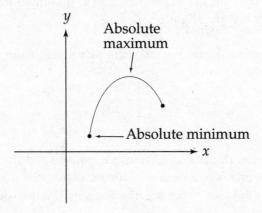

A typical word problem will ask you to find a maximum or a minimum value of a function, as it pertains to a certain situation. Sometimes you're given the equation; other times, you have to figure it out for yourself. Once you have the equation, you find its derivative and set it equal to zero. The values you get are called critical values. That is, if $f'(c) = 0$ or $f'(c)$ does not exist, then c is a critical value. Then, test these values to determine whether each value is a maximum or a minimum. The simplest way to do this is with the second derivative test.

If a function has a critical value at $x = c$, then that value is a relative maximum if $f''(c) < 0$ and it is a relative minimum if $f''(c) > 0$.

If the second derivative is also zero at $x = c$, then the point is neither a maximum nor a minimum but a point of inflection. More about that later.

It's time to do some examples.

Example 1: Find the minimum value on the curve $y = ax^2$, if $a > 0$.

Take the derivative and set it equal to zero.

$$\frac{dy}{dx} = 2ax = 0$$

The first derivative is equal to zero at $x = 0$. By plugging 0 back into the original equation, we can solve for the y-coordinate of the minimum (the y-coordinate is also 0, so the point is at the origin).

In order to determine if this is a maximum or a minimum, take the second derivative.

$$\frac{d^2 y}{dx^2} = 2a$$

Because a is positive, the second derivative is positive and the critical point we obtained from the first derivative is a minimum point. Had a been negative, the second derivative would have been negative and a maximum would have occurred at the critical point.

Example 2: A manufacturing company has determined that the total cost of producing an item can be determined from the equation $C = 8x^2 - 176x + 1,800$, where x is the number of units that the company makes. How many units should the company manufacture in order to minimize the cost?

Once again, take the derivative of the cost equation and set it equal to zero.

$$\frac{dC}{dx} = 16x - 176 = 0$$

$$x = 11$$

This tells us that 11 is a critical point of the equation. Now we need to figure out if this is a maximum or a minimum using the second derivative.

$$\frac{d^2 C}{dx^2} = 16$$

Because 16 is always positive, any critical value is going to be a minimum. Therefore, the company should manufacture 11 units in order to minimize its cost.

Example 3: A rocket is fired into the air, and its height in meters at any given time t can be calculated using the formula $h(t) = 1,600 + 196t - 4.9t^2$. Find the maximum height of the rocket and the time at which it occurs.

Take the derivative and set it equal to zero.

$$\frac{dh}{dt} = 196 - 9.8t$$

$$t = 20$$

Now that we know 20 is a critical point of the equation, use the second derivative test.

$$\frac{d^2h}{dt^2} = -9.8$$

This is always negative, so any critical value is a maximum. To determine the maximum height of the rocket, plug $t = 20$ into the equation.

$$h(20) = 1,600 + 196(20) - 4.9(20^2) = 3,560 \text{ meters}$$

The technique is always the same: (a) take the derivative of the equation; (b) set it equal to zero; and (c) use the second derivative test.

The hardest part of these word problems is when you have to set up the equation yourself. The following is a classic AP problem:

Example 4: Max wants to make a box with no lid from a rectangular sheet of cardboard that is 18 inches by 24 inches. The box is to be made by cutting a square of side x from each corner of the sheet and folding up the sides (see figure below). Find the value of x that maximizes the volume of the box.

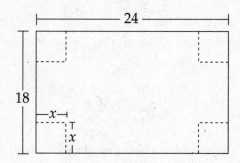

After we cut out the squares of side x and fold up the sides, the dimensions of the box will be

width: $18 - 2x$
length: $24 - 2x$
depth: x

Using the formula for the volume of a rectangular prism, we can get an equation for the volume in terms of x.

$$V = x(18 - 2x)(24 - 2x)$$

Multiply the terms together (and be careful with your algebra).

$$V = x(18 - 2x)(24 - 2x) = 4x^3 - 84x^2 + 432x$$

Now take the derivative.

$$\frac{dV}{dx} = 12x^2 - 168x + 432$$

Set the derivative equal to zero, and solve for x.

$$12x^2 - 168x + 432 = 0$$

$$x^2 - 14x + 36 = 0$$

$$x = \frac{14 \pm \sqrt{196 - 144}}{2} = 7 \pm \sqrt{13} \approx 3.4, 10.6$$

At the end of the day, no matter how complex the math might get, if a problem is based on a real world example, like this cardboard box, then the answer will make sense in reality.

Common sense tells us that you can't cut out two square pieces that measure 10.6 inches to a side (the sheet's only 18 inches wide!), so the maximizing value has to be 3.4 inches. Here's the second derivative test, just to be sure.

$$\frac{d^2V}{dx^2} = 24x - 168$$

At $x = 3.4$,

$$\frac{d^2V}{dx^2} = -86.4$$

So, the volume of the box will be maximized when $x = 3.4$.

Therefore, the dimensions of the box that maximize the volume are approximately: 11.2 in. × 17.2 in. × 3.4 in.

Sometimes, particularly when the domain of a function is restricted, you have to test the endpoints of the interval as well. This is because the highest or lowest value of a function may be at an endpoint of that interval; the critical value you obtained from the derivative might be just a local maximum or minimum. For the purposes of the AP exam, however, endpoints are considered separate from critical values.

Example 5: Find the absolute maximum and minimum values of $y = x^3 - x$ on the interval $[-3, 3]$.

Take the derivative and set it equal to zero.

$$\frac{dy}{dx} = 3x^2 - 1 = 0$$

Solve for x.

$$x = \pm\frac{1}{\sqrt{3}}$$

Test the critical points.

$$\frac{d^2y}{dx^2} = 6x$$

At $x = \dfrac{1}{\sqrt{3}}$, we have a minimum. At $x = -\dfrac{1}{\sqrt{3}}$, we have a maximum.

$$\text{At } x = -\frac{1}{\sqrt{3}}, \, y = -\frac{1}{3\sqrt{3}} + \frac{1}{\sqrt{3}} = \frac{2}{3\sqrt{3}} \approx 0.385$$

$$\text{At } x = \frac{1}{\sqrt{3}}, \, y = \frac{1}{3\sqrt{3}} - \frac{1}{\sqrt{3}} = -\frac{2}{3\sqrt{3}} \approx -0.385$$

Now it's time to check the endpoints of the interval.

$$\text{At } x = -3, \, y = -24$$

$$\text{At } x = 3, \, y = 24$$

We can see that the function actually has a *lower* value at $x = -3$ than at its "minimum" when $x = \dfrac{1}{\sqrt{3}}$. Similarly, the function has a *higher* value at $x = 3$ than at its "maximum" of $x = -\dfrac{1}{\sqrt{3}}$. This means that the function has a "local minimum" at

$x = \dfrac{1}{\sqrt{3}}$, and an "absolute minimum" when $x = -3$. And, the function has a "local

maximum" at $x = -\dfrac{1}{\sqrt{3}}$, and an "absolute maximum" at $x = 3$.

Example 6: A rectangle is to be inscribed in a semicircle with radius 4, with one side on the semicircle's diameter. What is the largest area this rectangle can have?

Let's look at this on the coordinate axes. The equation for a circle of radius 4, centered at the origin, is $x^2 + y^2 = 16$; a semicircle has the equation $y = \sqrt{16 - x^2}$. Our rectangle can then be expressed as a function of x, where the height is $\sqrt{16 - x^2}$ and the base is $2x$. See the following figure:

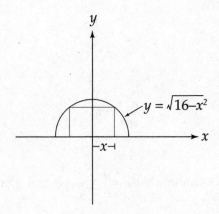

The area of the rectangle is: $A = 2x\sqrt{16 - x^2}$. Let's take the derivative of the area.

$$\frac{dA}{dx} = 2\sqrt{16 - x^2} - \frac{2x^2}{\sqrt{16 - x^2}}$$

The derivative is not defined at $x = \pm 4$. Setting the derivative equal to zero we get

If you're wondering why we don't use the negative root, it's because there is no such thing as a negative area.

$$2\sqrt{16 - x^2} - \frac{2x^2}{\sqrt{16 - x^2}} = 0$$

$$2\sqrt{16 - x^2} = \frac{2x^2}{\sqrt{16 - x^2}}$$

$$2(16 - x^2) = 2x^2$$

$$32 - 2x^2 = 2x^2$$

$$32 = 4x^2$$

$$x = \pm\sqrt{8}$$

Note that the domain of this function is $-4 \leq x \leq 4$, so these numbers serve as endpoints of the interval. Let's compare the critical values and the endpoints.

When $x = -4$, $y = 0$ and the area is 0.

When $x = 4$, $y = 0$ and the area is 0.

When $x = \sqrt{8}$, $y = \sqrt{8}$ and the area is 16.

Thus, the maximum area occurs when $x = \sqrt{8}$ and the area equals 16.

Try some of these solved problems on your own. As always, cover the answers as you work.

PROBLEM 1. A rectangular field, bounded on one side by a building, is to be fenced in on the other three sides. If 3,000 feet of fence is to be used, find the dimensions of the largest field that can be fenced in.

Answer: First, let's make a rough sketch of the situation.

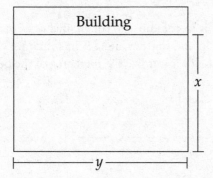

If we call the length of the field y and the width of the field x, the formula for the area of the field becomes

$$A = xy$$

The perimeter of the fencing is equal to the sum of two widths and the length.

$$2x + y = 3,000$$

Now solve this second equation for y.

$$y = 3,000 - 2x$$

When you plug this expression into the formula for the area, you get a formula for A in terms of x.

$$A = x(3{,}000 - 2x) = 3{,}000x - 2x^2$$

Next, take the derivative, set it equal to zero, and solve for x.

$$\frac{dA}{dx} = 3{,}000 - 4x = 0$$

$$x = 750$$

Let's check to make sure it's a maximum. Find the second derivative.

$$\frac{d^2 A}{dx^2} = -4$$

Because we have a negative result, $x = 750$ is a maximum. Finally, if we plug in $x = 750$ and solve for y, we find that $y = 1{,}500$. The largest field will measure 750 feet by 1,500 feet.

PROBLEM 2. A poster is to contain 100 square inches of picture surrounded by a 4-inch margin at the top and bottom and a 2-inch margin on each side. Find the overall dimensions that will minimize the total area of the poster.

Answer: First, make a sketch.

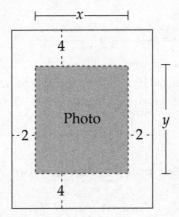

Let the area of the picture be $xy = 100$. The total area of the poster is $A = (x + 4)(y + 8)$. Then, expand the equation.

$$A = xy + 4y + 8x + 32$$

Substitute $xy = 100$ and $y = \dfrac{100}{x}$ into the area equation, and we get

$$A = 132 + \frac{400}{x} + 8x$$

Now take the derivative and set it equal to zero.

$$\frac{dA}{dx} = 8 - \frac{400}{x^2} = 0$$

Solving for x, we find that $x = \sqrt{50}$. Now solve for y by plugging $x = \sqrt{50}$ into the area equation: $y = 2\sqrt{50}$. Then check that these dimensions give us a minimum.

$$\frac{d^2 A}{dx^2} = \frac{800}{x^3}$$

This is positive when x is positive, so the minimum area occurs when $x = \sqrt{50}$. Thus, the overall dimensions of the poster are $4 + \sqrt{50}$ inches by $8 + 2\sqrt{50}$ inches.

PROBLEM 3. An open-top box with a square bottom and rectangular sides is to have a volume of 256 cubic inches. Find the dimensions that require the minimum amount of material.

Answer: First, make a sketch of the situation.

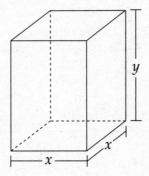

The amount of material necessary to make the box is equal to the surface area.

$$S = x^2 + 4xy$$

The formula for the volume of the box is $x^2 y = 256$.

If we solve the latter equation for y, $y = \dfrac{256}{x^2}$, and plug it into the former equation, we get

$$S = x^2 + 4x\frac{256}{x^2} = x^2 + \frac{1{,}024}{x}$$

Now take the derivative and set it equal to zero.

$$\frac{dS}{dx} = 2x - \frac{1{,}024}{x^2} = 0$$

If we solve this for x, we get $x^3 = 512$ and $x = 8$. Solving for y, we get $y = 4$.

Check that these dimensions give us a minimum.

$$\frac{d^2 A}{dx^2} = 2 + \frac{2{,}048}{x^3}$$

This is positive when x is positive, so the minimum surface area occurs when $x = 8$. The dimensions of the box should be 8 inches by 8 inches by 4 inches.

PROBLEM 4. Find the point on the curve $y = \sqrt{x}$ that is a minimum distance from the point (4, 0).

Answer: First, make that sketch.

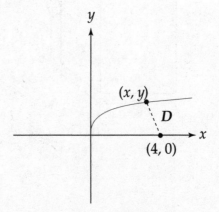

Using the distance formula, we get

$$D^2 = (x - 4)^2 + (y - 0)^2 = x^2 - 8x + 16 + y^2$$

Because $y = \sqrt{x}$,

$$D^2 = x^2 - 8x + 16 + x = x^2 - 7x + 16$$

Next, let $L = D^2$. We can do this because the minimum value of D^2 will occur at the same value of x as the minimum value of D. Therefore, it's simpler to minimize D^2 rather than D (because we won't have to take a square root!).

$$L = x^2 - 7x + 16$$

Now, take the derivative and set it equal to zero.

$$\frac{dL}{dx} = 2x - 7 = 0$$

$$x = \frac{7}{2}$$

Solving for y, we get $y = \sqrt{\dfrac{7}{2}}$.

Finally, because $\dfrac{d^2 L}{dx^2} = 2$, the point $\left(\dfrac{7}{2}, \sqrt{\dfrac{7}{2}} \right)$ is the minimum distance from the point (4, 0).

PRACTICE PROBLEM SET 10

Now try these problems on your own. The answers are in Chapter 19.

1. A rectangle has its base on the x-axis and its two upper corners on the parabola $y = 12 - x^2$. What is the largest possible area of the rectangle?

2. An open rectangular box is to be made from a 9 × 12 inch piece of tin by cutting squares of side x inches from the corners and folding up the sides. What should x be to maximize the volume of the box?

3. A 384-square-meter plot of land is to be enclosed by a fence and divided into two equal parts by another fence parallel to one pair of sides. What dimensions of the outer rectangle will minimize the amount of fence used?

4. What is the radius of a cylindrical soda can with volume of 512 cubic inches that will use the minimum material?

5. A swimmer is at a point 500 m from the closest point on a straight shoreline. She needs to reach a cottage located 1,800 m down shore from the closest point. If she swims at 4 m/s and she walks at 6 m/s, how far from the cottage should she come ashore so as to arrive at the cottage in the shortest time?

6. Find the closest point on the curve $x^2 + y^2 = 1$ to the point $(2, 1)$.

7. A window consists of an open rectangle topped by a semicircle and is to have a perimeter of 288 inches. Find the radius of the semicircle that will maximize the area of the window.

8. The range of a projectile is $R = \dfrac{v_0^2 \sin 2\theta}{g}$, where v_0 is its initial velocity, g is the acceleration due to gravity and is a constant, and θ is its firing angle. Find the angle that maximizes the projectile's range.

9. A computer company determines that its profit equation (in millions of dollars) is given by $P = x^3 - 48x^2 + 720x - 1,000$, where x is the number of thousands of units of software sold and $0 \le x \le 40$. Optimize the manufacturer's profit.

CURVE SKETCHING

Another topic on which students spend a lot of time in calculus is curve sketching. In the old days, whole courses (called "Analytic Geometry") were devoted to the subject, and students had to master a wide variety of techniques to learn how to sketch a curve accurately.

Fortunately (or unfortunately, depending on your point of view), students no longer need to be as good at analytic geometry. There are two reasons for this: (1) The AP exam tests only a few types of curves; and (2) you can use a graphing calculator. Because of the calculator, you can get an idea of the shape of the curve, and all you need to do is find important points to label the graph. We use calculus to find some of these points.

When it's time to sketch a curve, we'll show you a four-part analysis that'll give you all the information you need.

Step 1: Test the Function

Find where $f(x) = 0$. This tells you the function's x-intercepts (or roots). By setting $x = 0$, we can determine the y-intercepts. Then find any horizontal and/or vertical asymptotes.

Step 2: Test the First Derivative

Find where $f'(x) = 0$. This tells you the critical points. We can determine whether the curve is rising or falling, as well as where the maxima and minima are. It's also possible to determine if the curve has any points where it's nondifferentiable.

Step 3: Test the Second Derivative

Find where $f''(x) = 0$. This shows you where any points of inflection are. (These are points where the graph of a function changes concavity.) Then we can determine where the graph curves upward and where it curves downward.

Step 4: Test End Behavior

Look at what the general shape of the graph will be, based on the values of y for very large values of $\pm x$. Using this analysis, we can always come up with a sketch of a curve.

And now, the rules.

(1) When $f'(x) > 0$, the curve is rising; when $f'(x) < 0$, the curve is falling; when $f'(x) = 0$, the curve is at a critical point.

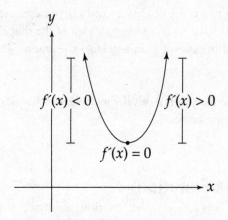

(2) When $f''(x) > 0$, the curve is "concave up"; when $f''(x) < 0$, the curve is "concave down"; and when $f''(x) = 0$, the curve is at a point of inflection.

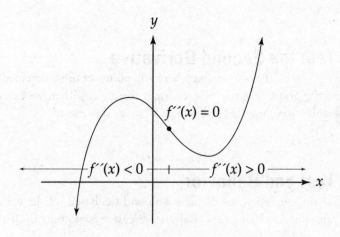

(3) The *y*-coordinates of each critical point are found by plugging the *x*-value into the original equation.

As always, this stuff will sink in better if we try a few examples.

Example 1: Sketch the equation $y = x^3 - 12x$.

Step 1: Find the *x*-intercepts.

$$x^3 - 12x = 0$$

$$x(x^2 - 12) = 0$$

$$x\left(x - \sqrt{12}\right)\left(x + \sqrt{12}\right) = 0$$

$$x = 0, \pm\sqrt{12}$$

The curve has *x*-intercepts at $\left(\sqrt{12},\ 0\right), \left(-\sqrt{12},\ 0\right)$, and (0, 0).

Next, find the *y*-intercepts.

$$y = (0)^3 - 12(0) = 0$$

The curve has a *y*-intercept at (0, 0).

There are no asymptotes, because there's no place where the curve is undefined (you won't have asymptotes for curves that are polynomials).

Step 2: Take the derivative of the function to find the critical points.

$$\frac{dy}{dx} = 3x^2 - 12$$

Set the derivative equal to zero, and solve for *x*.

$$3x^2 - 12 = 0$$

$$3(x^2 - 4) = 0$$

$$3(x - 2)(x + 2) = 0$$

so $x = 2, -2$.

Next, plug $x = 2, -2$ into the original equation to find the y-coordinates of the critical points.

$$y = (2)^3 - 12(2) = -16$$

$$y = (-2)^3 - 12(-2) = 16$$

Thus, we have critical points at $(2, -16)$ and $(-2, 16)$.

Step 3: Now, take the second derivative to find any points of inflection.

$$\frac{d^2y}{dx^2} = 6x$$

This equals zero at $x = 0$. We already know that when $x = 0$, $y = 0$, so the curve has a point of inflection at $(0, 0)$.

Now, plug the critical values into the second derivative to determine whether each is a maximum or a minimum. $f''(2) = 6(2) = 12$. This is positive, so the curve has a minimum at $(2, -16)$, and the curve is concave up at that point. $f''(-2) = 6(-2) = -12$. This value is negative, so the curve has a maximum at $(-2, 16)$ and the curve is concave down there.

Armed with this information, we can now plot the graph.

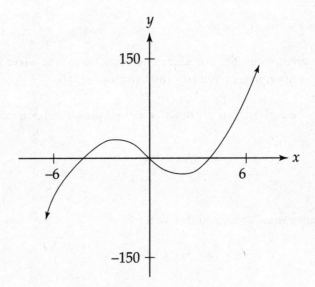

Example 2: Sketch the graph of $y = x^4 + 2x^3 - 2x^2 + 1$.

Step 1: First, let's find the x-intercepts.

$$x^4 + 2x^3 - 2x^2 + 1 = 0$$

If the equation doesn't factor easily, it's best not to bother to find the function's roots. Convenient, huh?

Next, let's find the y-intercepts.

$$y = (0)^4 + 2(0)^3 - 2(0)^2 + 1$$

The curve has a y-intercept at $(0, 1)$.

There are no vertical asymptotes because there is no place where the curve is undefined.

Step 2: Now we take the derivative to find the critical points.

$$\frac{dy}{dx} = 4x^3 + 6x^2 - 4x$$

The good news is that if the roots aren't easy to find, ETS won't ask you to find them, or you can find them with your calculator.

Set the derivative equal to zero.

$$4x^3 + 6x^2 - 4x = 0$$

$$2x(2x^2 + 3x - 2) = 0$$

$$2x(2x - 1)(x + 2) = 0$$

$$x = 0, \frac{1}{2}, -2$$

Next, plug these three values into the original equation to find the y-coordinates of the critical points. We already know that when $x = 0$, $y = 1$.

$$\text{When } x = \frac{1}{2}, y = \left(\frac{1}{2}\right)^4 + 2\left(\frac{1}{2}\right)^3 - 2\left(\frac{1}{2}\right)^2 + 1 = \frac{13}{16}$$

$$\text{When } x = -2, y = (-2)^4 + 2(-2)^3 - 2(-2)^2 + 1 = -7$$

Thus, we have critical points at $(0, 1)$, $\left(\frac{1}{2}, \frac{13}{16}\right)$, and $(-2, -7)$.

Step 3: Take the second derivative to find any points of inflection.

$$\frac{d^2 y}{dx^2} = 12x^2 + 12x - 4$$

Set this equal to zero.

$$12x^2 + 12x - 4 = 0$$

$$3x^2 + 3x - 1 = 0$$

$$x = \frac{-3 \pm \sqrt{21}}{6} \approx .26, -1.26$$

Therefore, the curve has points of inflection at $x = \dfrac{-3 \pm \sqrt{21}}{6}$.

Now solve for the y-coordinates.

$$(0.26, 0.90) \text{ and } (-1.26, -3.66)$$

We can now plug the critical values into the second derivative to determine whether each is a maximum or a minimum.

$$12(0)^2 + 12(0) - 4 = -4$$

This is negative, so the curve has a maximum at (0, 1); the curve is concave down there.

$$12\left(\frac{1}{2}\right)^2 + 12\left(\frac{1}{2}\right) - 4 = 5$$

This is positive, so the curve has a minimum at $\left(\dfrac{1}{2}, \dfrac{13}{16}\right)$; the curve is concave up there.

$$12(-2)^2 + 12(-2) - 4 = 20$$

This is positive, so the curve has a minimum at (–2, –7) and the curve is also concave up there.

We can now plot the graph.

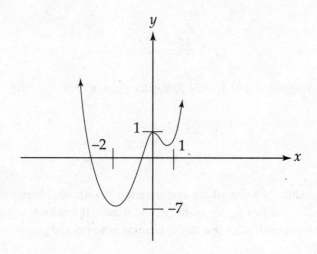

Finding a Cusp

If the derivative of a function approaches ∞ from one side of a point and $-\infty$ from the other, and if the function is continuous at that point, then the curve has a "cusp" at that point. In order to find a cusp, you need to look at points where the first derivative is undefined, as well as where it's zero.

Example 3: Sketch the graph of $y = 2 - x^{\frac{2}{3}}$.

Step 1: Find the *x*-intercepts.

$$2 - x^{\frac{2}{3}} = 0$$

$$x^{\frac{2}{3}} = 2 \qquad x = \pm 2^{\frac{3}{2}} = \pm 2\sqrt{2}$$

The *x*-intercepts are at $\left(\pm 2\sqrt{2},\ 0\right)$.

Next, find the *y*-intercepts.

$$y = 2 - (0)^{\frac{2}{3}} = 2$$

The curve has a *y*-intercept at (0, 2).

There are no asymptotes because there is no place where the curve is undefined.

Step 2: Now, take the derivative to find the critical points.

$$\frac{dy}{dx} = -\frac{2}{3}x^{-\frac{1}{3}}$$

What's next? You guessed it! Set the derivative equal to zero.

$$-\frac{2}{3}x^{-\frac{1}{3}} = 0$$

There are no values of x for which the equation is zero. But here's the new stuff to deal with: At $x = 0$, the derivative is undefined. If we look at the limit as x approaches 0 from both sides, we can determine whether the graph has a cusp.

$$\lim_{x \to 0^+} -\frac{2}{3}x^{-\frac{1}{3}} = -\infty \quad \text{and} \quad \lim_{x \to 0^-} -\frac{2}{3}x^{-\frac{1}{3}} = \infty$$

Therefore, the curve has a cusp at (0, 2).

There aren't any other critical points. But we can see that when $x < 0$, the derivative is positive (which means that the curve is rising to the left of zero), and when $x > 0$ the derivative is negative (which means that the curve is falling to the right of zero).

Step 3: Now, we take the second derivative to find any points of inflection.

$$\frac{d^2y}{dx^2} = \frac{2}{9}x^{-\frac{4}{3}}$$

Again, there's no x-value where this is zero. In fact, the second derivative is positive at all values of x except 0. Therefore, the graph is concave up everywhere.

Now it's time to graph this.

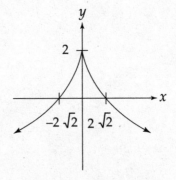

There's one other type of graph you should know about: a rational function. In order to graph a rational function, you need to know how to find that function's asymptotes.

How to Find Asymptotes

A line $y = c$ is a horizontal asymptote of the graph of $y = f(x)$ if

$$\lim_{x \to \infty} f(x) = c \text{ or if } \lim_{x \to -\infty} f(x) = c$$

A line $x = k$ is a vertical asymptote of the graph of $y = f(x)$ if

$$\lim_{x \to k^+} f(x) = \pm \infty \text{ or if } \lim_{x \to k^-} f(x) = \pm \infty$$

Example 4: Sketch the graph of $y = \dfrac{3x}{x + 2}$.

Step 1: Find the x-intercepts. A fraction can be equal to zero only when its numerator is equal to zero (provided that the denominator is not also zero there). All we have to do is set $3x = 0$, and you get $x = 0$. Thus, the graph has an x-intercept at $(0, 0)$. Note: This is also the y-intercept.

Next, look for asymptotes. The denominator is undefined at $x = -2$, and if we take the left- and right-hand limits of the function we see the following:

$$\lim_{x \to -2^+} \frac{3x}{x + 2} = -\infty \quad \text{and} \quad \lim_{x \to -2^-} \frac{3x}{x + 2} = \infty$$

The curve has a vertical asymptote at $x = -2$.

If we take $\lim_{x \to \infty} \dfrac{3x}{x + 2} = 3$ and $\lim_{x \to -\infty} \dfrac{3x}{x + 2} = 3$, the curve has a horizontal asymptote at $y = 3$.

Step 2: Now, take the derivative to figure out the critical points.

$$\frac{dy}{dx} = \frac{(x + 2)(3) - (3x)(1)}{(x + 2)^2} = \frac{6}{(x + 2)^2}$$

There are no values of x that make the derivative equal to zero. Because the numerator is 6 and the denominator is squared, the derivative will always be positive (the curve is always rising). You should note that the derivative is undefined at $x = -2$, but you already know that there's an asymptote at $x = -2$, so you don't need to examine this point further.

Step 3: Now, it's time for the second derivative.

$$\frac{d^2y}{dx^2} = \frac{-12}{(x+2)^3}$$

This is never equal to zero. The expression is positive when $x < -2$, so the graph is concave up when $x < -2$. The second derivative is negative when $x > -2$, so it's concave down when $x > -2$.

Now plot the graph.

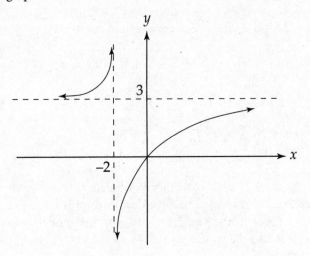

Now it's time to practice some problems. Do each problem, covering the answer first, then check your answer.

PROBLEM 1. Sketch the graph of $y = x^3 - 9x^2 + 24x - 10$. Plot all extrema, points of inflection, and asymptotes.

Answer: Follow the three steps.

First, see if the x-intercepts are easy to find. This is a cubic equation that isn't easily factored. So skip this step.

Next, find the y-intercepts by setting $x = 0$.

$$y = (0)^3 - 9(0)^2 + 24(0) - 10 = -10$$

The curve has a y-intercept at $(0, -10)$.

There are no asymptotes, because the curve is a simple polynomial.

Next, find the critical points using the first derivative.

$$\frac{dy}{dx} = 3x^2 - 18x + 24$$

Set the derivative equal to zero and solve for x.

$$3x^2 - 18x + 24 = 0$$

$$3(x^2 - 6x + 8) = 0$$

$$3(x - 4)(x - 2) = 0$$

$$x = 2, 4$$

Plug $x = 2$ and $x = 4$ into the original equation to find the y-coordinates of the critical points.

$$\text{When } x = 2, y = 10$$

$$\text{When } x = 4, y = 6$$

Thus, we have critical points at (2, 10) and (4, 6).

In our third step, the second derivative indicates any points of inflection.

$$\frac{d^2 y}{dx^2} = 6x - 18$$

This equals zero at $x = 3$.

Next, plug $x = 3$ into the original equation to find the y-coordinates of the point of inflection, which is at (3, 8). Plug the critical values into the second derivative to determine whether each is a maximum or a minimum.

$$6(2) - 18 = -6$$

This is negative, so the curve has a maximum at (2, 10), and the curve is concave down there.

$$6(4) - 18 = 6$$

This is positive, so the curve has a minimum at (4, 6), and the curve is concave up there.

It's graph-plotting time.

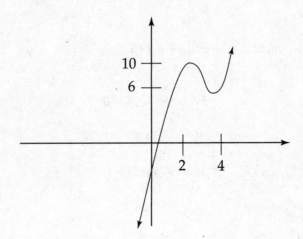

PROBLEM 2. Sketch the graph of $y = 8x^2 - 16x^4$. Plot all extrema, points of inflection, and asymptotes.

Answer: Factor the polynomial.

$$8x^2\left(1 - 2x^2\right) = 0$$

Solving for x, we get $x = 0$ (a double root), $x = \dfrac{1}{\sqrt{2}}$, and $x = -\dfrac{1}{\sqrt{2}}$.

Find the y-intercepts: when $x = 0$, $y = 0$.

There are no asymptotes, because the curve is a simple polynomial.

Find the critical points using the first derivative.

$$\frac{dy}{dx} = 16x - 64x^3$$

Set the derivative equal to zero and solve for x. You get $x = 0$, $x = \dfrac{1}{2}$, and $x = -\dfrac{1}{2}$.

Next, plug $x = 0$, $x = \dfrac{1}{2}$, and $x = -\dfrac{1}{2}$ into the original equation to find the y-coordinates of the critical points.

$$\text{When } x = 0, y = 0$$

$$\text{When } x = \frac{1}{2}, y = 1$$

$$\text{When } x = -\frac{1}{2}, y = 1$$

Thus, there are critical points at $(0, 0)$, $\left(\dfrac{1}{2}, 1\right)$, and $\left(-\dfrac{1}{2}, 1\right)$.

Take the second derivative to find any points of inflection.

$$\frac{d^2 y}{dx^2} = 16 - 192x^2$$

This equals zero at $x = \dfrac{1}{\sqrt{12}}$ and $x = -\dfrac{1}{\sqrt{12}}$.

Next, plug $x = \dfrac{1}{\sqrt{12}}$ and $x = -\dfrac{1}{\sqrt{12}}$ into the original equation to find the y-coordinates of the points of inflection, which are at $\left(\dfrac{1}{\sqrt{12}}, \dfrac{5}{9}\right)$ and $\left(-\dfrac{1}{\sqrt{12}}, \dfrac{5}{9}\right)$.

Now determine whether the points are maxima or minima.

At $x = 0$, we have a minimum; the curve is concave up there.

At $x = \dfrac{1}{2}$, it's a maximum, and the curve is concave down.

At $x = -\dfrac{1}{2}$, it's also a maximum (still concave down).

Now plot.

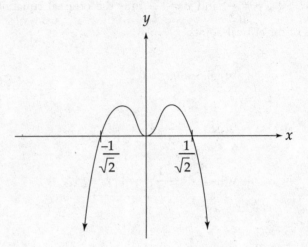

PROBLEM 3. Sketch the graph of $y = \left(\dfrac{x-4}{x+3}\right)^2$. Plot all extrema, points of inflection, and asymptotes.

Answer: This should seem rather routine by now.

Find the x-intercepts by setting the numerator equal to zero; $x = 4$. The graph has an x-intercept at $(4, 0)$. (It's a double root.)

Next, find the y-intercept by plugging in $x = 0$.

$$y = \frac{16}{9}$$

The denominator is undefined at $x = -3$, so there's a vertical asymptote at that point.

Look at the limits.

$$\lim_{x\to\infty}\left(\frac{x-4}{x+3}\right)^2 = 1 \text{ and } \lim_{x\to-\infty}\left(\frac{x-4}{x+3}\right)^2 = 1$$

The curve has a horizontal asymptote at $y = 1$.

It's time for the first derivative.

$$\frac{dy}{dx} = 2\left(\frac{x-4}{x+3}\right)\frac{(x+3)(1)-(x-4)(1)}{(x+3)^2} = \frac{14x-56}{(x+3)^3}$$

The derivative is zero when $x = 4$, and the derivative is undefined at $x = -3$. (There's an asymptote there, so we can ignore the point. If the curve were *defined* at $x = -3$, then it would be a critical point, as you'll see in the next example.)

Now for the second derivative.

$$\frac{d^2 y}{dx^2} = \frac{(x+3)^3 (14) - (14x - 56)3(x+3)^2}{(x+3)^6} = \frac{-28x + 210}{(x+3)^4}$$

This is zero when $x = \dfrac{15}{2}$. The second derivative is positive (and the graph is concave up) when $x < \dfrac{15}{2}$, and it's negative (and the graph is concave down) when $x > \dfrac{15}{2}$.

We can now plug $x = 4$ into the second derivative. It's positive there, so $(4, 0)$ is a minimum.

Your graph should look like the following:

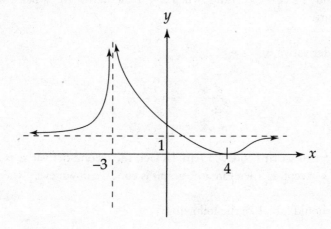

PROBLEM 4. Sketch the graph of $y = (x - 4)^{\frac{2}{3}}$. Plot all extrema, points of inflection, and asymptotes.

Answer: By inspection, the x-intercept is at $x = 4$.

Next, find the y-intercepts. When $x = 0$, $y = \sqrt[3]{16} \approx 2.52$.

No asymptotes exist because there's no place where the curve is undefined.

The first derivative is

$$\frac{dy}{dx} = \frac{2}{3}(x-4)^{-\frac{1}{3}}$$

Set it equal to zero.

$$\frac{2}{3}(x-4)^{-\frac{1}{3}} = 0$$

This can never equal zero. But, at $x = 4$ the derivative is undefined, so this is a critical point. If you look at the limit as x approaches 4 from both sides, you can see if there's a cusp.

$$\lim_{x \to 4^+} \frac{2}{3}(x-4)^{-\frac{1}{3}} = \infty \quad \text{and} \quad \lim_{x \to 4^-} \frac{2}{3}(x-4)^{-\frac{1}{3}} = -\infty$$

The curve has a cusp at $(4, 0)$.

There were no other critical points. But, we can see that when $x > 4$, the derivative is positive and the curve is rising; when $x < 4$ the derivative is negative, and the curve is falling.

The second derivative is

$$\frac{d^2 y}{dx^2} = -\frac{2}{9}(x-4)^{-\frac{4}{3}}$$

No value of x can set this equal to zero. In fact, the second derivative is negative at all values of x except 4. Therefore, the graph is concave down everywhere.

Your graph should look like the following:

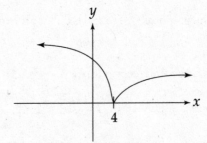

PRACTICE PROBLEM SET 11

It's time for you to try some of these on your own. Sketch each of the graphs below and check the answers in Chapter 19.

1. $y = x^3 - 9x - 6$

2. $y = -x^3 - 6x^2 - 9x - 4$

3. $y = \left(x^2 - 4\right)\left(9 - x^2\right)$

4. $y = \dfrac{x^4}{4} - 2x^2$

5. $y = \dfrac{x - 3}{x + 8}$

6. $y = \dfrac{x^2 - 4}{x - 3}$

7. $y = 3 + x^{\frac{2}{3}}$

8. $y = x^{\frac{2}{3}}\left(3 - 2x^{\frac{1}{3}}\right)$

9. $y = \dfrac{3x^2}{x^2 - 4}$

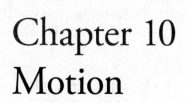

Chapter 10
Motion

This chapter deals with two different types of word problems that involve motion: related rates and the relationship between velocity and acceleration of a particle. The subject matter might seem arcane, but once you get the hang of them, you'll see that these aren't so hard, either. Besides, the AP exam tests only a few basic problem types.

RELATED RATES

The idea behind these problems is very simple. In a typical problem, you'll be given an equation relating two or more variables. These variables will change with respect to time, and you'll use derivatives to determine how the rates of change are related. (Hence the name: related rates.) Sounds easy, doesn't it?

Example 1: A circular pool of water is expanding at the rate of $16\pi \dfrac{\text{in}^2}{\text{sec}}$. At what rate is the radius expanding when the radius is 4 inches?

Note: The pool is expanding in square inches per second. We've been given the rate that the area is changing, and we need to find the rate of change of the radius. What equation relates the area of a circle to its radius? $A = \pi r^2$.

Step 1: Set up the equation and take the derivative of this equation with respect to t (time).

$$\frac{dA}{dt} = 2\pi r \frac{dr}{dt}$$

In this equation, $\dfrac{dA}{dt}$ represents the rate at which the area is changing, and $\dfrac{dr}{dt}$ is the rate at which the radius is changing. The simplest way to explain this is that whenever you have a variable in an equation (r, for example), the derivative with respect to time $\left(\dfrac{dr}{dt}\right)$ represents the rate at which that variable is increasing or decreasing.

Step 2: Now we can plug in the values for the rate of change of the area and for the radius. (Never plug in the values until after you have taken the derivative or you will get nonsense!)

$$16\pi = 2\pi\left(4\right)\frac{dr}{dt}$$

Solving for $\dfrac{dr}{dt}$, we get

$$16\pi = 8\pi \frac{dr}{dt} \text{ and } \frac{dr}{dt} = 2$$

The radius is changing at a rate of $2 \dfrac{\text{in}}{\text{sec}}$. It's important to note that this is the rate only when the radius is 4 inches. As the circle gets bigger and bigger, the radius will expand at a slower and slower rate.

Example 2: A 25-foot long ladder is leaning against a wall and sliding toward the floor. If the foot of the ladder is sliding away from the base of the wall at a rate of $15 \dfrac{\text{feet}}{\text{sec}}$, how fast is the top of the ladder sliding down the wall when the top of the ladder is 7 feet from the ground?

Here's another classic related rates problem. As always, a picture is worth 1,000 words.

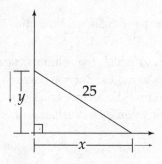

You can see that the ladder forms a right triangle with the wall. Let x stand for the distance from the foot of the ladder to the base of the wall, and let y represent the distance from the top of the ladder to the ground. What's our favorite theorem that deals with right triangles? The Pythagorean theorem tells us here that $x^2 + y^2 = 25^2$. Now we have an equation that relates the variables to each other.

Now take the derivative of the equation with respect to t.

$$2x\frac{dx}{dt} + 2y\frac{dy}{dt} = 0$$

Just plug in what you know and solve. Because we're looking for the rate at which the vertical distance is changing, we're going to solve for $\dfrac{dy}{dt}$.

Let's see what we know. We're given the rate at which the ladder is sliding away from the wall: $\dfrac{dx}{dt} = 15$. The distance from the ladder to the top of the wall is 7 feet ($y = 7$). To find x, use the Pythagorean theorem. If we plug in $y = 7$ to the equation $x^2 + y^2 = 25^2$, $x = 24$.

Now plug all this information into the derivative equation.

$$2(24)(15) + 2(7)\dfrac{dy}{dt} = 0$$

$$\dfrac{dy}{dt} = \dfrac{-360}{7}\dfrac{\text{feet}}{\text{sec}}$$

Example 3: A spherical balloon is expanding at a rate of $60\pi\,\dfrac{\text{in}^3}{\text{sec}}$. How fast is the surface area of the balloon expanding when the radius of the balloon is 4 in?

Step 1: You're given the rate at which the volume's expanding, and you know the equation that relates volume to radius. But you have to relate radius to surface area as well, because you have to find the surface area's rate of change. This means that you'll need the equations for volume and surface area of a sphere.

$$V = \dfrac{4}{3}\pi r^3$$

$$A = 4\pi r^2$$

You're trying to find $\dfrac{dA}{dt}$, but A is given in terms of r, so you have to get $\dfrac{dr}{dt}$ first. Because we know the volume, if we work with the equation that gives us volume in terms of radius, we can find $\dfrac{dr}{dt}$. From there, work with the other equation to find $\dfrac{dA}{dt}$. If we take the derivative of the equation with respect to t we get $\dfrac{dV}{dt} = 4\pi r^2 \dfrac{dr}{dt}$. Plugging in for $\dfrac{dV}{dt}$ and for r, we get $60\pi = 4\pi(4)^2\dfrac{dr}{dt}$.

Solving for $\dfrac{dr}{dt}$ we get

$$\frac{dr}{dt} = \frac{15}{16} \frac{\text{in}}{\text{sec}}$$

Step 2: Now we take the derivative of the other equation with respect to t.

$$\frac{dA}{dt} = 8\pi r \frac{dr}{dt}$$

We can plug in for r and $\dfrac{dr}{dt}$ from the previous step and we get

$$\frac{dA}{dt} = 8\pi (4) \frac{15}{16} = \frac{480\pi}{16} \frac{\text{in}^2}{\text{sec}} = 30\pi \frac{\text{in}^2}{\text{sec}}$$

One final example.

Example 4: An underground conical tank, standing on its vertex, is being filled with water at the rate of $18\pi \dfrac{\text{ft}^3}{\text{min}}$. If the tank has a height of 30 feet and a radius of 15 feet , how fast is the water level rising when the water is 12 feet deep?

This "cone" problem is also typical. The key point to getting these right is knowing that the ratio of the height of a right circular cone to its radius is constant. By telling us that the height of the cone is 30 and the radius is 15, we know that at any level, the height of the water will be twice its radius, or $h = 2r$.

You must find the rate at which the water is rising (the height is changing), or $\dfrac{dh}{dt}$.

Therefore, you want to eliminate the radius from the volume. By substituting $\dfrac{h}{2} = r$ into the equation for volume, we get

$$V = \frac{1}{3}\pi \left(\frac{h}{2}\right)^2 h = \frac{\pi h^3}{12}$$

Differentiate both sides with respect to t.

$$\frac{dV}{dt} = \frac{\pi}{12} 3h^2 \frac{dh}{dt}$$

The volume of a cone is $V = \dfrac{1}{3}\pi r^2 h$. (You'll learn to derive this formula through integration in Chapter 17.)

Now we can plug in and solve for $\dfrac{dh}{dt}$.

$$18\pi = \frac{\pi}{12}3(12)^2\frac{dh}{dt}$$

$$\frac{dh}{dt} = \frac{1}{2}\frac{\text{feet}}{\text{min}}$$

In order to solve related rates problems, you have to be good at determining relationships between variables. Once you figure that out, the rest is a piece of cake. Many of these problems involve geometric relationships, so review the formulas for the volumes and areas of cones, spheres, boxes, and other solids. Once you get the hang of setting up the problems, you'll see that these problems follow the same predictable patterns. Look through these sample problems.

PROBLEM 1. A circle is increasing in area at the rate of $16\pi\,\dfrac{\text{in}^2}{\text{s}}$. How fast is the radius increasing when the radius is 2 in?

Answer: Use the expression that relates the area of a circle to its radius: $A = \pi r^2$

Next, take the derivative of the expression with respect to t.

$$\frac{dA}{dt} = 2\pi r\frac{dr}{dt}$$

Now, plug in $\dfrac{dA}{dt} = 16\pi$ and $r = 2$.

$$16\pi = 2\pi(2)\frac{dr}{dt}$$

When you solve for $\dfrac{dr}{dt}$, you'll get $\dfrac{dr}{dt} = 4\,\dfrac{\text{in}}{\text{sec}}$.

PROBLEM 2. A rocket is rising vertically at a rate of 5,400 miles per hour. An observer on the ground is standing 20 miles from the rocket's launch point. How fast (in radians per second) is the angle of elevation between the ground and the observer's line of sight of the rocket increasing when the rocket is at an elevation of 40 miles?

Notice that velocity is given in miles per hour and the answer asks for radians per second. In situations like this one, you have to be sure to convert the units properly, or you'll get nailed.

Answer: First, draw a picture.

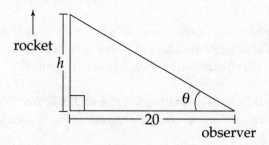

Now, find the equation that relates the angle of elevation to the rocket's altitude.

$$\tan\theta = \frac{h}{20}$$

If we take the derivative of both sides of this expression with respect to t, we get

$$\sec^2\theta \, \frac{d\theta}{dt} = \frac{1}{20} \frac{dh}{dt}$$

We know that $\frac{dh}{dt}$ = 5,400 miles per hour, but the problem asks for time in seconds, so we need to convert this number. There are 3,600 seconds in an hour, so $\frac{dh}{dt} = \frac{3}{2}$ miles per second. Next, we know that $\tan\theta = \frac{h}{20}$, so when h = 40, $\tan\theta = 2$. Because $1 + \tan^2\theta = \sec^2\theta$, we get $\sec^2\theta = 5$.

Plug in the following information:

$$5\frac{d\theta}{dt} = \frac{1}{20}\left(\frac{3}{2}\right) \text{ and } \frac{d\theta}{dt} = \frac{3}{200} \text{ radians per second}$$

PRACTICE PROBLEM SET 12

Now try these problems on your own. The answers are in Chapter 19.

1. Oil spilled from a tanker spreads in a circle whose circumference increases at a rate of 40 ft/sec. How fast is the area of the spill increasing when the circumference of the circle is 100π feet?

2. A spherical balloon is inflating at a rate of 27π in³/sec. How fast is the radius of the balloon increasing when the radius is 3 inches?

3. Cars A and B leave a town at the same time. Car A heads due south at a rate of 80 km/hr and car B heads due west at a rate of 60 km/hr. How fast is the distance between the cars increasing after three hours?

4. A cylindrical tank with a radius of 6 meters is filling with fluid at a rate of 108π m³/sec. How fast is the height increasing?

5. The sides of an equilateral triangle are increasing at the rate of 27 in/sec. How fast is the triangle's area increasing when the sides of the triangle are each 18 inches long?

6. An inverted conical container has a diameter of 42 inches and a depth of 15 inches. If water is flowing out of the vertex of the container at a rate of 35π in³/sec, how fast is the depth of the water dropping when the height is 5 in?

7. A boat is being pulled toward a dock by a rope attached to its bow through a pulley on the dock 7 feet above the bow. If the rope is hauled in at a rate of 4 ft/sec, how fast is the boat approaching the dock when 25 ft of rope is out?

8. A 6-foot-tall woman is walking at the rate of 4 ft/sec away from a street lamp that is 24 ft tall. How fast is the length of her shadow changing?

9. The voltage, V, in an electrical circuit is related to the current, I, and the resistance, R, by the equation $V = IR$. The current is decreasing at −4 amps/sec as the resistance increases at 20 ohms/sec. How fast is the voltage changing when the voltage is 100 volts and the current is 20 amps?

10. The minute hand of a clock is 6 inches long. Starting from noon, how fast is the area of the sector swept out by the minute hand increasing in in²/min at any instant?

POSITION, VELOCITY, AND ACCELERATION

Almost every AP exam has a question on position, velocity, or acceleration. It's one of the traditional areas of physics where calculus comes in handy. Some of these problems require the use of integral calculus, which we won't talk about until the second half of this book. So this unit is divided in half; you'll see the other half later.

If you have a function that gives you the position of an object (usually called a "particle") at a specified time, then the derivative of that function with respect to time is the velocity of the object, and the second derivative is the acceleration. These are usually represented by the following:

Position: $x(t)$ or sometimes $s(t)$

Velocity: $v(t)$, which is $x'(t)$

Acceleration: $a(t)$, which is $x''(t)$ or $v'(t)$

Please note that these equations are usually functions of time (t). Typically, t is greater than zero, but it doesn't have to be.

By the way, speed is the absolute value of velocity.

Example 1: If the position of a particle at a time t is given by the equation $x(t) = t^3 - 11t^2 + 24t$, find the velocity and the acceleration of the particle at time $t = 5$.

First, take the derivative of $x(t)$.

$$x'(t) = 3t^2 - 22t + 24 = v(t)$$

Second, plug in $t = 5$ to find the velocity at that time.

$$v(5) = 3(5^2) - 22(5) + 24 = -11$$

Third, take the derivative of $v(t)$ to find $a(t)$.

$$v'(t) = 6t - 22 = a(t)$$

Finally, plug in $t = 5$ to find the acceleration at that time.

$$a(5) = 6(5) - 22 = 8$$

See the negative velocity? The sign of the velocity is important because it indicates the direction of the particle. Make sure that you know the following:

> When the velocity is negative, the particle is moving to the left.
>
> When the velocity is positive, the particle is moving to the right.
>
> When the velocity and acceleration of the particle have the same signs, the particle's speed is increasing.
>
> When the velocity and acceleration of the particle have opposite signs, the particle's speed is decreasing (or slowing down).
>
> When the velocity is zero and the acceleration is not zero, the particle is momentarily stopped and changing direction.

Example 2: If the position of a particle is given by $x(t) = t^3 - 12t^2 + 36t + 18$, where $t > 0$, find the point at which the particle changes direction.

The derivative is

$$x'(t) = v(t) = 3t^2 - 24t + 36$$

Set it equal to zero and solve for t.

$$x'(t) = 3t^2 - 24t + 36 = 0$$

$$t^2 - 8t + 12 = 0$$

$$(t - 2)(t - 6) = 0$$

So we know that $t = 2$ or $t = 6$.

You need to check that the acceleration is not 0: $x''(t) = 6t - 24$. This equals 0 at $t = 4$. Therefore, the particle is changing direction at $t = 2$ and $t = 6$.

Example 3: Given the same position function as in Example 2, find the interval of time during which the particle is slowing down.

When $0 < t < 2$ and $t > 6$, the particle's velocity is positive; when $2 < t < 6$, the particle's velocity is negative. You can verify this by graphing the function and seeing when it's above or below the x-axis. Or, try some points in the regions between the roots and outside the roots. Now, we need to determine the same information about the acceleration.

$$a(t) = v'(t) = 6t - 24$$

So the acceleration will be negative when $t < 4$, and positive when $t > 4$.

So we have

Time	Velocity	Acceleration
$0 < t < 2$	Positive	Negative
$2 < t < 4$	Negative	Negative
$4 < t < 6$	Negative	Positive
$t > 6$	Positive	Positive

Whenever the velocity and acceleration have opposite signs, the particle is slowing down. Here the particle is slowing down during the first two seconds ($0 < t < 2$) and between the fourth and sixth seconds ($4 < t < 6$).

Another typical question you'll be asked is to find the distance a particle has traveled from one time to another. This is the distance that the particle has covered without regard to the sign, not just the displacement. In other words, if the particle had an odometer on it, what would it read? Usually, all you have to do is plug the two times into the position function and find the difference.

Example 4: How far does a particle travel between the eighth and tenth seconds if its position function is $x(t) = t^2 - 6t$?

Find $x(10) - x(8) = (100 - 60) - (64 - 48) = 24$.

Be careful about one very important thing: **If the velocity changes sign during the problem's time interval**, you'll get the wrong answer if you simply follow the method in the paragraph above. For example, suppose we had the same position function as above but we wanted to find the distance that the particle travels from $t = 2$ to $t = 4$.

$$x(4) - x(2) = (-8) - (-8) = 0$$

This is wrong. The particle travels from -8 back to -8, but it hasn't stood still. To fix this problem, divide the time interval into the time when the velocity is negative and the time when the velocity is positive, and add the absolute values of each distance. Here the velocity is $v(t) = 2t - 6$. The velocity is negative when $t < 3$ and positive when $t > 3$. So we find the absolute value of the distance traveled from $t = 2$ to $t = 3$, and add to that the absolute value of the distance traveled from $t = 3$ to $t = 4$.

Because $x(t) = t^2 - 6t$,

$$\left| x(3) - x(2) \right| + \left| x(4) - x(3) \right| = \left| -9 + 8 \right| + \left| -8 + 9 \right| = 2$$

This is the distance that the particle traveled.

Example 5: Given the position function $x(t) = t^4 - 8t^2$, find the distance that the particle travels from $t = 0$ to $t = 4$.

First, find the first derivative ($v(t) = 4t^3 - 16t$) and set it equal to zero.

$$4t^3 - 16t = 0 \quad 4t(t^2 - 4) = 0 \quad t = 0, 2, -2$$

So we need to divide the time interval into $t = 0$ to $t = 2$ and $t = 2$ to $t = 4$.

$$\left| x(2) - x(0) \right| + \left| x(4) - x(2) \right| = 16 + 144 = 160$$

Here are some solved problems. Do each problem, covering the answer first, then checking your answer.

PROBLEM 1. Find the velocity and acceleration of a particle whose position function is $x(t) = 2t^3 - 21t^2 + 60t + 3$, for $t > 0$.

Answer: Find the first two derivatives.

$$v(t) = 6t^2 - 42t + 60$$

$$a(t) = 12t - 42$$

PROBLEM 2. Given the position function in problem 1, find when the particle's speed is increasing.

Answer: First, set $v(t) = 0$.

$$6t^2 - 42t + 60 = 0$$

$$t^2 - 7t + 10 = 0$$

$$(t - 2)(t - 5) = 0$$

$$t = 2, t = 5$$

You should be able to determine that the velocity is positive from $0 < t < 2$, negative from $2 < t < 5$, and positive again from $t > 5$.

Now, set $a(t) = 0$.

$$12t - 42 = 0$$

$$t = \frac{7}{2}$$

You should be able to determine that the acceleration is negative from $0 < t < \dfrac{7}{2}$ and positive from $t > \dfrac{7}{2}$.

The intervals where the velocity and the acceleration have the same sign are $2 < t < \dfrac{7}{2}$ and $t > 5$.

PROBLEM 3. Given that the position of a particle is found by $x(t) = t^3 - 6t^2 + 1$, $t > 0$, find the distance that the particle travels from $t = 2$ to $t = 5$.

Answer: First, find $v(t)$.

$$v(t) = 3t^2 - 12t$$

Second, set $v(t) = 0$ and find the critical values.

$3t^2 - 12t = 0 \qquad\qquad 3t(t - 4) = 0 \qquad\qquad t = \{0, 4\}$

Because the particle changes direction after four seconds, you have to figure out two time intervals separately (from $t = 2$ to $t = 4$ and from $t = 4$ to $t = 5$) and add the absolute values of the distances.

$$\left|x(4) - x(2)\right| + \left|x(5) - x(4)\right| = \left|(-31) - (-15)\right| + \left|(-24) - (-31)\right| = 23$$

PRACTICE PROBLEM SET 13

Now try these problems. The answers are in Chapter 19.

1. Find the velocity and acceleration of a particle whose position function is $x(t) = t^3 - 9t^2 + 24t, \ t > 0$.

2. Find the velocity and acceleration of a particle whose position function is $x(t) = \sin(2t) + \cos(t)$.

3. If the position function of a particle is $x(t) = \dfrac{t}{t^2 + 9}, \ t > 0$, find when the particle is changing direction.

4. If the position function of a particle is $x(t) = \sin\left(\dfrac{t}{2}\right), \ 0 < t < 4\pi$, find when the particle is changing direction.

5. If the position function of a particle is $x(t) = 3t^2 + 2t + 4, \ t > 0$, find the distance that the particle travels from $t = 2$ to $t = 5$.

6. If the position function of a particle is $x(t) = t^2 + 8t, \ t > 0$, find the distance that the particle travels from $t = 0$ to $t = 4$.

7. If the position function of a particle is $x(t) = 2\sin^2 t + 2\cos^2 t, \ t > 0$, find the velocity and acceleration of the particle.

8. If the position function of a particle is $x(t) = t^3 + 8t^2 - 2t + 4, \ t > 0$, find when the particle is changing direction.

9. If the position function of a particle is $x(t) = 2t^3 - 6t^2 + 12t - 18$, $t > 0$, find when the particle is changing direction.

10. If the position function of a particle is $x(t) = \sin^2 2t, \ t > 0$, find the distance that the particle travels from $t = 0$ to $t = 2$.

Chapter 11
Exponential and Logarithmic Functions, Part One

As with trigonometric functions, you'll be expected to remember all of the logarithmic and exponential functions you've studied in the past. If you're not sure about any of this stuff, review the unit on Prerequisite Mathematics. Also, this is only part one of our treatment of exponents and logs. Much of what you need to know about these functions requires knowledge of integrals (the second half of the book), so we'll discuss them again later.

THE DERIVATIVE OF ln *x*

When you studied logs in the past, you probably concentrated on common logs (that is, those with a base of 10), and avoided natural logarithms (base *e*) as much as possible. Well, we have bad news for you: Most of what you'll see from now on involves natural logs. In fact, common logs almost never show up in calculus. But that's okay. All you have to do is memorize a bunch of rules, and you'll be fine.

Rule No. 1: If $y = \ln x$, then $\dfrac{dy}{dx} = \dfrac{1}{x}$

This rule has a corollary that incorporates the Chain Rule and is actually a more useful rule to memorize.

Rule No. 2: If $y = \ln u$, then $\dfrac{dy}{dx} = \dfrac{1}{u}\dfrac{du}{dx}$

Remember: *u* is a function of *x*, and $\dfrac{du}{dx}$ is its derivative.

You'll see how simple this rule is after we try a few examples.

Example 1: Find the derivative of $f(x) = \ln(x^3)$.

$$f'(x) = \frac{3x^2}{x^3} = \frac{3}{x}$$

If you recall your rules of logarithms, you could have done this another way.

$$\ln(x^3) = 3 \ln x$$

Therefore, $f'(x) = 3\left(\dfrac{1}{x}\right) = \dfrac{3}{x}$.

Example 2: Find the derivative of $f(x) = \ln(5x - 3x^6)$.

$$f'(x) = \frac{\left(5 - 18x^5\right)}{\left(5x - 3x^6\right)}$$

Example 3: Find the derivative of $f(x) = \ln(\cos x)$.

$$f'(x) = \frac{-\sin x}{\cos x} = -\tan x$$

Finding the derivative of a natural logarithm is just a matter of following a simple formula.

THE DERIVATIVE OF e^x

As you'll see in Rule No. 3, the derivative of e^x is probably the easiest thing that you'll ever have to do in calculus.

> Rule No. 3: If $y = e^x$, then $\dfrac{dy}{dx} = e^x$

That's not a typo. The derivative is the same as the original function! Incorporating the Chain Rule, we get a good formula for finding the derivative.

> Rule No. 4: If $y = e^u$, then $\dfrac{dy}{dx} = e^u \dfrac{du}{dx}$

And you were worried that all of this logarithm and exponential stuff was going to be hard!

Example 4: Find the derivative of $f(x) = e^{3x}$.

$$f'(x) = e^{3x}(3) = 3e^{3x}$$

Example 5: Find the derivative of $f(x) = e^{x^3}$.

$$f'(x) = e^{x^3}\left(3x^2\right) = 3x^2 e^{x^3}$$

Example 6: Find the derivative of $f(x) = e^{\tan x}$.

$$f'(x) = \left(\sec^2 x\right)e^{\tan x}$$

Example 7: Find the second derivative of $f(x) = e^{x^2}$.

$$f'(x) = 2xe^{x^2}$$
$$f''(x) = 2e^{x^2} + 4x^2 e^{x^2}$$

Once again, it's just a matter of following a formula.

THE DERIVATIVE OF $\log_a x$

This derivative is actually a little trickier than the derivative of a natural log. First, if you remember your logarithm rules about change of base, we can rewrite $\log_a x$.

$$\log_a x = \frac{\ln x}{\ln a}$$

Review the unit on Prerequisite Mathematics if this leaves you scratching your head. Anyway, because $\ln a$ is a constant, we can take the derivative.

$$\frac{1}{\ln a}\frac{1}{x}$$

This leads us to our next rule.

Rule No. 5: If $y = \log_a x$, then $\dfrac{dy}{dx} = \dfrac{1}{x \ln a}$

Once again, incorporating the Chain Rule gives us a more useful formula.

Rule No. 6: If $y = \log_a u$, then $\dfrac{dy}{dx} = \dfrac{1}{u \ln a} \dfrac{du}{dx}$

Example 8: Find the derivative of $f(x) = \log_{10} x$.

$$f'(x) = \frac{1}{x \ln 10}$$

Note: We refer to the $\log_{10} x$ as $\log x$.

Example 9: Find the derivative of $f(x) = \log_8 (x^2 + x)$.

$$f'(x) = \frac{2x + 1}{(x^2 + x) \ln 8}$$

Example 10: Find the derivative of $f(x) = \log_e x$.

$$f'(x) = \frac{1}{x \ln e} = \frac{1}{x}$$

You can expect this result from Rules 1 and 2 involving natural logs.

THE DERIVATIVE OF a^x

You should recall from your precalculus days that we can rewrite a^x as $e^{x \ln a}$. Keep in mind that $\ln a$ is just a constant, which gives us the next rule.

$$\text{Rule No. 7: If } y = a^x, \text{then } \frac{dy}{dx} = \left(e^{x\ln a}\right)\ln a = a^x\left(\ln a\right)$$

Given the pattern of this chapter, you can guess what's coming: another rule that incorporates the Chain Rule.

$$\text{Rule No. 8: If } y = a^u, \text{ then } \frac{dy}{dx} = a^u\left(\ln a\right)\frac{du}{dx}$$

And now, some examples.

Example 11: Find the derivative of $f(x) = 3^x$.

$$f'(x) = 3^x \ln 3$$

Example 12: Find the derivative of $f(x) = 8^{4x^5}$.

$$f'(x) = 8^{4x^5}\left(20x^4\right)\ln 8$$

Example 13: Find the derivative of $f(x) = \pi^{\sin x}$.

$$f'(x) = \pi^{\sin x}\left(\cos x\right)\ln \pi$$

Finally, here's every nasty teacher's favorite exponential derivative.

Example 14: Find the derivative of $f(x) = x^x$.

First, rewrite this as $f(x) = e^{x\ln x}$. Then, take the derivative.

$$f'(x) = e^{x\ln x}\left(\ln x + \frac{x}{x}\right) = e^{x\ln x}\left(\ln x + 1\right) = x^x\left(\ln x + 1\right)$$

Would you have thought of that? Remember this trick. It might come in handy! Okay. Ready for some practice? Here are some more solved problems. Cover the solutions and get cracking.

PROBLEM 1. Find the derivative of $y = 3\ln(5x^2 + 4x)$.

Answer: Use Rule No. 2.

$$\frac{dy}{dx} = 3\frac{10x+4}{5x^2+4x} = \frac{30x+12}{5x^2+4x}$$

PROBLEM 2. Find the derivative of $f(x) = \ln\left(\sin\left(x^5\right)\right)$.

Answer: Use Rule No. 2 in addition to the Chain Rule.

$$f'(x) = \frac{5x^4\cos\left(x^5\right)}{\sin\left(x^5\right)} = 5x^4\cot\left(x^5\right)$$

PROBLEM 3. Find the derivative of $f(x) = e^{3x^7-4x^2}$.

Answer: Use Rule No. 4.

$$f'(x) = \left(21x^6 - 8x\right)e^{3x^7-4x^2}$$

PROBLEM 4. Find the derivative of $f(x) = \log_4(\tan x)$.

Answer: Use Rule No. 6.

$$f'(x) = \frac{1}{\ln 4}\frac{\sec^2 x}{\tan x}$$

PROBLEM 5. Find the derivative of $y = \log_8\sqrt{\dfrac{x^3}{1+x^2}}$.

Answer: First, use the rules of logarithms to rewrite the equation.

$$y = \frac{1}{2}\left[3\log_8 x - \log_8\left(1+x^2\right)\right]$$

Now it's much easier to find the derivative.

$$\frac{dy}{dx} = \frac{1}{2}\left[3\frac{1}{x\ln 8}\right] - \frac{1}{2}\left[\frac{2x}{\ln 8\left(1+x^2\right)}\right] = \frac{1}{2\ln 8}\left[\frac{3}{x} - \frac{2x}{\left(1+x^2\right)}\right]$$

PROBLEM 6. Find the derivative of $y = 5^{\sqrt{x}}$.

Answer: Use Rule No. 8.

$$\frac{dy}{dx} = 5^{\sqrt{x}} \frac{1}{2\sqrt{x}} \ln 5 = \frac{5^{\sqrt{x}} \ln 5}{2\sqrt{x}}$$

PROBLEM 7. Find the derivative of $y = \dfrac{e^{x^3}}{5^{\cos x}}$.

Answer: Here, you need to use the Quotient Rule and Rules Nos. 4 and 8.

$$\frac{dy}{dx} = \frac{5^{\cos x}\left(3x^2 e^{x^3}\right) - e^{x^3}\left(5^{\cos x} \ln 5 (-\sin x)\right)}{\left(5^{\cos x}\right)^2} = 5^{\cos x} e^{x^3} \frac{\left(3x^2\right) + \left(\sin x \ln 5\right)}{5^{2\cos x}} = e^{x^3} \frac{3x^2 + \sin x \ln 5}{5^{\cos x}}$$

PRACTICE PROBLEM SET 14

Now find the derivative of each of the following functions. The answers are in Chapter 19.

1. $f(x) = \ln\left(x^4 + 8\right)$

2. $f(x) = \ln\left(3x\sqrt{3+x}\right)$

3. $f(x) = \ln\left(\cot x - \csc x\right)$

4. $f(x) = x \ln \cos 3x - x^3$

5. $f(x) = \ln\left(\dfrac{5x^2}{\sqrt{5+x^2}}\right)$

6. $f(x) = e^{x \cos x}$

7. $f(x) = e^{-3x} \sin 5x$

8. $f(x) = \dfrac{e^{\tan 4x}}{4x}$

9. $f(x) = e^{\pi x} - \ln e^{\pi x}$

10. $f(x) = \log_{12}(x^3)$

11. $f(x) = \log_6(3x \tan x)$

12. $f(x) = \dfrac{\log_4 x}{e^{4x}}$

13. $f(x) = \log \sqrt{10^{3x}}$

14. $f(x) = \ln x \log x$

15. $f(x) = e^{3x} - 3^{ex}$

16. $f(x) = 10^{\sin x}$

17. $f(x) = 5^{\tan x}$

18. $f(x) = \ln(10^x)$

19. $f(x) = x^5 5^x$

Chapter 12
Other Topics
in Differential
Calculus

This chapter is devoted to other topics involving differential calculus that don't fit into a specific category.

THE DERIVATIVE OF AN INVERSE FUNCTION

ETS occasionally asks a question about finding the derivative of an inverse function. To do this, you need to learn only this simple formula.

Suppose we have a function $x = f(y)$ that is defined and differentiable at $y = a$ where $x = c$. Suppose we also know that the $f^{-1}(x)$ exists at $x = c$. Thus, $f(a) = c$ and $f^{-1}(c) = a$. Then, because $\dfrac{dy}{dx} = \dfrac{1}{\dfrac{dx}{dy}}$,

$$\frac{d}{dx} f^{-1}(x) \bigg|_{x=c} = \frac{1}{\left[\dfrac{d}{dy} f(y) \right]_{y=a}}$$

The short translation of this is: We can find the derivative of a function's inverse at a particular point by taking the reciprocal of the derivative at that point's corresponding y-value. These examples should help clear up any confusion.

Example 1: If $f(x) = x^2$, find a derivative of $f^{-1}(x)$ at $x = 9$.

First, notice that $f(3) = 9$. One of the most confusing parts of finding the derivative of an inverse function is that when you're asked to find the derivative at a value of x, they're *really* asking you for the derivative of the inverse of the function at the value that *corresponds* to $f(x) = 9$. This is because x-values of the inverse correspond to $f(x)$-values of the original function.

The rule is very simple: When you're asked to find the derivative of $f^{-1}(x)$ at $x = c$, you take the reciprocal of the derivative of $f(x)$ at $x = a$, where $f(a) = c$.

We know that $\dfrac{d}{dx} f(x) = 2x$. This means that we're going to plug $x = 3$ into the formula (because $f(3) = 9$). This gives us

$$\frac{1}{2x} \bigg|_{x=3} = \frac{1}{6}$$

We can verify this by finding the inverse of the function first and then taking the derivative. The inverse of the function $f(x) = x^2$ is the function $f^{-1}(x) = \sqrt{x}$. Now we find the derivative and evaluate it at $f(3) = 9$.

$$\frac{d}{dx}\sqrt{x} = \left.\frac{1}{2\sqrt{x}}\right|_{x=9} = \frac{1}{6}$$

Remember the rule: Find the value, a, of $f(x)$ that gives you the value of x that the problem asks for. Then plug that value, a, into the reciprocal of the derivative of the *inverse* function.

Example 2: Find a derivative of the inverse of $y = x^3 - 1$ when $y = 7$.

First, we need to find the x-value that corresponds to $y = 7$. A little algebra tells us that this is $x = 2$. Then,

$$\frac{dy}{dx} = 3x^2 \text{ and } \frac{1}{\dfrac{dy}{dx}} = \frac{1}{3x^2}$$

Therefore, the derivative of the inverse is

$$\left.\frac{1}{3x^2}\right|_{x=2} = \frac{1}{12}$$

Verify it: The inverse of the function $y = x^3 - 1$ is the function $y = \sqrt[3]{x+1}$. The derivative of this latter function is

$$\left.\frac{1}{3\sqrt[3]{(x+1)^2}}\right|_{x=7} = \frac{1}{12}$$

Let's do one more.

Example 3: Find a derivative of the inverse of $y = x^2 + 4$ when $y = 29$.

At $y = 29$, $x = 5$, the derivative of the function is

$$\frac{dy}{dx} = 2x$$

So, a derivative of the inverse is

$$\left.\frac{1}{2x}\right|_{x=5} = \frac{1}{10}$$

Note that $x = -5$ also gives us $y = 29$, so $-\dfrac{1}{10}$ is also a derivative. It's not that hard, once you get the hang of it.

This is all you'll be required to know involving derivatives of inverses. Naturally, there are ways to create harder problems, but the AP exam stays away from them and sticks to simpler stuff.

Here are some solved problems. Do each problem, cover the answer first, and then check your answer.

PROBLEM 1. Find a derivative of the inverse of $f(x) = 2x^3 + 5x + 1$ at $y = 8$.

Answer: First, we take the derivative of $f(x)$.

$$f'(x) = 6x^2 + 5$$

A possible value of x is $x = 1$.

Then, we use the formula to find the derivative of the inverse.

$$\frac{1}{f'(1)} = \frac{1}{11}$$

PROBLEM 2. Find a derivative of the inverse of $f(x) = 3x^3 - x + 7$ at $y = 9$.

Answer: First, take the derivative of $f(x)$.

$$f'(x) = 9x^2 - 1$$

A possible value of x is $x = 1$.

Then, use the formula to find the derivative of the inverse.

$$\frac{1}{f'(1)} = \frac{1}{8}$$

PROBLEM 3. Find a derivative of the inverse of $y = \dfrac{8}{x^3}$ at $y = 1$.

Answer: Take the derivative of y.

$$y' = -\frac{24}{x^4}$$

Find the value of x where $y = 1$.

$$1 = \frac{8}{x^3}$$
$$x = 2$$

Use the formula.

$$\left.\frac{1}{\dfrac{dy}{dx}}\right|_{x=2} = \frac{1}{\left.\left(-\dfrac{24}{x^4}\right)\right|_{x=2}} = -\frac{2}{3}$$

Here's one more.

PROBLEM 4. Find a derivative of the inverse of $y = 2x - x^3$ at $y = 1$.

Answer: The derivative of the function is

$$\frac{dy}{dx} = 2 - 3x^2$$

Next, find the value of x where $y = 1$. By inspection, $y = 1$ when $x = 1$.

Then, we use the formula to find the derivative of the inverse.

$$\left.\frac{1}{\dfrac{dy}{dx}}\right|_{x=1} = \frac{1}{\left.\left(2 - 3x^2\right)\right|_{x=1}} = -1$$

PRACTICE PROBLEM SET 15

Find a derivative of the inverse of each of the following functions. The answers are in Chapter 19.

1. $y = x + \dfrac{1}{x}$ at $y = \dfrac{17}{4}$; where $x > 1$

2. $y = 3x - 5x^3$ at $y = 2$

3. $y = e^x$ at $y = e$

4. $f(x) = x^7 - 2x^5 + 2x^3$ at $f(x) = 1$

5. $y = x + x^3$ at $y = -2$

6. $y = 4x - x^3$ at $y = 3$

7. $y = \ln x$ at $y = 0$

8. $y = x^{\frac{1}{3}} + x^{\frac{1}{5}}$ at $y = 2$

DIFFERENTIALS

Sometimes this is called "linearization." A differential is a very small quantity that corresponds to a change in a number. We use the symbol Δx to denote a differential. What are differentials used for? The AP exam mostly wants you to use them to approximate the value of a function or to find the error of an approximation.

Recall the formula for the definition of the derivative.

$$f'(x) = \lim_{h \to 0} \frac{f(x+h) - f(x)}{h}$$

Replace h with Δx, which also stands for a very small increment of x, and get rid of the limit.

$$f'(x) \approx \frac{f(x + \Delta x) - f(x)}{\Delta x}$$

Notice that this is no longer equal to the derivative, but an approximation of it. If Δx is kept small, the approximation remains fairly accurate. Next, rearrange the equation as follows:

$$f(x + \Delta x) \approx f(x) + f'(x)\Delta x$$

This is our formula for differentials. It says that "the value of a function (at x plus a little bit) equals the value of the function (at x) plus the product of the derivative of the function (at x) and the little bit."

Example 1: Use differentials to approximate $\sqrt{9.01}$.

You can start by letting $x = 9$, $\Delta x = +0.01$, $f(x) = \sqrt{x}$. Next, we need to find $f'(x)$.

$$f'(x) = \frac{1}{2\sqrt{x}}$$

Now, plug in to the formula.

$$f(x + \Delta x) \approx f(x) + f'(x)\Delta x$$
$$\sqrt{x + \Delta x} \approx \sqrt{x} + \frac{1}{2\sqrt{x}}\Delta x$$

Now, if we plug in $x = 9$ and $\Delta x = + 0.01$.

$$\sqrt{9.01} \approx \sqrt{9} + \frac{1}{2\sqrt{9}}(0.01) \approx 3.001666666$$

If you enter $\sqrt{9.01}$ into your calculator, you get: 3.001666204. As you can see, our answer is a pretty good approximation. It's not so good, however, when Δx is too big. How big is too big? Good question.

Example 2: Use differentials to approximate $\sqrt{9.5}$.

Let $x = 9$, $\Delta x = +.5$, $f(x) = \sqrt{x}$ and plug in to what you found in Example 1.

$$\sqrt{9.5} \approx \sqrt{9} + \frac{1}{2\sqrt{9}}(.5) \approx 3.083333333$$

However, $\sqrt{9.5}$ equals 3.082207001 on a calculator. This is good to only two decimal places. As the ratio of $\frac{\Delta x}{x}$ grows larger, the approximation gets less accurate, and we start to get away from the actual value.

There's another approximation formula that you'll need to know for the AP exam. This formula is used to estimate the error in a measurement, or to find the effect on a formula when a small change in measurement is made. The formula is:

$$dy = f'(x)dx$$

This notation may look a little confusing. It says that the change in a measurement dy, due to a differential dx, is found by multiplying the derivative of the equation for y by the differential. Let's do an example.

Note that this equation is simply a rearrangement of $\frac{dy}{dx} = f'(x)$.

Example 3: The radius of a circle is increased from 3 to 3.04. Estimate the change in area.

Let $A = \pi r^2$. Then our formula says that $dA = A'dr$, where A' is the derivative of the area with respect to r, and $dr = 0.04$ (the change). First, find the derivative of the area: $A' = 2\pi r$. Now, plug in to the formula.

$$dA = 2\pi r\, dr = 2\pi(3)(0.04) = 0.754$$

The actual change in the area is from 9π to 9.2416π, which is approximately 0.759. As you can see, this approximation formula is pretty accurate.

Here are some sample problems involving this differential formula. Try them out, then check your work against the answers directly beneath.

PROBLEM 1. Use differentials to approximate $(3.98)^4$.

Answer: Let $f(x) = x^4$, $x = 4$, and $\Delta x = -0.02$. Next, find $f'(x)$, which is: $f'(x) = 4x^3$.

Now, plug in to the formula.

$$f(x + \Delta x) \approx f(x) + f'(x)\Delta x$$
$$(x + \Delta x)^4 \approx x^4 + 4x^3\Delta x$$

If you plug in $x = 4$ and $\Delta x = -0.02$, you get

$$(3.98)^4 \approx 4^4 + 4(4)^3(-0.02) \approx 250.88$$

Check $(3.98)^4$ by using your calculator; you should get 250.9182722. Not a bad approximation.

You're probably asking yourself, why can't I just use my calculator every time? Because most math teachers are dedicated to teaching you several complicated ways to calculate things without your calculator.

PROBLEM 2. Use differentials to approximate $\sin 46°$.

Answer: This is a tricky question. The formula doesn't work if you use degrees. Here's why: Let $f(x) = \sin x$, $x = 45°$, and $\Delta x = 1°$. The derivative is $f'(x) = \cos x$.

If you plug this information into the formula, you get: $\sin 46° \approx \sin 45° + \cos 45°(1°) = \sqrt{2}$. You should recognize that this is nonsense for two reasons: (1) the sine of any angle is between −1 and 1; and (2) the answer should be close to $\sin 45° = \dfrac{1}{\sqrt{2}}$.

What went wrong? You have to use radians! As we mentioned before, angles in calculus problems are measured in radians, not degrees.

Let $f(x) = \sin x$, $x = \dfrac{\pi}{4}$, and $\Delta x = \dfrac{\pi}{180}$. Now plug in to the formula.

$$\sin\left(\frac{46\pi}{180}\right) \approx \sin\frac{\pi}{4} + \left(\cos\frac{\pi}{4}\right)\left(\frac{\pi}{180}\right) = 0.7194$$

PROBLEM 3. The radius of a sphere is measured to be 4 cm with an error of ±0.01 cm. Use differentials to approximate the error in the surface area.

Answer: Now it's time for the other differential formula. The formula for the surface area of a sphere is

$$S = 4\pi r^2$$

The formula says that $dS = S'\,dr$, so first, we find the derivative of the surface area, $(S' = 8\pi r)$ and plug away.

$$dS = 8\pi r\,dr = 8\pi(4)(\pm 0.01) = \pm 1.0053$$

This looks like a big error, but given that the surface area of a sphere with radius 4 is approximately 201 cm², the error is quite small.

PRACTICE PROBLEM SET 16

Use the differential formulas in this chapter to solve these problems. The answers are in Chapter 19.

1. Approximate $\sqrt{25.02}$.

2. Approximate $\sqrt[3]{63.97}$.

3. Approximate $\tan 61°$.

4. Approximate $(9.99)^3$.

5. The side of a cube is measured to be 6 in. with an error of ± 0.02 in. Estimate the error in the volume of the cube.

6. When a spherical ball bearing is heated, its radius increases by 0.01 mm. Estimate the change in volume of the ball bearing when the radius is 5 mm.

7. A side of an equilateral triangle is measured to be 10 cm. Estimate the change in the area of the triangle when the side shrinks to 9.8 cm.

8. A cylindrical tank is constructed to have a diameter of 5 meters and a height of 20 meters. Find the error in the volume if

   (a) the diameter is exact, but the height is 20.1 meters; and

   (b) the height is exact, but the diameter is 5.1 meters.

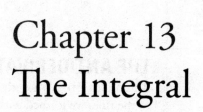

Chapter 13
The Integral

Welcome to the other half of calculus! This, unfortunately, is the more difficult half, but don't worry. We'll get you through it. In differential calculus, you learned all of the fun things that you can do with the derivative. Now you'll learn to do the reverse: how to take an integral. As you might imagine, there's a bunch of new fun things that you can do with integrals, too.

It's also time for a new symbol $\int$, which stands for integration. An integral actually serves several different purposes, but the first, and most basic, is that of the antiderivative.

THE ANTIDERIVATIVE

An antiderivative is a derivative in reverse. Therefore, we're going to reverse some of the rules we learned with derivatives and apply them to integrals. For example, we know that the derivative of x^2 is $2x$. If we are given the derivative of a function and have to figure out the original function, we use antidifferentiation. Thus, the antiderivative of $2x$ is x^2. (Actually, the answer is slightly more complicated than that, but we'll get into that in a few moments.)

Now we need to add some info here to make sure that you get this absolutely correct. First, as far as notation goes, it is traditional to write the antiderivative of a function using its uppercase letter, so the antiderivative of $f(x)$ is $F(x)$, the antiderivative of $g(x)$ is $G(x)$, and so on.

The second idea is very important: Each function has more than one antiderivative. In fact, there are an infinite number of antiderivatives of a function. Let's go back to our example to help illustrate this.

Remember that the antiderivative of $2x$ is x^2? Well, consider: If you take the derivative of $x^2 + 1$, you get $2x$. The same is true for $x^2 + 2$, $x^2 - 1$, and so on. In fact, if *any* constant is added to x^2, the derivative is still $2x$ because the derivative of a constant is zero.

Because of this, we write the antiderivative of $2x$ as $x^2 + C$, where C stands for any constant.

Just remember that you must always use the *dx* symbol, and teachers love to take points off for forgetting the *dx*. Don't ask why, but they do!

Finally, whenever you take the integral (or antiderivative) of a function of x, you always add the term *dx* (or *dy* if it's a function of y, etc.) to the integrand (the thing inside the integral). You'll learn why later.

Here is the **Power Rule** for antiderivatives.

$$\text{If } f(x) = x^n, \text{ then } \int f(x)dx = \frac{x^{n+1}}{n+1} + C \ \ (\text{except when } n = -1).$$

Example 1: Find $\int x^3 dx$.

Using the Power Rule, we get

$$\int x^3 dx = \frac{x^4}{4} + C$$

Don't forget the constant C, or your teachers will take points off for that, too!

Example 2: Find $\int x^{-3} dx$.

The Power Rule works with negative exponents, too.

$$\int x^{-3} dx = \frac{x^{-2}}{-2} + C$$

Think of a constant as something that is always there, and you won't forget it.

Not terribly hard, is it? Now it's time for a few more rules that look remarkably similar to the rules for derivatives that we saw in Chapter 6.

$$\int kf(x)dx = k \int f(x)dx$$

$$\int [f(x) + g(x)]dx = \int f(x)dx + \int g(x)dx$$

$$\int k\,dx = kx + C$$

Here are a few more examples to make you an expert.

Example 3: $\int 5\,dx = 5x + C$

Example 4: $\int 7x^3 dx = \frac{7x^4}{4} + C$

Example 5: $\int \left(3x^2 + 2x\right) dx = x^3 + x^2 + C$

Example 6: $\int \sqrt{x}\; dx = \dfrac{x^{\frac{3}{2}}}{\frac{3}{2}} + C = \dfrac{2x^{\frac{3}{2}}}{3} + C$

INTEGRALS OF TRIG FUNCTIONS

The integrals of some trigonometric functions follow directly from the derivative formulas in Chapter 6.

$$\int \sin\; ax\; dx = -\frac{\cos ax}{a} + C$$

$$\int \cos ax\; dx = \frac{\sin ax}{a} + C$$

$$\int \sec ax \tan ax\; dx = \frac{\sec ax}{a} + C$$

$$\int \sec^2 ax\; dx = \frac{\tan ax}{a} + C$$

$$\int \csc ax \cot ax\; dx = -\frac{\csc ax}{a} + C$$

$$\int \csc^2 ax\; dx = -\frac{\cot ax}{a} + C$$

We didn't mention the integrals of tangent, cotangent, secant, and cosecant, because you need to know some rules about logarithms to figure them out. We'll get to them in a few chapters. Notice also that each of the answers is divided by a constant. This is to account for the Chain Rule. Let's do some examples.

Example 7: Check the integral $\int \sin 5x\; dx = -\dfrac{\cos 5x}{5} + C$ by differentiating the answer.

$$\frac{d}{dx}\left[-\frac{\cos 5x}{5} + C \right] = -\frac{1}{5}\left(-\sin 5x\right)(5) = \sin 5x$$

Notice how the constant is accounted for in the answer?

Example 8: $\int \sec^2 3x\; dx = \dfrac{\tan 3x}{3} + C$

Example 9: $\int \cos \pi x\; dx = \dfrac{\sin \pi x}{\pi} + C$

Example 10: $\int \sec\left(\dfrac{x}{2}\right)\tan\left(\dfrac{x}{2}\right) dx = 2\sec\left(\dfrac{x}{2}\right) + C$

If you're not sure if you have the correct answer when you take an integral, you can always check by differentiating the answer and seeing if you get what you started with. Try to get in the habit of doing that at the beginning, because it'll help you build confidence in your ability to find integrals properly. You'll see that, although you can differentiate just about any expression that you'll normally encounter, you won't be able to integrate many of the functions you see.

ADDITION AND SUBTRACTION

By using the rules for addition and subtraction, we can integrate most polynomials.

Example 11: Find $\int \left(x^3 + x^2 - x \right) dx$.

We can break this into separate integrals, which gives us

$$\int x^3 dx + \int x^2 dx - \int x \ dx$$

Now you can integrate each of these individually.

$$\frac{x^4}{4} + C + \frac{x^3}{3} + C - \frac{x^2}{2} + C$$

You can combine the constants into one constant (it doesn't matter how many C's we use, because their sum is one collective constant whose derivative is zero).

$$\frac{x^4}{4} + \frac{x^3}{3} - \frac{x^2}{2} + C$$

Sometimes you'll be given information about the function you're seeking that will enable you to solve for the constant. Often, this is an "initial value," which is the value of the function when the variable is zero. As we've seen, normally there are an infinite number of solutions for an integral, but when we solve for the constant, there's only one.

Example 12: Find the equation of y where $\dfrac{dy}{dx} = 3x + 5$ and $y = 6$ when $x = 0$.

Let's put this in integral form.

$$y = \int (3x + 5) \ dx$$

Integrating, we get

$$y = \frac{3x^2}{2} + 5x + C$$

Now we can solve for the constant because we know that $y = 6$ when $x = 0$.

$$6 = \frac{3(0)^2}{2} + 5(0) + C$$

Therefore, $C = 6$ and the equation is

$$y = \frac{3x^2}{2} + 5x + 6$$

Example 13: Find $f(x)$ if $f'(x) = \sin x - \cos x$ and $f(\pi) = 3$.

Integrate $f'(x)$.

$$f(x) = \int (\sin x - \cos x)\ dx = -\cos x - \sin x + C$$

Now solve for the constant.

$$3 = -\cos(\pi) - \sin(\pi) + C$$

$$C = 2$$

Therefore, the equation becomes

$$f(x) = -\cos x - \sin x + 2$$

Now we've covered the basics of integration. However, integration is a very sophisticated topic and there are many types of integrals that will cause you trouble. We will need several techniques to learn how to evaluate these integrals. The first and most important is called *u*-substitution, which we will cover in the second half of this chapter.

In the meantime, here are some solved problems. Do each problem, covering the answer first, then check your answer.

PROBLEM 1. Evaluate $\int x^{\frac{3}{5}} dx$.

Answer: Here's the Power Rule again.

$$\int x^n dx = \frac{x^{n+1}}{n+1} + C$$

Using the rule,

$$\int x^{\frac{3}{5}} dx = \frac{x^{\frac{8}{5}}}{\frac{8}{5}} + C$$

You can rewrite it as $\dfrac{5x^{\frac{8}{5}}}{8}+C$.

PROBLEM 2. Evaluate $\displaystyle\int \left(5x^3 + x^2 - 6x + 4\right) dx$.

Answer: We can break this up into several integrals.

$$\int \left(5x^3 + x^2 - 6x + 4\right) dx = \int 5x^3\, dx + \int x^2\, dx - \int 6x\, dx + 4\int dx$$

Each of these can be integrated according to the Power Rule.

$$\frac{5x^4}{4} + C + \frac{x^3}{3} + C - \frac{6x^2}{2} + C + 4x + C$$

This can be rewritten as

$$\frac{5x^4}{4} + \frac{x^3}{3} - 3x^2 + 4x + C$$

Notice that we combine the constant terms into one constant term C.

PROBLEM 3. Evaluate $\displaystyle\int \left(3 - x^2\right)^2 dx$.

Answer: First, expand the integrand.

$$\int \left(9 - 6x^2 + x^4\right) dx$$

Break this up into several integrals.

$$\int 9\, dx - 6\int x^2\, dx + \int x^4\, dx$$

And integrate according to the Power Rule.

$$9x - 2x^3 + \frac{x^5}{5} + C$$

PROBLEM 4. Evaluate $\displaystyle\int \left(4 \sin x - 3 \cos x\right) dx$.

Answer: Break this problem into two integrals.

$$4\int \sin x\, dx - 3\int \cos x\, dx$$

Each of these trig integrals can be evaluated according to its rule.

$$-4 \cos x - 3 \sin x + C$$

PROBLEM 5. Evaluate $\int \left(2 \sec^2 x - 5 \csc^2 x\right) dx$.

Answer: Break the integral in two.

$$2 \int \sec^2 x \, dx - 5 \int \csc^2 x \, dx$$

Each of these trig integrals can be evaluated according to its rule.

$$2 \tan x + 5 \cot x + C$$

PRACTICE PROBLEM SET 17

Now evaluate the following integrals. The answers are in Chapter 19.

1. $\int \dfrac{1}{x^4} dx$

2. $\int \dfrac{5}{\sqrt{x}} dx$

3. $\int \dfrac{x^5 + 7}{x^2} dx$

4. $\int \left(5x^4 - 3x^2 + 2x + 6\right) dx$

5. $\int \left(3x^{-3} - 2x^{-2} + x^4 + 16x^7\right) dx$

6. $\int \left(1 + x^2\right)(x - 2) dx$

7. $\int x^{\frac{1}{3}} (2 + x) dx$

8. $\displaystyle\int \left(x^3 + x\right)^2 dx$

9. $\displaystyle\int \frac{x^6 - 2x^4 + 1}{x^2}\, dx$

10. $\displaystyle\int x\left(x - 1\right)^3 dx$

11. $\displaystyle\int \left(\cos x - 5\sin x\right) dx$

12. $\displaystyle\int \sec x\left(\sec x + \tan x\right) dx$

13. $\displaystyle\int \left(\sec^2 x + x\right) dx$

14. $\displaystyle\int \frac{\sin x}{\cos^2 x}\, dx$

15. $\displaystyle\int \frac{\cos^3 x + 4}{\cos^2 x}\, dx$

16. $\displaystyle\int \frac{\sin 2x}{\cos x}\, dx$

17. $\displaystyle\int \left(1 + \cos^2 x \sec x\right) dx$

18. $\displaystyle\int \left(\tan^2 x\right) dx$

19. $\displaystyle\int \frac{1}{\csc x}\, dx$

20. $\displaystyle\int \left(x - \frac{2}{\cos^2 x}\right) dx$

u-SUBSTITUTION

When we discussed differentiation, one of the most important techniques we mastered was the Chain Rule. Now, you'll learn the integration corollary of the Chain Rule (called *u*-substitution), which we use when the integrand is a composite function. All you do is replace the function with *u*, and then you can integrate the simpler function using the Power Rule (as shown below).

$$\int u^n \, du = \frac{u^{n+1}}{n+1} + C$$

Suppose you have to integrate $\int (x-4)^{10} \, dx$. You could expand out this function and integrate each term, but that'll take a while. Instead, you can follow these four steps.

Step 1: Let $u = x - 4$. Then $\dfrac{du}{dx} = 1$ (rearrange this to get $du = dx$).

Step 2: Substitute $u = x - 4$ and $du = dx$ into the integrand.

$$\int u^{10} \, du$$

Step 3: Integrate.

$$\int u^{10} \, du = \frac{u^{11}}{11} + C$$

Step 4: Substitute back for *u*.

$$\frac{(x-4)^{11}}{11} + C$$

That's *u*-substitution. The main difficulty you'll have will be picking the appropriate function to set equal to *u*. The best way to get better is to practice. The object is to pick a function and replace it with *u*, then take the derivative of *u* to find *du*. If we can't replace *all* of the terms in the integrand, we *can't* do the substitution.

Let's do some examples.

Example 1: $\int 10x\left(5x^2-3\right)^6 dx =$

Once again, you could expand this out and integrate each term, but that would be difficult. Use *u*-substitution.

Let $u = 5x^2 - 3$. Then $\dfrac{du}{dx} = 10x$ and $du = 10x\, dx$. Now you can substitute.

$$\int u^6\, du$$

And integrate.

$$\int u^6\, du = \frac{u^7}{7} + C$$

Substituting back gives you

$$\frac{\left(5x^2-3\right)^7}{7} + C$$

Confirm that this is the integral by differentiating $\dfrac{\left(5x^2-3\right)^7}{7} + C$.

$$\frac{d}{dx}\left[\frac{\left(5x^2-3\right)^7}{7} + C\right] = \frac{7\left(5x^2-3\right)^6}{7}(10x) = \left(5x^2-3\right)^6(10x)$$

Example 2: $\int 2x\sqrt{x^2-5}\ dx =$

If $u = x^2 - 5$, then $\dfrac{du}{dx} = 2x$ and $du = 2x\, dx$. Substitute u into the integrand.

$$\int u^{\frac{1}{2}}\ du$$

Integrate.

$$\int u^{\frac{1}{2}}\ du = \frac{u^{\frac{3}{2}}}{\frac{3}{2}} + C = \frac{2u^{\frac{3}{2}}}{3} + C$$

And substitute back.

$$\frac{2\left(x^2-5\right)^{\frac{3}{2}}}{3}+C$$

Note: From now on, we're not going to rearrange $\frac{du}{dx}$; we'll go directly to "$du=$" format. You should be able to do that step on your own.

Example 3: $\displaystyle\int 3\sin\left(3x-1\right)dx=$

Let $u=3x-1$. Then $du=3dx$. Substitute the u in the integral.

$$\int \sin u\; du$$

Figure out the integral.

$$\int \sin u\; du=-\cos u+C$$

And throw the x's back in.

$$-\cos(3x-1)+C$$

So far, this is only the simplest kind of u-substitution; naturally, the process can get worse when the substitution isn't as easy. Usually, you'll have to insert a constant term to put the integrand into a workable form.

Example 4: $\displaystyle\int \left(5x+7\right)^{20}dx=$

Let $u=5x+7$. Then $du=5\;dx$. Notice that we can't do the substitution immediately because we need to substitute for dx and we have $5\;dx$. No problem: Because 5 is a constant, just solve for dx.

$$\frac{1}{5}du=dx$$

Now you can substitute.

$$\int \left(5x+7\right)^{20}dx=\int u^{20}\left(\frac{1}{5}\right)du$$

Rearrange the integral and solve.

$$\int u^{20}\left(\frac{1}{5}\right)du = \frac{1}{5}\int u^{20}\,du = \frac{1}{5}\frac{u^{21}}{21}+C = \frac{u^{21}}{105}+C$$

And now it's time to substitute back.

$$\frac{u^{21}}{105}+C = \frac{(5x+7)^{21}}{105}+C$$

Example 5: $\int x\cos\left(3x^2+1\right)dx =$

Let $u = 3x^2 + 1$. Then $du = 6x\,dx$. We need to substitute for $x\,dx$, so we can rearrange the du term.

$$\frac{1}{6}du = x\,dx$$

Now substitute.

$$\int \frac{1}{6}\cos u\,du$$

Evaluate the integral.

$$\int \frac{1}{6}\cos u\,du = \frac{1}{6}\sin u + C$$

And substitute back.

$$\frac{1}{6}\sin u + C = \frac{1}{6}\sin\left(3x^2+1\right)+C$$

Example 6: $\int x\sec^2\left(x^2\right)dx =$

Let $u = x^2$. Then $du = 2x\,dx$ and $\frac{1}{2}du = x\,dx$.

Substitute.

$$\int \frac{1}{2}\sec^2 u\,du$$

Evaluate the integral.

$$\int \frac{1}{2}\sec^2 u \ du = \frac{1}{2}\tan u + C$$

Now the original function goes back in.

$$\frac{1}{2}\tan\left(x^2\right)+C$$

This is a good technique to master, so practice on the following solved problems. Do each problem, covering the answer first, then check your answer.

PROBLEM 1. Evaluate $\int \sec^2 3x \ dx$.

Answer: Let $u = 3x$ and $du = 3dx$. Then $\frac{1}{3}du = dx$.

Substitute and integrate.

$$\frac{1}{3}\int \sec^2 u \ du = \frac{1}{3}\tan u + C$$

Then substitute back.

$$\frac{1}{3}\tan 3x + C$$

PROBLEM 2. Evaluate $\int \sqrt{5x-4} \ dx$.

Answer: Let $u = 5x - 4$ and $du = 5dx$. Then $\frac{1}{5}du = dx$.

Substitute and integrate.

$$\frac{1}{5}\int u^{\frac{1}{2}} \ du = \frac{2}{15}u^{\frac{3}{2}} + C$$

Then substitute back.

$$\frac{2}{15}(5x-4)^{\frac{3}{2}} + C$$

PROBLEM 3. Evaluate $\int x\left(4x^2 - 7\right)^{10} dx$.

Answer: Let $u = 4x^2 - 7$ and $du = 8x\, dx$. Then $\dfrac{1}{8} du = x\, dx$.

Substitute and integrate.

$$\frac{1}{8}\int u^{10}\, du = \frac{1}{88} u^{11} + C$$

Then substitute back.

$$\frac{1}{88}\left(4x^2 - 7\right)^{11} + C$$

PROBLEM 4. Evaluate $\int \tan\dfrac{x}{3}\,\sec^2\dfrac{x}{3} dx$.

Answer: Let $u = \tan\dfrac{x}{3}$ and $du = \dfrac{1}{3}\sec^2\dfrac{x}{3} dx$. Then $3du = \sec^2\dfrac{x}{3} dx$.

Substituting, we get

$$3\int u\, du = \frac{3}{2} u^2 + C$$

Then substitute back.

$$\frac{3}{2}\tan^2\frac{x}{3} + C$$

PRACTICE PROBLEM SET 18

Now evaluate the following integrals. The answers are in Chapter 19.

1. $\int \sin 2x \cos 2x \, dx$

2. $\int \dfrac{3x \, dx}{\sqrt[3]{10 - x^2}}$

3. $\int x^3 \sqrt{5x^4 + 20} \, dx$

4. $\int \dfrac{dx}{(x-1)^2}$

5. $\int (x^2 + 1)(x^3 + 3x)^{-5} \, dx$

6. $\int \dfrac{1}{\sqrt{x}} \sin \sqrt{x} \, dx$

7. $\int x^2 \sec^2 x^3 \, dx$

8. $\int \dfrac{\cos\left(\dfrac{3}{x}\right)}{x^2} \, dx$

9. $\int \dfrac{\sin 2x}{(1 - \cos 2x)^3} \, dx$

10. $\int \sin(\sin x) \cos x \, dx$

Chapter 14
Definite Integrals

AREA UNDER A CURVE

It's time to learn one of the most important uses of the integral. We've already discussed how integration can be used to "antidifferentiate" a function; now you'll see that you can also use it to find the area under a curve. First, here's a little background about how to find the area without using integration.

Suppose you have to find the area under the curve $y = x^2 + 2$ from $x = 1$ to $x = 3$. The graph of the curve looks like the following:

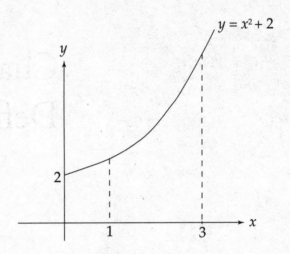

Don't panic yet. Nothing you've learned in geometry thus far has taught you how to find the area of something like this. You have learned how to find the area of a rectangle, though, and we're going to use rectangles to approximate the area between the curve and the x-axis.

Let's divide the region into two rectangles, one from $x = 1$ to $x = 2$ and one from $x = 2$ to $x = 3$, where the top of each rectangle comes just under the curve. It looks like the following:

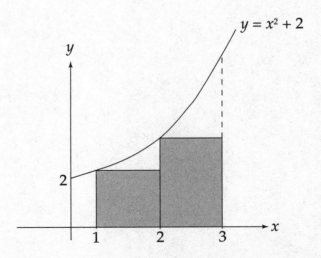

Notice that the width of each rectangle is 1. The height of the left rectangle is found by plugging 1 into the equation $y = x^2 + 2$ (yielding 3); the height of the right rectangle is found by plugging 2 into the same equation (yielding 6). The combined area of the two rectangles is $(1)(3) + (1)(6) = 9$. So we could say that the area under the curve is approximately 9 square units.

Naturally, this is a pretty rough approximation that significantly underestimates the area. Look at how much of the area we missed by using two rectangles! How do you suppose we could make the approximation better? Divide the region into more, thinner rectangles.

This time, cut up the region into four rectangles, each with a width of $\frac{1}{2}$. It looks like the following:

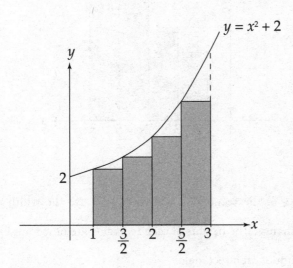

Now find the height of each rectangle the same way as before. Notice that the values we use are the left endpoints of each rectangle. The heights of the rectangles are, respectively,

$$(1)^2 + 2 = 3; \ \left(\frac{3}{2}\right)^2 + 2 = \frac{17}{4}; \ (2)^2 + 2 = 6 \text{ and } \left(\frac{5}{2}\right)^2 + 2 = \frac{33}{4}$$

Now, multiply each height by the width of $\frac{1}{2}$ and add up the areas.

$$\left(\frac{1}{2}\right)(3) + \left(\frac{1}{2}\right)\left(\frac{17}{4}\right) + \left(\frac{1}{2}\right)(6) + \left(\frac{1}{2}\right)\left(\frac{33}{4}\right) = \frac{43}{4}$$

This is a much better approximation of the area, but there's still a lot of space that isn't accounted for. We're still underestimating the area. The rectangles need to be thinner. But before we do that, let's do something else.

Notice how each of the rectangles is inscribed in the region. Suppose we used circumscribed rectangles instead—that is, we could determine the height of each rectangle by the higher of the two y-values, not the lower. The following graph shows what the region would look like:

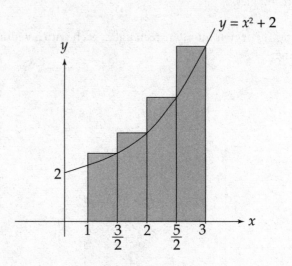

To find the area of the rectangles, we would still use the width of $\frac{1}{2}$, but the heights would change. The heights of each rectangle are now found by plugging in the right endpoint of each rectangle.

$$\left(\frac{3}{2}\right)^2 + 2 = \frac{17}{4}; (2)^2 + 2 = 6; \left(\frac{5}{2}\right)^2 + 2 = \frac{33}{4}; \text{ and } (3)^2 + 2 = 11$$

Once again, multiply each height by the width of $\frac{1}{2}$ and add up the areas.

$$\left(\frac{1}{2}\right)\left(\frac{17}{4}\right) + \left(\frac{1}{2}\right)(6) + \left(\frac{1}{2}\right)\left(\frac{33}{4}\right) + \left(\frac{1}{2}\right)(11) = \frac{59}{4}$$

The area under the curve using four left endpoint rectangles is $\frac{43}{4}$, and the area using four right endpoint rectangles is $\frac{59}{4}$, so why not average the two? This gives us $\frac{51}{4}$, which is a better approximation of the area.

Now that we've found the area using the rectangles a few times, let's turn the method into a formula. Call the left endpoint of the interval a and the right endpoint of the interval b, and set the number of rectangles we use equal to n. So the width of each rectangle is $\frac{b-a}{n}$. The height of the first inscribed rectangle is y_0, the height of the second rectangle is y_1, the height of the third rectangle is y_2, and so on, up to the last rectangle, which is y_{n-1}. If we use the left endpoint of each rectangle, the area under the curve is

$$\left(\frac{b-a}{n}\right)\left[y_0 + y_1 + y_2 + y_3 \ldots + y_{n-1}\right]$$

If we use the right endpoint of each rectangle, then the formula is

$$\left(\frac{b-a}{n}\right)\left[y_1 + y_2 + y_3 \ldots + y_n\right]$$

Now for the fun part. Remember how we said that we could make the approximation better by making more, thinner rectangles? By letting n approach infinity, we create an infinite number of rectangles that are infinitesimally thin. The formula for "left-endpoint" rectangles becomes

$$\lim_{n \to \infty}\left(\frac{b-a}{n}\right)\left[y_0 + y_1 + y_2 + y_3 \ldots + y_{n-1}\right]$$

For "right-endpoint" rectangles, the formula becomes

$$\lim_{n \to \infty} \left(\frac{b-a}{n} \right) \left[y_1 + y_2 + y_3 \ldots + y_n \right]$$

We could also find the area using the midpoint of each interval. Let's once again use the above example, dividing the region into four rectangles. The region would look like the following:

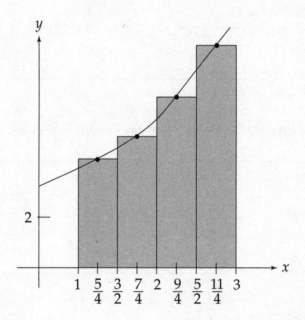

To find the area of the rectangles, we would still use the width of $\frac{1}{2}$, but the heights would now be found by plugging the midpoint of each interval into the equation.

$$\left(\frac{5}{4} \right)^2 + 2 = \frac{57}{16}; \left(\frac{7}{4} \right)^2 + 2 = \frac{81}{16}; \left(\frac{9}{4} \right)^2 + 2 = \frac{113}{16}; \left(\frac{11}{4} \right)^2 + 2 = \frac{153}{16}$$

Once again, multiply each height by the width of $\frac{1}{2}$ and add up the areas.

$$\left(\frac{1}{2}\right)\left(\frac{57}{16}\right)+\left(\frac{1}{2}\right)\left(\frac{81}{16}\right)+\left(\frac{1}{2}\right)\left(\frac{113}{16}\right)+\left(\frac{1}{2}\right)\left(\frac{153}{16}\right)=\frac{404}{32}=\frac{101}{8}=12.625$$

The general formula for approximating the area under a curve using midpoints is

$$\left(\frac{b-a}{n}\right)\left[y_{\frac{1}{2}}+y_{\frac{3}{2}}+y_{\frac{5}{2}}+\ldots+y_{\frac{2n-1}{2}}\right]$$

(Note: The fractional subscript means to evaluate the function at the number halfway between each integral pair of values of n.)

When you take the limit of this infinite sum, you get the integral. (You knew the integral had to show up sometime, didn't you?) Actually, we write the integral as

$$\int_{1}^{3}\left(x^2+2\right)dx$$

It is called a **definite integral**, and it means that we're finding the area under the curve x^2+2 from $x=1$ to $x=3$. (We'll discuss how to evaluate this in a moment.) On the AP exam, you'll only be asked to divide the region into a small number of rectangles, so it won't be very hard. Let's do an example.

Example 1: Approximate the area under the curve $y=x^3$ from $x=2$ to $x=3$ using four left-endpoint rectangles.

Draw four rectangles that look like the following:

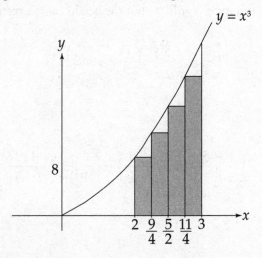

The width of each rectangle is $\frac{1}{4}$. The heights of the rectangles are

$$2^3, \left(\frac{9}{4}\right)^3, \left(\frac{5}{2}\right)^3, \text{ and } \left(\frac{11}{4}\right)^3$$

Therefore, the area is

$$\left(\frac{1}{4}\right)\left(2^3\right)+\left(\frac{1}{4}\right)\left(\frac{9}{4}\right)^3+\left(\frac{1}{4}\right)\left(\frac{5}{2}\right)^3+\left(\frac{1}{4}\right)\left(\frac{11}{4}\right)^3 = \frac{893}{64} \approx 13.953$$

Example 2: Repeat Example 1 using four right-endpoint rectangles.

Now draw four rectangles that look like the following:

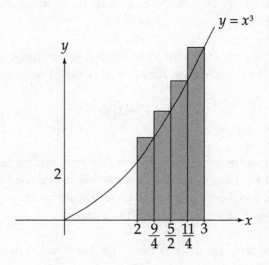

The width of each rectangle is still $\frac{1}{4}$, but the heights of the rectangles are now

$$\left(\frac{9}{4}\right)^3, \left(\frac{5}{2}\right)^3, \left(\frac{11}{4}\right)^3, \text{ and } \left(3^3\right)$$

The area is now

$$\left(\frac{1}{4}\right)\left(\frac{9}{4}\right)^3+\left(\frac{1}{4}\right)\left(\frac{5}{2}\right)^3+\left(\frac{1}{4}\right)\left(\frac{11}{4}\right)^3+\left(\frac{1}{4}\right)\left(3^3\right) = \frac{1197}{64} \approx 18.703$$

Example 3: Repeat Example 1 using four midpoint rectangles.

Now draw four rectangles that look like the following:

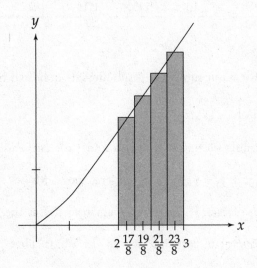

The width of each rectangle is still $\dfrac{1}{4}$, but the heights of the rectangles are now

$$\left(\frac{17}{8}\right)^3, \left(\frac{19}{8}\right)^3, \left(\frac{21}{8}\right)^3, \text{ and } \left(\frac{23}{8}\right)^3$$

The area is now

$$\left(\frac{1}{4}\right)\left(\frac{17}{8}\right)^3 + \left(\frac{1}{4}\right)\left(\frac{19}{8}\right)^3 + \left(\frac{1}{4}\right)\left(\frac{21}{8}\right)^3 + \left(\frac{1}{4}\right)\left(\frac{23}{8}\right)^3 = \frac{2075}{128} \approx 16.211$$

TABULAR RIEMANN SUMS

Sometimes the AP exam will ask you to find a Riemann sum, or to approximate an integral (same thing, right?), but won't give you a function to work with. Instead, they will give you a table of values for x and $f(x)$. These are quite simple to evaluate. All you do is use the right-hand or left-hand sum formula, plugging in the appropriate values for $f(x)$. One thing you should watch out for is that sometimes the x-values are not evenly spaced, so make sure that you use the correct values for the widths of the rectangles. Let's do an example.

Example 4: Suppose we are given the following table of values for x and $f(x)$.

x	2	4	6	8	10	12
$f(x)$	10	13	15	14	9	3

Use a *right-hand* Riemann sum with 5 subintervals indicated by the data in the table to approximate $\int_{2}^{12} f(x)dx$.

Recall that the formula for finding the area under the curve using the right endpoints is: $\left(\dfrac{b-a}{n}\right)[y_1 + y_2 + y_3 + \ldots + y_n]$. Here, the width of each rectangle is 2. We find the height of each rectangle by evaluating $f(x)$ at the appropriate value of x, the right endpoint of each interval on the x-axis. Here, $y_1 = 13$, $y_2 = 15$, $y_3 = 14$, $y_4 = 9$, and $y_5 = 3$. Therefore, we can approximate the integral with:

$$\int_{2}^{12} f(x)dx = (2)(13) + (2)(15) + (2)(14) + (2)(9) + (2)(3) = 108.$$

Let's do another example but this time the values of x will not be evenly spaced on the x-axis.

Example 5: Given the following table of values for x and $f(x)$:

x	0	2	5	11	19	22	23
$f(x)$	4	6	16	18	22	29	50

Use a *left-hand* Riemann sum with 6 subintervals indicated by the data in the table to approximate $\int_{0}^{23} f(x)dx$.

Recall that the formula for finding the area under the curve using the left endpoints is: $\left(\dfrac{b-a}{n}\right)[y_0 + y_1 + y_2 + \ldots + y_{n-1}]$. This formula assumes that the x-values are evenly spaced but they aren't here, so we will replace the values of $\left(\dfrac{b-a}{n}\right)$ with the appropriate widths of each rectangle. The width of the first rectangle is $2 - 0 = 2$; the second width is $5 - 2 = 3$; the third is $11 - 5 = 6$; the fourth is $19 - 11 = 8$; the fifth is $22 - 19 = 3$; and the sixth is $23 - 22 = 1$. We find the height of each rectangle by evaluating $f(x)$ at the appropriate value of x, the left endpoint of each interval on the x-axis. Here, $y_0 = 4$, $y_1 = 6$, $y_2 = 16$, $y_3 = 18$, $y_4 = 22$, and $y_5 = 29$. Therefore, we can approximate the integral with:

$$\int_0^{23} f(x)dx = (2)(4) + (3)(6) + (6)(16) + (8)(18) + (3)(22) + (1)(29) = 361.$$

That's all there is to approximating the area under a curve using rectangles. Now let's learn how to find the area exactly. In order to evaluate this, you'll need to know. . .

The Fundamental Theorem of Calculus

Before, we said that if you create an infinite number of infinitely thin rectangles, you'll get the area under the curve, which is an integral. For the example above, the integral is

$$\int_2^3 x^3 dx$$

There is a rule for evaluating an integral like this. The rule is called the **Fundamental Theorem of Calculus**, and it says

$$\int_a^b f(x)dx = F(b) - F(a)\,;\text{ where } F(x) \text{ is the antiderivative of } f(x).$$

Using this rule, you can find $\int_2^3 x^3 dx$ by integrating it, and we get $\dfrac{x^4}{4}$. Now all you do is plug in 3 and 2 and take the difference. We use the following notation to symbolize this

$$\left.\frac{x^4}{4}\right|_2^3$$

Thus, we have

$$\frac{3^4}{4} - \frac{2^4}{4} = \frac{81}{4} - \frac{16}{4} = \frac{65}{4}$$

Because $\dfrac{65}{4} = 16.25$, you can see how close we were with our three earlier approximations.

Example 6: Find $\int_1^3 (x^2 + 2) dx$.

The Fundamental Theorem of Calculus yields the following:

$$\int_1^3 (x^2 + 2) dx = \left(\frac{x^3}{3} + 2x \right)\Bigg|_1^3$$

If we evaluate this, we get the following:

$$\left(\frac{3^3}{3} + 2(3) \right) - \left(\frac{1^3}{3} + 2(1) \right) = \frac{38}{3}$$

This is the first function for which we found the approximate area by using inscribed rectangles. Our final estimate, where we averaged the inscribed and circumscribed rectangles, was $\frac{51}{4}$, and as you can see, that was very close (off by $\frac{1}{12}$). When we used the midpoints, we were off by $\frac{1}{24}$.

We're going to do only a few approximations using rectangles, because it's not a big part of the AP exam. On the other hand, definite integrals are a huge part of the rest of this book.

Example 7: $\int_1^5 (x^2 - x) dx = \left(\frac{x^3}{3} - \frac{x^2}{2} \right)\Bigg|_1^5 = \left(\frac{125}{3} - \frac{25}{2} \right) - \left(\frac{1}{3} - \frac{1}{2} \right) = \frac{88}{3}$

Example 8: $\int_0^{\frac{\pi}{2}} \sin x \, dx = (-\cos x)\Bigg|_0^{\frac{\pi}{2}} = \left(-\cos \frac{\pi}{2} \right) - (-\cos 0) = 1$

Example 9: $\int_0^{\frac{\pi}{4}} \sec^2 x \, dx = \tan x\Bigg|_0^{\frac{\pi}{4}} = \tan \frac{\pi}{4} - \tan 0 = 1$

The Trapezoid Rule

There's another approximation method that's even better than the rectangle method. Essentially, all you do is divide the region into trapezoids instead of rectangles. Let's use the problem that we did at the beginning of the chapter.

We get a picture that looks like the following:

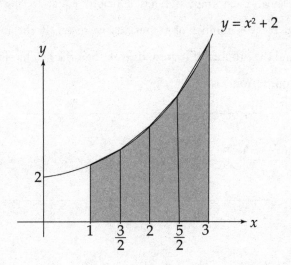

As you should recall from geometry, the formula for the area of a trapezoid is

$$\frac{1}{2}\left(b_1 + b_2\right)h$$

(Note: b_1 and b_2 are the two bases of the trapezoid.) Notice that each of the shapes is a trapezoid on its side, so the height of each trapezoid is the length of the interval $\frac{1}{2}$, and the bases are the y-values that correspond to each x-value. We found these earlier in the rectangle example; they are, in order: $3, \frac{17}{4}, 6, \frac{33}{4}$, and 11. We can find the area of each trapezoid and add them up.

$$\frac{1}{2}\left(3 + \frac{17}{4}\right)\left(\frac{1}{2}\right) + \frac{1}{2}\left(\frac{17}{4} + 6\right)\left(\frac{1}{2}\right) + \frac{1}{2}\left(6 + \frac{33}{4}\right)\left(\frac{1}{2}\right) + \frac{1}{2}\left(\frac{33}{4} + 11\right)\left(\frac{1}{2}\right) = \frac{51}{4},$$

or 12.75

Recall that the actual value of the area is $\dfrac{38}{3}$ or 12.67; the Trapezoid Rule gives a pretty good approximation.

Notice how each trapezoid shares a base with the trapezoid next to it, except for the end ones. This enables us to simplify the formula for using the Trapezoid Rule. Each trapezoid has a height equal to the length of the interval divided by the number of trapezoids we use. If the interval is from $x = a$ to $x = b$, and the number of trapezoids is n, then the height of each trapezoid is $\dfrac{b-a}{n}$. Then our formula is

$$\left(\frac{1}{2}\right)\left(\frac{b-a}{n}\right)\left[y_0 + 2y_1 + 2y_2 + 2y_3 \ldots + 2y_{n-2} + 2y_{n-1} + y_n\right]$$

This is all you need to know about the Trapezoid Rule. Just follow the formula, and you won't have any problems. Let's do one more example.

Example 10: Approximate the area under the curve $y = x^3$ from $x = 2$ to $x = 3$ using four inscribed trapezoids.

Following the rule, the height of each trapezoid is $\dfrac{3-2}{4} = \dfrac{1}{4}$. Thus, the approximate area is

$$\left(\frac{1}{2}\right)\left(\frac{1}{4}\right)\left[2^3 + 2\left(\frac{9}{4}\right)^3 + 2\left(\frac{5}{2}\right)^3 + 2\left(\frac{11}{4}\right)^3 + 3^3\right] = \frac{1,045}{64}$$

Compare this answer to the actual value we found earlier—it's pretty close!

PROBLEM 1. Approximate the area under the curve $y = 4 - x^2$ from $x = -1$ to $x = 1$ with $n = 4$ inscribed rectangles.

Answer: Draw four rectangles that look like this.

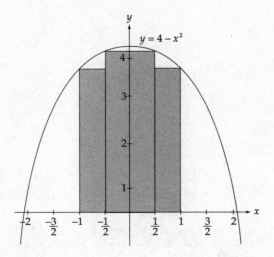

The width of each rectangle is $\dfrac{1}{2}$. The heights of the rectangles are found by evaluating $y = 4 - x^2$ at the appropriate endpoints.

$$\left(4 - (-1)^2\right), \left(4 - \left(-\frac{1}{2}\right)^2\right), \left(4 - \left(\frac{1}{2}\right)^2\right), \text{ and } \left(4 - (1)^2\right)$$

These can be simplified to $3, \dfrac{15}{4}, \dfrac{15}{4}$, and 3. Therefore, the area is

$$\left(\frac{1}{2}\right)(3) + \left(\frac{1}{2}\right)\left(\frac{15}{4}\right) + \left(\frac{1}{2}\right)\left(\frac{15}{4}\right) + \left(\frac{1}{2}\right)(3) = \frac{27}{4}$$

PROBLEM 2. Find the area under the curve $y = 4 - x^2$ from $x = -1$ to $x = 1$ with $n = 4$ circumscribed rectangles.

Answer: Draw four rectangles that look like the following:

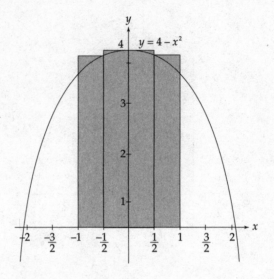

The width of each rectangle is $\dfrac{1}{2}$. The heights of the rectangles are found by evaluating $y = 4 - x^2$ at the appropriate endpoints.

$$\left(4 - \left(-\frac{1}{2}\right)^2\right), \left(4 - (0)^2\right), \left(4 - (0)^2\right), \text{ and } \left(4 - \left(\frac{1}{2}\right)^2\right)$$

These can be simplified to $\dfrac{15}{4}$, 4, 4, and $\dfrac{15}{4}$. Therefore, the area is

$$\left(\frac{1}{2}\right)\left(\frac{15}{4}\right) + \left(\frac{1}{2}\right)(4) + \left(\frac{1}{2}\right)(4) + \left(\frac{1}{2}\right)\left(\frac{15}{4}\right) = \frac{31}{4}$$

PROBLEM 3. Find the area under the curve $y = 4 - x^2$ from $x = -1$ to $x = 1$ using the Trapezoid Rule with $n = 4$.

Answer: Draw four trapezoids that look like the following:

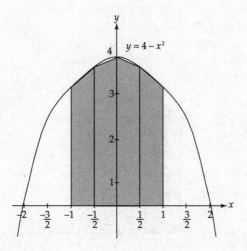

The width of each trapezoid is $\dfrac{1}{2}$. Evaluate the bases of the trapezoid by calculating $y = 4 - x^2$ at the appropriate endpoints. Following the rule, we get that the area is approximately

$$\left(\frac{1}{2}\right)\left(\frac{1}{2}\right)\left[\left(4 - (-1)^2\right) + 2\left(4 - \left(-\frac{1}{2}\right)^2\right) + 2\left(4 - (0)^2\right) + 2\left(4 - \left(\frac{1}{2}\right)^2\right) + \left(4 - (1)^2\right)\right] =$$
$$\left(\frac{1}{2}\right)\left(\frac{1}{2}\right)\left[3 + \frac{15}{2} + 8 + \frac{15}{2} + 3\right] = \frac{29}{4}$$

PROBLEM 4. Find the area under the curve $y = 4 - x^2$ from $x = -1$ to $x = 1$ using the Midpoint Formula with $n = 4$.

Answer: Draw four rectangles that look like the following:

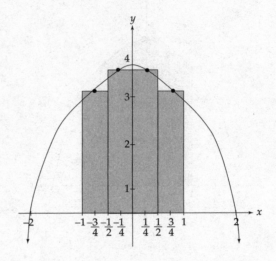

The width of each rectangle is $\dfrac{1}{2}$. The heights of the rectangles are found by evaluating $y = 4 - x^2$ at the appropriate points.

$$4 - \left(-\frac{3}{4}\right)^2 = \frac{55}{16}; \ 4 - \left(-\frac{1}{4}\right)^2 = \frac{63}{16}; \ 4 - \left(\frac{1}{4}\right)^2 = \frac{63}{16}; \ 4 - \left(\frac{3}{4}\right)^2 = \frac{55}{16}$$

Multiply each height by the width of $\dfrac{1}{2}$, and add up the areas.

$$\left(\frac{1}{2}\right)\left(\frac{55}{16}\right) + \left(\frac{1}{2}\right)\left(\frac{63}{16}\right) + \left(\frac{1}{2}\right)\left(\frac{63}{16}\right) + \left(\frac{1}{2}\right)\left(\frac{55}{16}\right) = \frac{59}{8}$$

PROBLEM 5. Find the area under the curve $y = 4 - x^2$ from $x = -1$ to $x = 1$.

Answer: Now we can use the definite integral by evaluating.

$$\int_{-1}^{1} \left(4 - x^2\right) dx$$

This can be rewritten as

$$\left(4x - \frac{x^3}{3}\right)\Bigg|_{-1}^{1}$$

Follow the Fundamental Theorem of Calculus.

$$\left(4 - \frac{1}{3}\right) - \left(4(-1) - \frac{(-1)^3}{3}\right) = \frac{22}{3}$$

PROBLEM 6. Given the following table of values for t and $f(t)$:

t	0	2	4	7	11	13	14
$f(t)$	5	6	10	15	20	26	30

Use a *right-hand* Riemann sum with 6 subintervals indicated by the data in the table to approximate $\int_0^{14} f(t)\, dt$.

Answer: The width of the first rectangle is 2 − 0 = 2; the second width is 4 − 2 = 2; the third is 7 − 4 = 3; the fourth is 11 − 7 = 4; the fifth is 13 − 11 = 2; and the sixth is 14 − 13 = 1. We find the height of each rectangle by evaluating $f(t)$ at the appropriate value of t, the right endpoint of each interval on the t-axis. Here $y_1 = 6$, $y_2 = 10$, $y_3 = 15$ $y_4 = 20$, $y_5 = 26$, and $y_6 = 30$. Therefore, we can approximate the integral with

$$\int_0^{14} f(t)\, dt = (2)(6) + (2)(10) + (3)(15) + (4)(20) + (2)(26) + (1)(30) = 239$$

PRACTICE PROBLEM SET 19

Here's a great opportunity to practice finding the area beneath a curve and evaluating integrals. The answers are in Chapter 19.

1. Find the area under the curve $y = 2x - x^2$ from $x = 1$ to $x = 2$ with $n = 4$ left-endpoint rectangles.

2. Find the area under the curve $y = 2x - x^2$ from $x = 1$ to $x = 2$ with $n = 4$ right-endpoint rectangles.

3. Find the area under the curve $y = 2x - x^2$ from $x = 1$ to $x = 2$ using the Trapezoid Rule with $n = 4$.

4. Find the area under the curve $y = 2x - x^2$ from $x = 1$ to $x = 2$ using the Midpoint Formula with $n = 4$.

5. Find the area under the curve $y = 2x - x^2$ from $x = 1$ to $x = 2$.

6. Evaluate $\int_{-\frac{\pi}{2}}^{\frac{\pi}{2}} \cos x \, dx$.

7. Evaluate $\int_{1}^{9} 2x\sqrt{x} \, dx$.

8. Evaluate $\int_{0}^{1} \left(x^4 - 5x^3 + 3x^2 - 4x - 6\right) dx$.

9. Evaluate $\int_{-4}^{4} |x| \, dx$.

10. Evaluate $\int_{-\frac{\pi}{2}}^{\frac{\pi}{2}} \sin x \, dx$.

11. Suppose we are given the following table of values for x and $g(x)$:

x	0	1	3	5	9	14
$g(x)$	10	8	11	17	20	23

Use a left-hand Riemann sum with 5 subintervals indicated by the data in the table to approximate $\int_{0}^{14} g(x) dx$.

THE MEAN VALUE THEOREM FOR INTEGRALS

As you recall, we did the Mean Value Theorem once before, in Chapter 8, but this time we'll apply it to integrals, not derivatives. In fact, some books refer to it as the "Mean Value Theorem for Integrals" or MVTI. The most important aspect of the MVTI is that it enables you to find the average value of a function. In fact, the AP exam will often ask you to find the average value of a function, which is just its way of testing your knowledge of the MVTI.

Here's the theorem.

If $f(x)$ is continuous on a closed interval $[a, b]$, then at some point c in the interval $[a, b]$ the following is true:

$$\int_a^b f(x)dx = f(c)(b - a)$$

This tells you that the area under the curve of $f(x)$ on the interval $[a, b]$ is equal to the value of the function at some value c (between a and b) times the length of the interval. If you look at this graphically, you can see that you're finding the area of a rectangle whose base is the interval and whose height is some value of $f(x)$ that creates a rectangle with the same area as the area under the curve.

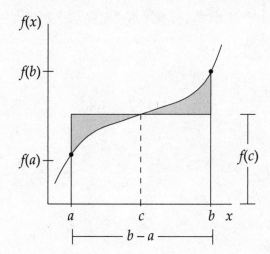

The number $f(c)$ gives us the average value of f on $[a, b]$. Thus, if we rearrange the theorem, we get the formula for finding the average value of $f(x)$ on $[a, b]$.

$$f(c) = \frac{1}{b-a}\int_a^b f(x)dx$$

There's all you need to know about finding average values. Try some examples.

Example 1: Find the average value of $f(x) = x^2$ from $x = 2$ to $x = 4$.

Evaluate the integral $\dfrac{1}{4-2}\displaystyle\int_2^4 x^2 dx$.

$$\frac{1}{4-2}\int_2^4 x^2 dx = \frac{1}{2}\left[\frac{x^3}{3}\bigg|_2^4\right] = \frac{1}{2}\left(\frac{64}{3} - \frac{8}{3}\right) = \frac{28}{3}$$

Example 2: Find the average value of $f(x) = \sin x$ on $[0, \pi]$.

Evaluate $\dfrac{1}{\pi - 0}\displaystyle\int_0^\pi \sin x \, dx$.

$$\frac{1}{\pi - 0}\int_0^\pi \sin x \, dx = \frac{1}{\pi}\left(-\cos x\right)\bigg|_0^\pi = \frac{1}{\pi}\left(-\cos\pi + \cos 0\right) = \frac{2}{\pi}$$

The Second Fundamental Theorem of Calculus

As you saw in the last chapter, we've only half-learned the theorem. It has two parts, often referred to as the **First and Second Fundamental Theorems of Calculus.**

The First Fundamental Theorem of Calculus (which you've already seen):

If $f(x)$ is continuous at every point of $[a, b]$, and $F(x)$ is an antiderivative of $f(x)$ on $[a, b]$, then $\int_a^b f(x)dx = F(b) - F(a)$.

The Second Fundamental Theorem of Calculus:

If $f(x)$ is continuous on $[a, b]$, then the derivative of the function $F(x) = \int_a^x f(t)dt$ is

$$\frac{dF}{dx} = \frac{d}{dx}\int_a^x f(t)dt = f(x)$$

These theorems are sometimes taught in reverse order.

We've already made use of the first theorem in evaluating definite integrals. In fact, we use the first Fundamental Theorem every time we evaluate a definite integral, so we're not going to give you any examples of that here. There is one aspect of the first Fundamental Theorem, however, that involves the area between curves (we'll discuss that in Chapter 16).

But for now, you should know the following:

If we have a point c in the interval $[a, b]$, then

$$\int_a^c f(x)dx + \int_c^b f(x)dx = \int_a^b f(x)dx$$

In other words, we can divide up the region into parts, add them up, and find the area of the region based on the result. We'll get back to this in the chapter on the area between two curves.

The second theorem tells us how to find the derivative of an integral.

Example 3: Find $\dfrac{d}{dx}\displaystyle\int_1^x \cos t\, dt$.

The second Fundamental Theorem says that the derivative of this integral is just $\cos x$.

Example 4: Find $\dfrac{d}{dx}\displaystyle\int_2^x \left(1-t^3\right) dt$.

Here, the theorem says that the derivative of this integral is just $(1 - x^3)$.

Isn't this easy? Let's add a couple of nuances. First, the constant term in the limits of integration is a "dummy term." Any constant will give the same answer. For example,

$$\frac{d}{dx}\int_2^x (1-t^3)\,dt = \frac{d}{dx}\int_{-2}^x (1-t^3)\,dt = \frac{d}{dx}\int_\pi^x (1-t^3)\,dt = 1-x^3$$

In other words, all we're concerned with is the variable term.

Second, if the upper limit is a function of x, instead of just plain x, we multiply the answer by the derivative of that term. For example,

$$\frac{d}{dx}\int_2^{x^2} \left(1-t^3\right) dt = \left[1-\left(x^2\right)^3\right](2x) = \left(1-x^6\right)(2x)$$

Example 5: Find $\dfrac{d}{dx}\displaystyle\int_0^{3x^4} \left(t+4t^2\right) dt = \left[3x^4 + 4(3x^4)^2\right](12x^3)$.

Try these solved problems on your own. You know the drill.

PROBLEM 1. Find the average value of $f(x) = \dfrac{1}{x^2}$ on the interval $[1, 3]$.

Answer: According to the Mean Value Theorem, the average value is found by evaluating

$$\frac{1}{3-1}\int_1^3 \frac{dx}{x^2}$$

Your result should be

$$\frac{1}{2}\left(-\frac{1}{x}\right)\Big|_1^3 = \frac{1}{2}\left(-\frac{1}{3}+1\right) = \frac{1}{3}$$

PROBLEM 2. Find the average value of $f(x) = \sin x$ on the interval $[-\pi, \pi]$.

Answer: According to the Mean Value Theorem, the average value is found by evaluating

$$\frac{1}{2\pi}\left(\int_{-\pi}^{\pi} \sin x \; dx\right)$$

Integrating, we get

$$\frac{1}{2\pi}\left(-\cos x\right)\Big|_{-\pi}^{\pi} = \frac{1}{2\pi}\left(-\cos\pi + \cos(-\pi)\right) = 0$$

PROBLEM 3. Find $\dfrac{d}{dx}\displaystyle\int_1^x \dfrac{dt}{1-\sqrt[3]{t}}$.

Answer: According to the Second Fundamental Theorem of Calculus,

$$\frac{d}{dx}\int_1^x \frac{dt}{1-\sqrt[3]{t}} = \frac{1}{1-\sqrt[3]{x}}$$

PROBLEM 4. Find $\dfrac{d}{dx}\displaystyle\int_1^{x^2} \dfrac{t\;dt}{\sin t}$.

Answer: According to the Second Fundamental Theorem of Calculus,

$$\frac{d}{dx}\int_1^{x^2} \frac{t\;dt}{\sin t} = \frac{x^2}{\sin(x^2)}(2x) = \frac{2x^3}{\sin x^2}$$

Accumulation Functions

The AP exam will also have problems that deal with **accumulation functions.** These are simply functions of the form $F(x) = \displaystyle\int_0^x f(t)\;dt$. These are called accumulation functions because the value of the integral is the area under the curve from the constant to the value x, and as x gets bigger, so does the area (it "accumulates"). In these functions, t is a dummy variable that is used as the variable of integration.

Let's do an example.

Example 1: Suppose we have the function $F(x) = \int_0^x t\, dt$. Let's evaluate this at different values of x. First, let's find $F(1)$. Graphically, we are looking for the area under the curve $y = t$ from $t = 0$ to $t = 1$. It looks like the following:

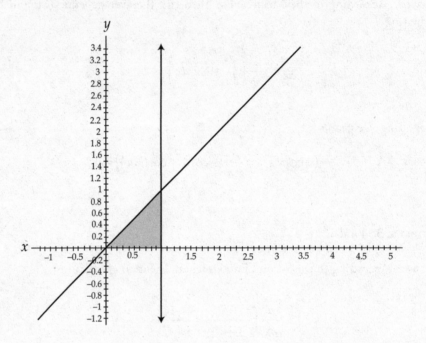

$F(1)$ is just the area of the triangle $A = \dfrac{1}{2}(1)(1) = \dfrac{1}{2}$. If we evaluate the integral, we get: $F(1) = \int_0^1 t\, dt = \left(\dfrac{t^2}{2}\right)\Bigg|_0^1 = \dfrac{1}{2} - 0 = \dfrac{1}{2}$. Let's now find $F(2)$. This is the area under the curve $y = t$ from $t = 0$ to $t = 2$.

It looks like the following:

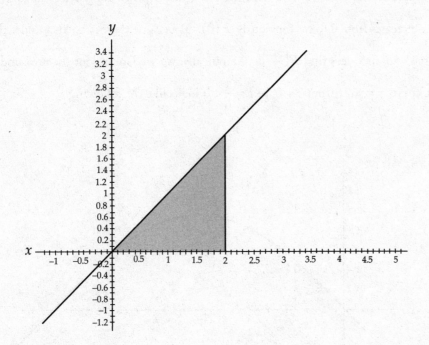

$F(2)$ is just the area of the triangle: $A = \dfrac{1}{2}(2)(2) = 2$. If we evaluate the integral, we get: $F(2) = \displaystyle\int_{0}^{2} t\,dt = \left(\dfrac{t^{2}}{2}\right)\Bigg|_{0}^{2} = 2 - 0 = 2$.

We can see that, as x increases, the function increases.

This was a fairly simple example. Let's do another one.

Example 2: Suppose we have the function $F(x) = \int_0^x \sin t \, dt$. Let's evaluate this as x increases from 0 to π. Obviously $F(0) = 0$ because there is no area under the curve. So, first, let's find $F\left(\dfrac{\pi}{6}\right)$. Graphically, we are looking for the area under the curve $y = \sin t$ from $t = 0$ to $t = \dfrac{\pi}{6}$. It looks like the following:

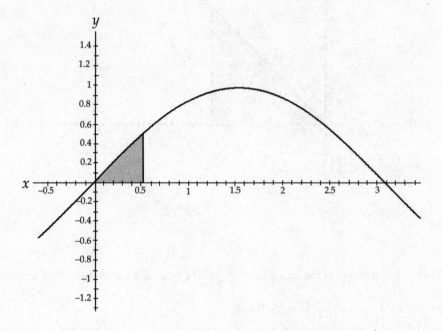

If we evaluate the integral, we get

$$F\left(\frac{\pi}{6}\right) = \int_0^{\frac{\pi}{6}} \sin t \, dt = \left(-\cos t\right)\Big|_0^{\frac{\pi}{6}} = -\cos\frac{\pi}{6} + \cos 0 = 1 - \frac{\sqrt{3}}{2} \approx 0.134.$$

Now let's find $F\left(\dfrac{\pi}{4}\right)$. It looks like the following:

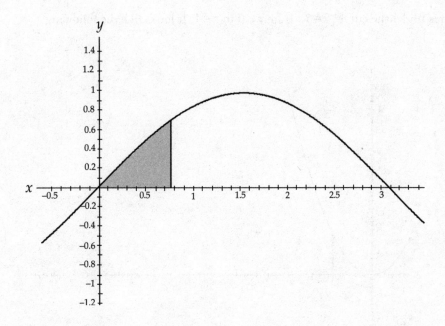

If we evaluate the integral, we get

$$F\left(\frac{\pi}{4}\right) = \int_0^{\frac{\pi}{4}} \sin t \, dt = \left(-\cos t\right)\Big|_0^{\frac{\pi}{4}} = -\cos\frac{\pi}{4} + \cos 0 = 1 - \frac{\sqrt{2}}{2} \approx 0.293.$$

Let's make a table of values of the accumulation function for different values of x.

x	0	$\dfrac{\pi}{6}$	$\dfrac{\pi}{4}$	$\dfrac{\pi}{3}$	$\dfrac{\pi}{2}$	π
$F(x)$	0	$1 - \dfrac{\sqrt{3}}{2}$	$1 - \dfrac{\sqrt{2}}{2}$	$\dfrac{1}{2}$	1	2

We can see that the area is accumulating under the curve as x increases from 0 to π. Naturally, the values will shrink as we move from π to 2π because the values of $\sin x$ are negative. But, with accumulation functions, we are usually concerned only with positive areas.

Let's do one more example.

Example 3: Suppose we have the function $F(x) = \int_0^x t^2 \, dt$. Let's evaluate this as x increases from 0 to 4. First, let's find $F(1)$. Graphically, we are looking for the area under the curve $y = t^2$ from $t = 0$ to $t = 1$. It looks like the following:

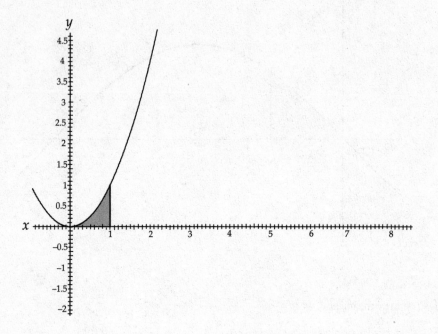

If we evaluate the integral, we get $F(1) = \int_0^1 t^2 \, dt = \left(\dfrac{t^3}{3} \right) \Bigg|_0^1 = \dfrac{1}{3}$.

As in the previous example, let's make a table of values of the accumulation function for different values of x.

x	1	2	3	4
$F(x)$	$\dfrac{1}{3}$	$\dfrac{8}{3}$	9	$\dfrac{64}{3}$

We can see that the values of $F(x)$ will increase as x increases.

PRACTICE PROBLEM SET 20

Now try these problems. The answers are in Chapter 19.

1. Find the average value of $f(x) = 4x \cos x^2$ on the interval $\left[0, \sqrt{\dfrac{\pi}{2}}\right]$.

2. Find the average value of $f(x) = \sqrt{x}$ on the interval $[0, 16]$.

3. Find the average value of $f(x) = \sqrt{1-x}$ on the interval $[-1, 1]$.

4. Find the average value of $f(x) = 2|x|$ on the interval $[-1, 1]$.

5. Find $\dfrac{d}{dx}\displaystyle\int_1^x \sin^2 t \; dt$.

6. Find $\dfrac{d}{dx}\displaystyle\int_1^{3x} \left(t^2 - t\right) dt$.

7. Find $\dfrac{d}{dx}\displaystyle\int_0^{x^2} |t| \; dt$.

8. Find $\dfrac{d}{dx}\displaystyle\int_1^x -2\cos t \; dt$.

Chapter 15
Exponential and Logarithmic Functions, Part Two

You've learned how to integrate polynomials and some of the trig functions (there are more of them to come), and you have the first technique of integration: u-substitution. Now it's time to learn how to integrate some other functions—namely, exponential and logarithmic functions. Yes, the long-awaited second part of Chapter 11. The first integral is the natural logarithm.

$$\int \frac{du}{u} = \ln|u| + C$$

Notice the absolute value in the logarithm. This ensures that you aren't taking the logarithm of a negative number. If you know that the term you're taking the log of is positive (for example, $x^2 + 1$), we can dispense with the absolute value marks. Let's do some examples.

Example 1: Find $\int \frac{5dx}{x+3}$.

Whenever an integrand contains a fraction, check to see if the integral is a logarithm. Usually, the process involves u-substitution. Let $u = x + 3$ and $du = dx$. Then,

$$\int \frac{5dx}{x+3} = 5\int \frac{du}{u} = 5\ln|u| + C$$

Substituting back, the final result is

$$5 \ln |x + 3| + C$$

Example 2: Find $\int \frac{2x\,dx}{x^2+1}$.

Let $u = x^2 + 1$ $du = 2x\,dx$ and substitute into the integrand.

$$\int \frac{2x\,dx}{x^2+1} = \int \frac{du}{u} = \ln|u| + C$$

Then substitute back.

$$\ln (x^2 + 1) + C$$

MORE INTEGRALS OF TRIG FUNCTIONS

Remember when we started antiderivatives and we didn't do the integral of tangent, cotangent, secant, or cosecant? Well, their time has come.

Example 3: Find $\int \tan x \, dx$.

First, rewrite this integral as

$$\int \frac{\sin x}{\cos x} \, dx$$

Now, we let $u = \cos x$ and $du = -\sin x \, dx$ and substitute.

$$\int -\frac{du}{u}$$

Now, integrate and re-substitute.

$$\int -\frac{du}{u} = -\ln|u| = -\ln|\cos x| + C$$

Thus, $\int \tan x \, dx = -\ln|\cos x| + C$.

Example 4: Find $\int \cot x \, dx$.

Just as before, rewrite this integral in terms of sine and cosine.

$$\int \frac{\cos x}{\sin x} \, dx$$

Now we let $u = \sin x$ and $du = \cos x \, dx$ and substitute.

$$\int \frac{du}{u}$$

Now, integrate.

$$\ln|u| + C = \ln|\sin x| + C$$

Therefore, $\int \cot x \, dx = \ln \left| \sin x \right| + C.$

This looks a lot like the previous example, doesn't it?

Example 5: Find $\int \sec x \, dx.$

You could rewrite this integral as

$$\int \frac{1}{\cos x} \, dx$$

However, if you try u-substitution at this point, it won't work. So what should you do? You'll probably never guess, so we'll show you: Multiply the sec x by $\dfrac{\sec x + \tan x}{\sec x + \tan x}$. This gives you

$$\int \sec x \left(\frac{\sec x + \tan x}{\sec x + \tan x} \right) dx = \int \frac{\sec^2 x + \sec x \tan x}{\sec x + \tan x} \, dx$$

Now you can do u-substitution. Let $u = \sec x + \tan x \, du = (\sec x \tan x + \sec^2 x) \, dx$.

Then rewrite the integral as

$$\int \frac{du}{u}$$

Pretty slick, huh?

The rest goes according to plan as you integrate.

$$\int \frac{du}{u} = \ln |u| + C = \ln |\sec x + \tan x| + C$$

Therefore, $\int \sec x \, dx = \ln |\sec x + \tan x| + C.$

Example 6: Find $\int \csc x \, dx$.

You guessed it! Multiply $\csc x$ by $\dfrac{\csc x + \cot x}{\csc x + \cot x} \, dx$. This gives you

$$\int \csc x \left(\frac{\csc x + \cot x}{\csc x + \cot x} \right) dx = \int \frac{\csc^2 x + \csc x \cot x}{\csc x + \cot x} \, dx$$

Let $u = \csc x + \cot x$ and $du = (-\csc x \cot x - \csc^2 x) \, dx$. And, just as in Example 5, you can rewrite the integral as

$$\int -\frac{du}{u}$$

And integrate.

$$\int -\frac{du}{u} = -\ln |u| + C = -\ln |\csc x + \cot x| + C$$

> Therefore, $\int \csc x \, dx = -\ln |\csc x + \cot x| + C$.

As we do more integrals, the natural log will turn up over and over. It's important that you get good at recognizing when integrating requires the use of the natural log.

INTEGRATING e^x AND a^x

Now let's learn how to find the integral of e^x. Remember that $\dfrac{d}{dx} e^x = e^x$? Well, you should be able to predict the following formula:

> $$\int e^u \, du = e^u + C$$

As with the natural logarithm, most of these integrals use u-substitution.

Example 7: Find $\int e^{7x} dx$.

Let $u = 7x$, $du = 7dx$, and $\frac{1}{7} du = dx$. Then you have

$$\int e^{7x} dx = \frac{1}{7}\int e^u du = \frac{1}{7} e^u + C$$

Substituting back, you get

$$\frac{1}{7} e^{7x} + C$$

In fact, whenever you see $\int e^{kx} dx$, where k is a constant, the integral is

$$\int e^{kx} dx = \frac{1}{k} e^{kx} + C$$

Example 8: Find $\int xe^{3x^2+1} dx$.

Let $u = 3x^2 + 1$, $du = 6x\, dx$, and $\frac{1}{6} du = x\, dx$. The result is

$$\int xe^{3x^2+1} dx = \frac{1}{6}\int e^u\, du = \frac{1}{6} e^u + C$$

Now it's time to put the x's back in.

$$\frac{1}{6} e^{3x^2+1} + C$$

Example 9: Find $\int e^{\sin x} \cos x\, dx$.

Let $u = \sin x$ and $du = \cos x\, dx$. The substitution here couldn't be simpler.

$$\int e^u du = e^u + C = e^{\sin x} + C$$

As you can see, these integrals are pretty straightforward. The key is to use *u*-substitution to transform nasty-looking integrals into simple ones.

There's another type of exponential function whose integral you'll have to find occasionally.

$$\int a^u \, du$$

As you should recall from your rules of logarithms and exponents, the term a^u can be written as $e^{u \ln a}$. Because $\ln a$ is a constant, we can transform $\int a^u \, du$ into $\int e^{u \ln a} \, du$. If you integrate this, you'll get

$$\int e^{u \ln a} \, du = \frac{1}{\ln a} e^{u \ln a} + C$$

Now substituting back a^u for $e^{u \ln a}$,

$$\int a^u \, du = \frac{1}{\ln a} a^u + C$$

Example 10: Find $\int 5^x \, dx$.

Follow the rule we just derived.

$$\int 5^x \, dx = \frac{1}{\ln 5} 5^x + C$$

Because these integrals don't show up too often on the AP exam, this is the last you'll see of them in this book. You should, however, be able to integrate them using the rule, or by converting them into a form of $\int e^u \, du$.

Try these on your own. Do each problem with the answers covered, and then check your answer.

PROBLEM 1. Evaluate $\int \dfrac{dx}{3x}$.

Answer: Move the constant term outside of the integral, like this.

$$\frac{1}{3} \int \frac{dx}{x}$$

Now you can integrate.

$$\frac{1}{3} \int \frac{dx}{x} = \frac{1}{3} \ln|x| + C$$

PROBLEM 2. Evaluate $\int \dfrac{3x^2}{x^3 - 1} \, dx$.

Answer: Let $u = x^3 - 1$ and $du = 3x^2 \, dx$, and substitute.

$$\int \frac{du}{u}$$

Now integrate.

$$\ln|u| + C$$

And substitute back.

$$\ln|x^3 - 1| + C$$

PROBLEM 3. Evaluate $\int e^{5x} \, dx$.

Answer: Let $u = 5x$ and $du = 5dx$. Then $\dfrac{1}{5} du = dx$. Substitute in.

$$\frac{1}{5} \int e^u \, du$$

Integrate.

$$\frac{1}{5} e^u + C$$

And substitute back.

$$\frac{1}{5} e^{5x} + C$$

PROBLEM 4. Evaluate $\int 2^{3x}\,dx$.

Answer: Let $u = 3x$ and $du = 3dx$. Then $\dfrac{1}{3}\,du = dx$. Make the substitution.

$$\frac{1}{3}\int 2^u\,du$$

Integrate according to the rule. Your result should be

$$\frac{1}{3\ln 2}2^u + C$$

Now get back to the expression as a function of x.

$$\frac{1}{3\ln 2}2^{3x} + C = \frac{1}{\ln 8}2^{3x} + C$$

PRACTICE PROBLEM SET 21

Evaluate the following integrals. The answers are in Chapter 19.

1. $\displaystyle\int \frac{\sec^2 x}{\tan x}\,dx$

2. $\displaystyle\int \frac{\cos x}{1 - \sin x}\,dx$

3. $\displaystyle\int \frac{1}{x\ln x}\,dx$

4. $\displaystyle\int \frac{1}{x}\cos(\ln x)\,dx$

5. $\displaystyle\int \frac{\sin x - \cos x}{\cos x}\, dx$

6. $\displaystyle\int \frac{dx}{\sqrt{x}(1 + 2\sqrt{x})}$

7. $\displaystyle\int \frac{e^x\, dx}{1 + e^x}$

8. $\displaystyle\int xe^{5x^2 - 1}\, dx$

9. $\displaystyle\int e^x \cos(2 + e^x)\, dx$

10. $\displaystyle\int \frac{e^x + e^{-x}}{e^x - e^{-x}}\, dx$

11. $\displaystyle\int x 4^{-x^2}\, dx$

12. $\displaystyle\int 7^{\sin x} \cos x\, dx$

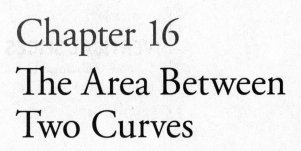

Chapter 16
The Area Between
Two Curves

These next two units discuss some of the most difficult topics you'll encounter in AP calculus. For some reason, students have terrible trouble setting up these problems. Fortunately, the AP exam asks only relatively simple versions of these problems on the exam.

Unfortunately, this unit and the next are always on the AP exam. We'll try to make them as simple as possible. You've already learned that if you want to find the area under a curve, you can integrate the function of the curve by using the endpoints as limits. So far, though, we've talked only about the area between a curve and the x-axis. What if you have to find the area between two curves?

VERTICAL SLICES

Suppose you wanted to find the area between the curve $y = x$ and the curve $y = x^2$ from $x = 2$ to $x = 4$. First, sketch the curves.

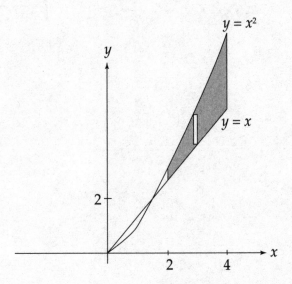

You can find the area by slicing up the region vertically, into a bunch of infinitely thin strips, and adding up the areas of all the strips. The height of each strip is $x^2 - x$, and the width of each strip is dx. Add up all the strips by using the integral.

$$\int_2^4 (x^2 - x)\, dx$$

Then, evaluate it.

$$\left(\frac{x^3}{3} - \frac{x^2}{2} \right)\Bigg|_2^4 = \left(\frac{64}{3} - \frac{16}{2} \right) - \left(\frac{8}{3} - \frac{4}{2} \right) = \frac{38}{3}$$

That wasn't so hard, was it? Don't worry. The process gets more complicated, but the idea remains the same. Now let's generalize this and come up with a rule.

If a region is bounded by $f(x)$ above and $g(x)$ below at all points of the interval $[a, b]$, then the area of the region is given by

$$\int_a^b \left[f(x) - g(x)\right] dx$$

Example 1: Find the area of the region between the parabola $y = 1 - x^2$ and the line $y = 1 - x$.

First, make a sketch of the region.

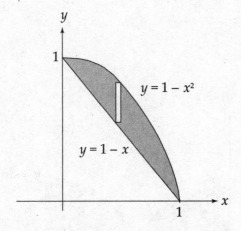

To find the points of intersection of the graphs, set the two equations equal to each other and solve for x.

$$1 - x^2 = 1 - x$$
$$x^2 - x = 0$$
$$x(x-1) = 0$$
$$x = 0, 1$$

The left-hand edge of the region is $x = 0$ and the right-hand edge is $x = 1$, so the limits of integration are from 0 to 1.

Next, note that the top curve is always $y = 1 - x^2$, and the bottom curve is always $y = 1 - x$. (If the region has a place where the top and bottom curve switch, you need to make two integrals, one for each region. Fortunately, that's not the case here.) Thus, we need to evaluate the following:

$$\int_0^1 \left[\left(1 - x^2 \right) - \left(1 - x \right) \right] dx$$

$$\int_0^1 \left[\left(1 - x^2 \right) - \left(1 - x \right) \right] dx = \int_0^1 \left(-x^2 + x \right) dx = \left(-\frac{x^3}{3} + \frac{x^2}{2} \right) \Bigg|_0^1 = \frac{1}{6}$$

Sometimes you're given the endpoints of the region; sometimes you have to find them on your own.

Example 2: Find the area of the region between the curve $y = \sin x$ and the curve $y = \cos x$ from 0 to $\dfrac{\pi}{2}$.

First, sketch the region.

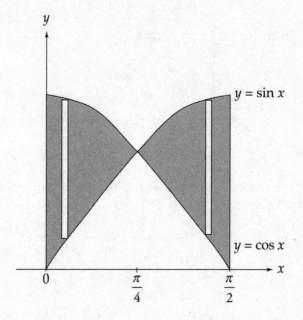

Notice that $\cos x$ is on top between 0 and $\dfrac{\pi}{4}$, then $\sin x$ is on top between $\dfrac{\pi}{4}$ and $\dfrac{\pi}{2}$. The point where they cross is $\dfrac{\pi}{4}$, so you have to divide the area into two integrals: one from 0 to $\dfrac{\pi}{4}$, and the other from $\dfrac{\pi}{4}$ to $\dfrac{\pi}{2}$. In the first region, $\cos x$ is above $\sin x$, so the integral to evaluate is

$$\int_0^{\frac{\pi}{4}} \left(\cos x - \sin x \right) dx$$

The integral of the second region is a little different, because $\sin x$ is above $\cos x$.

$$\int_{\frac{\pi}{4}}^{\frac{\pi}{2}} \left(\sin x - \cos x \right) dx$$

If you add the two integrals, you'll get the area of the whole region.

$$\int_0^{\frac{\pi}{4}} \left(\cos x - \sin x \right) dx = \left(\sin x + \cos x \right) \Big|_0^{\frac{\pi}{4}} = \sqrt{2} - 1$$

$$\int_{\frac{\pi}{4}}^{\frac{\pi}{2}} \left(\sin x - \cos x \right) dx = \left(-\cos x - \sin x \right) \Big|_{\frac{\pi}{4}}^{\frac{\pi}{2}} = \sqrt{2} - 1$$

Adding these, we get that the area is $2\sqrt{2} - 2$.

HORIZONTAL SLICES

Now for the fun part. We can slice a region vertically when one function is at the top of our section and a different function is at the bottom. But what if the same function is both the top and the bottom of the slice (what we call a double-valued function)? You have to slice the region horizontally.

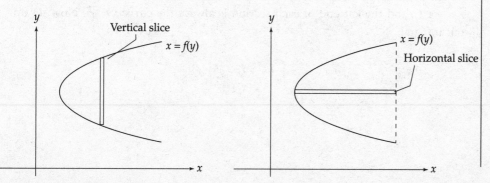

If we were to slice vertically, as in the left-hand picture, we'd have a problem. But if we were to slice horizontally, as in the right-hand picture, we don't have a problem. Instead of integrating an equation $f(x)$ with respect to x, we need to integrate an equation $f(y)$ with respect to y. As a result, our area formula changes a little.

If a region is bounded by $f(y)$ on the right and $g(y)$ on the left at all points of the interval $[c, d]$, then the area of the region is given by

$$\int_c^d \left[f(y) - g(y)\right] dy$$

Example 3: Find the area of the region between the curve $x = y^2$ and the curve $x = y + 6$ from $y = 0$ to $y = 3$.

First, sketch the region.

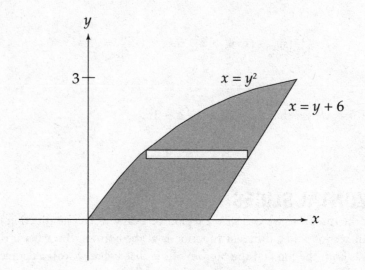

When you slice up the area horizontally, the right end of each section is the curve $x = y + 6$, and the left end of each section is always the curve $x = y^2$. Now set up our integral.

$$\int_0^3 (y + 6 - y^2)\, dy$$

Evaluating this gives us the area.

$$\int_0^3 (y + 6 - y^2)\, dy = \left(\frac{y^2}{2} + 6y - \frac{y^3}{3} \right)\bigg|_0^3 = \frac{27}{2}$$

Example 4: Find the area between the curve $y = \sqrt{x + 3}$ and the curve $y = \sqrt{3 - x}$ and the x-axis from $x = -3$ to $x = 3$.

First, sketch the curves.

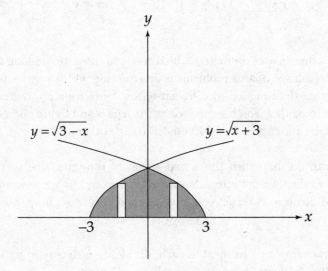

From $x = -3$ to $x = 0$, if you slice the region vertically, the curve $y = \sqrt{x + 3}$ is on top, and the x-axis is on the bottom; from $x = 0$ to $x = 3$, the curve $y = \sqrt{3 - x}$ is on top and the x-axis is on the bottom. Therefore, you can find the area by evaluating two integrals.

$$\int_{-3}^0 (\sqrt{x + 3} - 0)\, dx \text{ and } \int_0^3 (\sqrt{3 - x} - 0)\, dx$$

Your results should be

$$\frac{2}{3}(x + 3)^{\frac{3}{2}}\bigg|_{-3}^0 + \left(-\frac{2}{3}(3 - x)^{\frac{3}{2}} \right)\bigg|_0^3 = 4\sqrt{3}$$

Let's suppose you sliced the region horizontally instead. The curve $y = \sqrt{x + 3}$ is always on the left, and the curve $y = \sqrt{3 - x}$ is always on the right. If you solve each equation for x in terms of y, you save some time by using only one integral instead of two.

The two equations are $x = y^2 - 3$ and $x = 3 - y^2$. We also have to change the limits of integration from x-limits to y-limits. The two curves intersect at $y = \sqrt{3}$, so our limits of integration are from $y = 0$ to $y = \sqrt{3}$. The new integral is

$$\int_0^{\sqrt{3}} \left[(3 - y^2) - (y^2 - 3) \right] dy = \int_0^{\sqrt{3}} (6 - 2y^2)\, dy = 6y - \frac{2y^3}{3} \Big|_0^{\sqrt{3}} = 4\sqrt{3}$$

You get the same answer no matter which way you integrate (as long as you do it right!). The challenge of area problems is determining which way to integrate and then converting the equation to different terms. Unfortunately, there's no simple rule for how to do this. You have to look at the region and figure out its endpoints, as well as where the curves are with respect to each other.

Once you can do that, then the actual set-up of the integral(s) isn't that hard. Sometimes, evaluating the integrals isn't easy; however, if the integral of an AP question is difficult to evaluate, you'll be required only to set it up, not to evaluate it.

Here are some sample problems. On each, decide the best way to set up the integrals, and then evaluate them. Then check your answer.

Problem 1. Find the area of the region between the curve $y = 3 - x^2$ and the line $y = 1 - x$ from $x = 0$ to $x = 2$.

Answer: First, make a sketch.

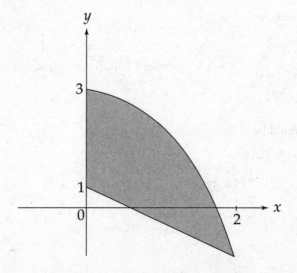

Because the curve $y = 3 - x^2$ is always above $y = 1 - x$ within the interval, you have to evaluate the following integral:

$$\int_0^2 \Big[(3 - x^2) - (1 - x) \Big] dx = \int_0^2 (2 + x - x^2) dx$$

Therefore, the area of the region is

$$\left(2x + \frac{x^2}{2} - \frac{x^3}{3} \right) \Bigg|_0^2 = \frac{10}{3}$$

PROBLEM 2. Find the area between the x-axis and the curve $y = 2 - x^2$ from $x = 0$ to $x = 2$.

Answer: First, sketch the graph over the interval.

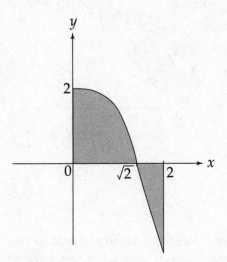

Because the curve crosses the x-axis at $\sqrt{2}$, you have to divide the region into two parts: from $x = 0$ to $x = \sqrt{2}$ and from $x = \sqrt{2}$ to $x = 2$. In the latter region, you'll need to integrate $y = -\left(2 - x^2\right) = x^2 - 2$ to adjust for the region's being below the x-axis. Therefore, we can find the area by evaluating

$$\int_0^{\sqrt{2}} (2 - x^2) \, dx + \int_{\sqrt{2}}^2 (x^2 - 2) \, dx$$

Integrating, we get

$$\left(2x + - \frac{x^3}{3}\right)\Bigg|_0^{\sqrt{2}} + \left(\frac{x^3}{3} - 2x\right)\Bigg|_{\sqrt{2}}^2 = \left(2\sqrt{2} - \frac{2\sqrt{2}}{3}\right) - 0 + \left(\frac{8}{3} - 4\right) - \left(\frac{2\sqrt{2}}{3} - 2\sqrt{2}\right) = \frac{8\sqrt{2} - 4}{3}$$

PROBLEM 3. Find the area of the region between the curve $x = y^2 - 4y$ and the line $x = y$.

Answer: First, sketch the graph over the interval.

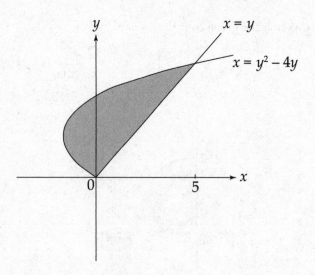

You don't have the endpoints this time, so you need to find where the two curves intersect. If you set them equal to each other, they intersect at $y = 0$ and at $y = 5$. The curve $x = y^2 - 4y$ is always to the left of $x = y$ over the interval we just found, so we can evaluate the following integral:

$$\int_0^5 \left[y - (y^2 - 4y)\right] dy = \int_0^5 (5y - y^2) dy$$

The result of the integration should be

$$\left(\frac{5y^2}{2} - \frac{y^3}{3}\right)\Bigg|_0^5 = \frac{125}{6}$$

PROBLEM 4. Find the area between the curve $x = y^3 - y$ and the line $x = 0$ (the y-axis).

Answer: First, sketch the graph over the interval.

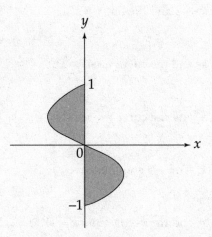

Next, find where the two curves intersect. By setting $y^3 - y = 0$, you'll find that they intersect at $y = -1$, $y = 0$, and $y = 1$. Notice that the curve is to the right of the y-axis from $y = -1$ to $y = 0$ and to the left of the y-axis from $y = 0$ to $y = 1$. Thus, the region must be divided into two parts: from $y = -1$ to $y = 0$ and from $y = 0$ to $y = 1$.

Set up the two integrals.

$$\int_{-1}^{0} (y^3 - y)\,dy + \int_{0}^{1} (y - y^3)\,dy$$

And integrate them.

$$\left(\frac{y^4}{4} - \frac{y^2}{2} \right)\Bigg|_{-1}^{0} + \left(\frac{y^2}{2} - \frac{y^4}{4} \right)\Bigg|_{0}^{1} = \frac{1}{2}$$

PRACTICE PROBLEM SET 22

Find the area of the region between the two curves in each problem, and be sure to sketch each one. (We gave you only endpoints in one of them.) The answers are in Chapter 19.

1. The curve $y = x^2 - 2$ and the line $y = 2$.

2. The curve $y = x^2$ and the curve $y = 4x - x^2$.

3. The curve $y = x^3$ and the curve $y = 3x^2 - 4$.

4. The curve $y = x^2 - 4x - 5$ and the curve $y = 2x - 5$.

5. The curve $y = x^3$ and the x-axis, from $x = -1$ to $x = 2$.

6. The curve $x = y^2$ and the line $x = y + 2$.

7. The curve $x = y^2$ and the curve $x = 3 - 2y^2$.

8. The curve $x = y^3 - y^2$ and the line $x = 2y$.

9. The curve $x = y^2 - 4y + 2$ and the line $x = y - 2$.

10. The curve $x = y^{\frac{2}{3}}$ and the curve $x = 2 - y^4$.

Chapter 17
The Volume of a
Solid of Revolution

Does the chapter title leave you in a cold sweat? Don't worry. You're not alone. This chapter covers a topic widely seen as one of the most difficult on the AP exam. There is *always* a volume question on the test. The good news is that you're almost never asked to evaluate the integral—you usually only have to set it up. The difficulty with this chapter, as with Chapter 16, is that there aren't any simple rules to follow. You have to draw the picture and figure it out.

In this chapter, we're going to take the region between two curves, rotate it around a line (usually the *x*- or *y*-axis), and find the volume of the region. There are two methods of doing this: the **washers method** and the **cylindrical shells method**. Sometimes you'll hear the washers method called the **disk method**, but a disk is only a washer without a hole in the middle.

WASHERS AND DISKS

Let's look at the region between the curve $y = \sqrt{x}$ and the *x*-axis (the curve $y = 0$), from $x = 0$ to $x = 1$, and revolve it about the *x*-axis. The picture looks like the following:

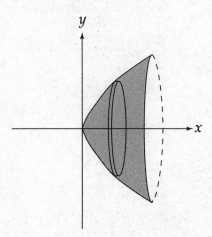

If you slice the resulting solid perpendicular to the *x*-axis, each cross-section of the solid is a circle, or disk (hence the phrase "disk method"). The radii of the disks vary from one value of *x* to the next, but you can find each radius by plugging it into the equation for *y*: Each radius is $\sqrt{x}$. Therefore, the area of each disk is

$$\pi \left(\sqrt{x} \right)^2 = \pi x$$

Each disk is infinitesimally thin, so its thickness is dx; if you add up the volumes of all the disks, you'll get the entire volume. The way to add these up is by using the integral, with the endpoints of the interval as the limits of integration. Therefore, to find the volume, evaluate the integral.

$$\int_0^1 \pi x \, dx = \left. \frac{\pi x^2}{2} \right|_0^1 = \frac{\pi}{2}$$

Now let's generalize this. If you have a region whose area is bounded by the curve $y = f(x)$ and the x-axis on the interval $[a, b]$, each disk has a radius of $f(x)$, and the area of each disk will be

$$\pi \left[f(x) \right]^2$$

To find the volume, evaluate the integral.

$$\pi \int_a^b \left[f(x) \right]^2 dx$$

This is the formula for finding the volume using disks.

Example 1: Find the volume of the solid that results when the region between the curve $y = x$ and the x-axis, from $x = 0$ to $x = 1$, is revolved about the x-axis.

As always, sketch the region to get a better look at the problem.

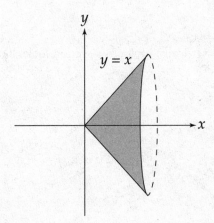

When you slice vertically, the top curve is $y = x$ and the limits of integration are from $x = 0$ to $x = 1$. Using our formula, we evaluate the integral.

$$\pi \int_0^1 x^2 \, dx$$

The result is

$$\pi \int_0^1 x^2 \, dx = \pi \left. \frac{x^3}{3} \right|_0^1 = \frac{\pi}{3}$$

By the way, did you notice that the solid in the problem is a cone with a height and radius of 1? The formula for the volume of a cone is $\frac{1}{3}\pi r^2 h$, so you should expect to get $\frac{\pi}{3}$.

Now let's figure out how to find the volume of the solid that results when we revolve a region that does not touch the x-axis. Consider the region bounded above by the curve $y = x^3$ and below by the curve $y = x^2$, from $x = 2$ to $x = 4$, which is revolved about the x-axis. Sketch the region first.

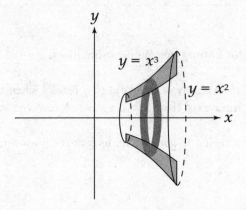

If you slice this region vertically, each cross-section looks like a washer (hence the phrase "washer method").

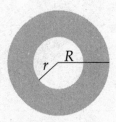

The outer radius is $R = x^3$ and the inner radius is $r = x^2$. To find the area of the region between the two circles, take the area of the outer circle, πR^2, and subtract the area of the inner circle, πr^2.

We can simplify this to

$$\pi R^2 - \pi r^2 = \pi\left(R^2 - r^2\right)$$

Because the outer radius is $R = x^3$ and the inner radius is $r = x^2$, the area of each region is $\pi\left(x^6 - x^4\right)$. You can sum up these regions using the integral.

$$\pi\int_2^4 \left(x^6 - x^4\right) dx = \frac{74{,}336\pi}{35}$$

Here's the general idea: In a region whose area is bounded above by the curve $y = f(x)$ and below by the curve $y = g(x)$, on the interval $[a, b]$, then each washer will have an area of

$$\pi\left[f(x)^2 - g(x)^2\right]$$

To find the volume, evaluate the integral.

$$\pi\int_a^b \left[f(x)^2 - g(x)^2\right] dx$$

This is the formula for finding the volume using washers when the region is rotated around the x-axis.

Example 2: Find the volume of the solid that results when the region bounded by $y = x$ and $y = x^2$, from $x = 0$ to $x = 1$, is revolved about the x-axis.

Sketch it first.

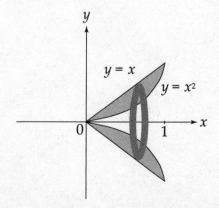

The top curve is $y = x$ and the bottom curve is $y = x^2$ throughout the region. Then our formula tells us that we evaluate the integral.

$$\pi \int_0^1 \left(x^2 - x^4 \right) dx$$

The result is

$$\pi \int_0^1 \left(x^2 - x^4 \right) dx = \pi \left(\frac{x^3}{3} - \frac{x^5}{5} \right) \Bigg|_0^1 = \frac{2\pi}{15}$$

Suppose the region we're interested in is revolved around the y-axis instead of the x-axis. Now, to find the volume, you have to slice the region horizontally instead of vertically. We discussed how to do this in the previous unit on area.

Now, if you have a region whose area is bounded on the right by the curve $x = f(y)$ and on the left by the curve $x = g(y)$, on the interval $[c, d]$, then each washer has an area of

$$\pi \left[f(y)^2 - g(y)^2 \right]$$

To find the volume, evaluate the integral.

$$\pi \int_c^d \left[f(y)^2 - g(y)^2 \right] dy$$

This is the formula for finding the volume using washers when the region is rotated about the y-axis.

Example 3: Find the volume of the solid that results when the region bounded by the curve $x = y^2$ and the curve $x = y^3$, from $y = 0$ to $y = 1$, is revolved about the y-axis.

Sketch away.

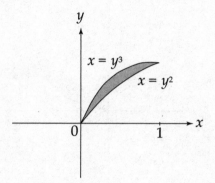

Because $x = y^2$ is always on the outside and $x = y^3$ is always on the inside, you have to evaluate the integral.

$$\pi \int_0^1 \left(y^4 - y^6 \right) dy$$

You should get the following:

$$\pi \int_0^1 \left(y^4 - y^6 \right) dx = \pi \left[\frac{y^5}{5} - \frac{y^7}{7} \right]_0^1 = \frac{2\pi}{35}$$

There's only one more nuance to cover. Sometimes you'll have to revolve the region about a line instead of one of the axes. If so, this will affect the radii of the washers; you'll have to adjust the integral to reflect the shift. Once you draw a picture, it usually isn't too hard to see the difference.

Example 4: Find the volume of the solid that results when the area bounded by the curve $y = x^2$ and the curve $y = 4x$ is revolved about the line $y = -2$. <u>Set up but do not evaluate the integral.</u> (This is how the AP exam will say it!)

You're not given the limits of integration here, so you need to find where the two curves intersect by setting the equations equal to each other.

$$x^2 = 4x$$
$$x^2 - 4x = 0$$
$$x = 0,\ 4$$

These will be our limits of integration. Next, sketch the curve.

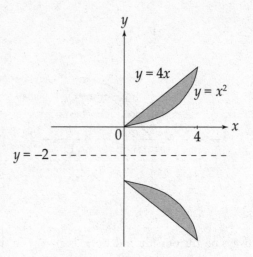

Notice that the distance from the axis of revolution is no longer found by just using each equation. Now, you need to add 2 to each equation to account for the shift in the axis. Thus, the radii are $x^2 + 2$ and $4x + 2$. This means that we need to evaluate the integral.

$$\pi \int_0^4 \left[\left(4x + 2\right)^2 - \left(x^2 + 2\right)^2 \right] dx$$

Suppose instead that the region was revolved about the line $x = -2$. Sketch the region again.

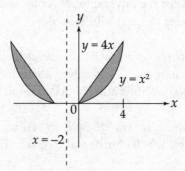

You'll have to slice the region horizontally this time; this means you're going to solve each equation for x in terms of y: $x = \sqrt{y}$ and $x = \dfrac{y}{4}$. We also need to find the y-coordinates of the intersection of the two curves: $y = 0, 16$.

Notice also that, again, each radius is going to be increased by 2 to reflect the shift in the axis of revolution. Thus, we will have to evaluate the integral.

$$\pi \int_0^{16} \left[\left(\sqrt{y} + 2 \right)^2 - \left(\frac{y}{4} + 2 \right)^2 \right] dy$$

Finding the volumes isn't that hard, once you've drawn a picture, figured out whether you need to slice vertically or horizontally, and determined whether the axis of revolution has been shifted. Sometimes, though, there will be times when you want to slice vertically yet revolve around the y-axis (or slice horizontally yet revolve about the x-axis). Here's the method for finding volumes in this way.

CYLINDRICAL SHELLS

Let's examine the region bounded above by the curve $y = 2 - x^2$ and below by the curve $y = x^2$, from $x = 0$ to $x = 1$. Suppose you had to revolve the region about the y-axis instead of the x-axis.

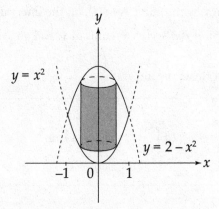

If you slice the region vertically and revolve the slice, you won't get a washer; you'll get a cylinder instead. Because each slice is an infinitesimally thin rectangle, the cylinder's "thickness" is also very, very thin, but real nonetheless. Thus, if you find the surface area of each cylinder and add them up, you'll get the volume of the region.

Why work in the dark? Just as you spend time practicing formulas in order to memorize them, make sure you actually work on drawing these examples. If you can't visualize the problem, you won't be able to set up the integral.

The formula for the surface area of a cylinder is $2\pi rh$. The height of the cylinder is the length of the vertical slice, $(2 - x^2) - x^2 = 2 - 2x^2$, and the radius of the slice is x. Thus, evaluate the integral

$$2\pi \int_0^1 x(2 - 2x^2)\,dx$$

The math goes like the following:

$$2\pi \int_0^1 x(2 - 2x^2)\,dx = 2\pi \int_0^1 (2x - 2x^3)\,dx = 2\pi \left(x^2 - \frac{x^4}{2} \right)\Bigg|_0^1 = \pi$$

Suppose you tried to slice the region horizontally and use washers. You'd have to convert each equation and find the new limits of integration. Because the region is not bounded by the same pair of curves throughout, you would have to evaluate the region using several integrals. The cylindrical shells method was invented precisely so you can avoid this.

From a general standpoint: If we have a region whose area is bounded above by the curve $y = f(x)$ and below by the curve $y = g(x)$, on the interval $[a, b]$, then each cylinder will have a height of $f(x) - g(x)$, a radius of x, and an area of $2\pi x[f(x) - g(x)]$.

To find the volume, evaluate the integral.

$$2\pi \int_a^b x[(f(x) - g(x)]\,dx$$

This is the formula for finding the volume using cylindrical shells when the region is rotated around the y-axis.

Example 5: Find the volume of the region that results when the region bounded by the curve $y = \sqrt{x}$, the x-axis, and the line $x = 9$ is revolved about the y-axis. Set up but do not evaluate the integral.

Your sketch should look like the following:

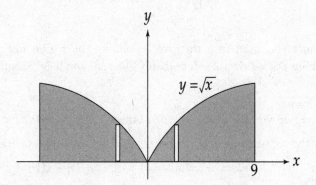

Notice that the limits of integration are from $x = 0$ to $x = 9$, and that each vertical slice is bounded from above by the curve $y = \sqrt{x}$ and from below by the x-axis ($y = 0$). We need to evaluate the integral.

$$2\pi \int_0^9 x(\sqrt{x} - 0)\,dx = 2\pi \int_0^9 x(\sqrt{x})\,dx$$

Example 6:   Find the volume that results when the region in Example 5 is revolved about the line $x = -1$. <u>Set up but do not evaluate the integral.</u>

Sketch the figure.

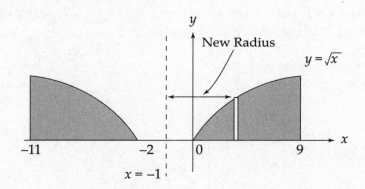

If you slice the region vertically, the height of the shell doesn't change because of the shift in axis of revolution, but you have to add 1 to each radius.

Our integral thus becomes

$$2\pi \int_0^9 (x+1)(\sqrt{x})\,dx$$

The last formula you need to learn involves slicing the region horizontally and revolving it about the *x*-axis. As you probably guessed, you'll get a cylindrical shell.

If you have a region whose area is bounded on the right by the curve $x = f(y)$ and on the left by the curve $x = g(y)$, on the interval $[c, d]$, then each cylinder will have a height of $f(y) - g(y)$, a radius of *y*, and an area of $2\pi y[f(y) - g(y)]$.

To find the volume, evaluate the integral.

$$2\pi \int_c^d y[(f(y) - g(y)]\,dy$$

This is the formula for finding the volume using cylindrical shells when the region is rotated around the *x*-axis.

Example 7: Find the volume of the region that results when the region bounded by the curve $x = y^3$ and the line $x = y$, from $y = 0$ to $y = 1$, is rotated about the *x*-axis. <u>Set up but do not evaluate the integral.</u>

Let your sketch be your guide.

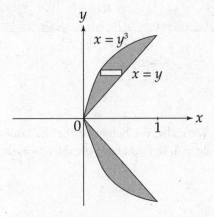

Each horizontal slice is bounded on the right by the curve $x = y$ and on the left by the line $x = y^3$. The integral to evaluate is

$$2\pi \int_0^1 y(y - y^3)\,dy$$

Suppose that you had to revolve this region about the line $y = -1$ instead. Now the region looks like the following:

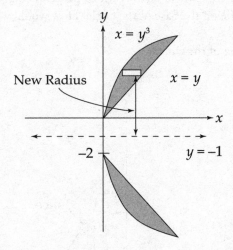

The radius of each cylinder is increased by 1 because of the shift in the axis of revolution, so the integral looks like the following:

$$2\pi \int_0^1 (y + 1)(y - y^3)\,dy$$

Wasn't this fun? Volumes of Solids of Revolution require you to sketch the region carefully and to decide whether it'll be easier to slice the region vertically or horizontally. Once you figure out the slices' boundaries and the limits of integration (and you've adjusted for an axis of revolution, if necessary), it's just a matter of plugging into the integral. Usually, you won't be asked to evaluate the integral unless it's a simple one. Once you've conquered this topic, you're ready for anything.

VOLUMES OF SOLIDS WITH KNOWN CROSS-SECTIONS

There is one other type of volume that you need to be able to find. Sometimes, you will be given an object where you know the shape of the base and where perpendicular cross-sections are all the same regular, planar geometric shape. These sound hard, but are actually quite straightforward. This is easiest to explain through an example.

Example 8: Suppose we are asked to find the volume of a solid whose base is the circle $x^2 + y^2 = 4$, and where cross-sections perpendicular to the x-axis are all squares whose sides lie on the base of the circle. How would we find the volume?

First, make a drawing of the circle.

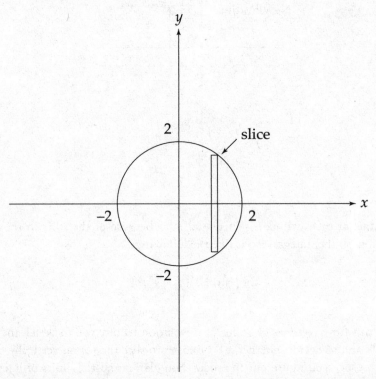

What this problem is telling us is that every time we make a vertical slice, the slice is the length of the base of a square. If we want to find the volume of the solid, all we have to do is integrate the area of the square, from one endpoint of the circle to the other.

The side of the square is the vertical slice whose length is $2y$, which we can find by solving the equation of the circle for y and multiplying by 2. We get $y = \sqrt{4 - x^2}$. Then the length of a side of the square is $2\sqrt{4 - x^2}$. Because the area of a square is $side^2$, we can find the volume by: $\int_{-2}^{2} (16 - 4x^2)\,dx$.

Let's perform the integration, although on some problems you will be permitted to find the answer with a calculator.

$$\int_{-2}^{2} (16 - 4x^2)\,dx = \left(16x - \frac{4x^3}{3} \right)\Bigg|_{-2}^{2} = \left(32 - \frac{32}{3} \right) - \left(32 + \frac{32}{3} \right)$$

$$= 64 - \frac{64}{3} = \frac{128}{3}$$

As you can see, the technique is very simple. First, you find the side of the cross-section in terms of y. This will involve a vertical slice. Then, you plug the side into the equation for the area of the cross-section. Then, integrate the area from one endpoint of the base to the other. On the AP exam, cross-sections will be squares, equilateral triangles, circles, or semi-circles, or maybe isosceles right triangles. So here are some handy formulas to know.

Given the side of an equilateral triangle, the area is $A = (side)^2 \, \frac{\sqrt{3}}{4}$.

Given the diameter of a semi-circle, the area is $A = (diameter)^2 \frac{\pi}{8}$.

Given the hypotenuse of an isosceles right triangle, the area is $A = \frac{(hypotenuse)^2}{4}$.

Example 9: Use the same base as Example 8, except this time the cross-sections are equilateral triangles. We find the side of the triangle just as we did above. It is $2y$, which is $2\sqrt{4 - x^2}$. Now, because the area of an equilateral triangle is $(side)^2 \, \frac{\sqrt{3}}{4}$, we can find the volume by evaluating the integral

$$\frac{\sqrt{3}}{4} \int_{-2}^{2} (4)(4 - x^2)\,dx = \sqrt{3} \int_{-2}^{2} (4 - x^2)\,dx.$$

We get $\sqrt{3} \int_{-2}^{2} (4 - x^2)\,dx = \sqrt{3} \left(4x - \frac{x^3}{3} \right)\Bigg|_{-2}^{2} = \frac{32\sqrt{3}}{3}$

Example 10: Use the same base as Example 8, except this time the cross-sections are semi-circles whose diameters lie on the base. We find the side of the semi-circle just as we did above. It is $2y$, which is $2\sqrt{4-x^2}$. Now because the area of a semi-circle is $(diameter)^2\dfrac{\pi}{8}$, we can find the volume by evaluating the integral

$$\frac{\pi}{8}\int_{-2}^{2}(4)(4-x^2)\,dx = \frac{\pi}{2}\int_{-2}^{2}(4-x^2)\,dx\,.$$

We get $\dfrac{\pi}{2}\displaystyle\int_{-2}^{2}(4-x^2)\,dx = \dfrac{\pi}{2}\left(4x-\dfrac{x^3}{3}\right)\Bigg|_{-2}^{2} = \dfrac{16\pi}{3}\,.$

Here are some solved problems. Do each problem, covering the answer first, then check your answer.

PROBLEM 1. Find the volume of the solid that results when the region bounded by the curve $y = 16 - x^2$ and the curve $y = 16 - 4x$ is rotated about the x-axis. Use the washer method and <u>set up but do not evaluate the integral</u>.

Answer: First, sketch the region.

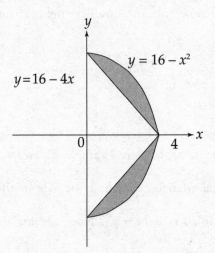

Next, find where the curves intersect by setting the two equations equal to each other.

$$16 - x^2 = 16 - 4x$$
$$x^2 = 4x$$
$$x^2 - 4x = 0$$
$$x = 0,\ 4$$

Slicing vertically, the top curve is always $y = 16 - x^2$ and the bottom is always $y = 16 - 4x$, so the integral looks like the following:

$$\pi \int_0^4 \left[(16 - x^2)^2 - (16 - 4x)^2 \right] dx$$

PROBLEM 2. Repeat Problem 1, but revolve the region about the y-axis and use the cylindrical shells method. <u>Set up but do not evaluate the integral</u>.

Answer: Sketch the situation.

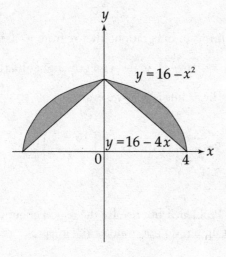

Slicing vertically, the top of each cylinder is $y = 16 - x^2$, the bottom is $y = 16 - 4x$, and the radius is x. Therefore, you should set up the following:

$$2\pi \int_0^4 x \left[(16 - x^2) - \left(16 - 4x \right) \right] dx$$

PROBLEM 3. Repeat Problem 1 but revolve the region about the x-axis and use the cylindrical shells method. <u>Set up but do not evaluate the integral</u>.

Answer: Sketch the situation.

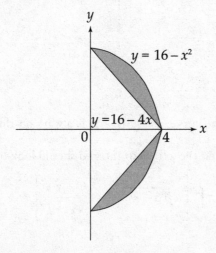

To slice horizontally, you have to solve each equation for x in terms of y and find the limits of integration with respect to y. First, solve for x in terms of y.

$$y = 16 - x^2 \text{ becomes } x = \sqrt{16 - y}$$

and

$$y = 16 - 4x \text{ becomes } x = \frac{16 - y}{4}$$

Next, determine the limits of integration: They're from $y = 0$ to $y = 16$. Slicing horizontally, the curve $x = \sqrt{16 - y}$ is always on the right and the curve $x = \frac{16 - y}{4}$ is always on the left. The radius is y, so we evaluate

$$2\pi \int_0^{16} y \left[\left(\sqrt{16 - y}\right) - \left(\frac{16 - y}{4} \right) \right] dy$$

PROBLEM 4. Repeat Problem 1 but revolve the region about the y-axis and use the washers method. <u>Set up but do not evaluate the integral</u>.

Answer: Sketch.

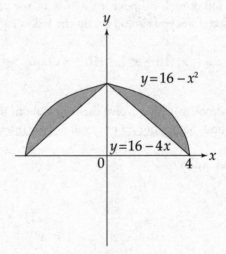

Slicing horizontally, the curve $x = \sqrt{16 - y}$ is always on the right and the curve $x = \frac{16 - y}{4}$ is always on the left. Your integral should look like the following:

$$\pi \int_0^{16} \left[\left(\sqrt{16 - y}\right)^2 - \left(\frac{16 - y}{4} \right)^2 \right] dy$$

PROBLEM 5. Repeat Problem 1, but revolve the region about the line $y = -3$. You may use either method. <u>Set up but do not evaluate the integral</u>.

Answer: Your sketch should resemble the one below (note that it's not drawn exactly to scale).

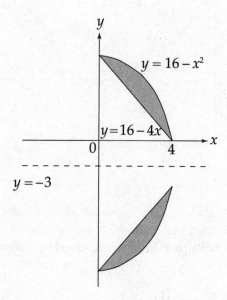

If you were to slice the region vertically, you would use washers. You'll need to add 3 to each radius to adjust for the axis of revolution. The integral to evaluate is

$$\pi \int_0^4 \left[(16 - x^2 + 3)^2 - \left(16 - 4x + 3 \right)^2 \right] dx$$

To slice the region horizontally, use cylindrical shells. The radius of each shell would increase by 3, and you would evaluate

$$2\pi \int_0^{16} (y+3) \left[(\sqrt{16 - y}) - \left(\frac{16 - y}{4} \right) \right] dy$$

PROBLEM 6. Repeat Problem 1, but revolve the region about the line $x = 8$. You may use either method. <u>Set up but do not evaluate the integral</u>.

Warning! This one is tricky!

Answer: First, sketch the region.

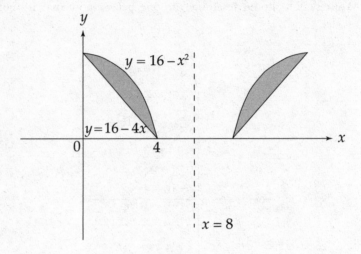

If you choose cylindrical shells, slice the region vertically; you'll need to adjust for the axis of revolution. Each radius can be found by subtracting x from 8. (Not 8 from x. That was the tricky part, in case you missed it.) The integral to evaluate is

$$2\pi \int_0^4 (8-x)\left[(16-x^2) - \left(16-4x\right)\right] dx$$

If you choose washers, slice the region horizontally. The radius of each washer is found by subtracting each equation from 8. Notice also that the curve $x = \dfrac{16-y}{4}$ is now the outer radius of the washer, and the curve $x = \sqrt{16-y}$ is the inner radius. The integral looks like the following:

$$\pi \int_0^{16}\left[\left(8 - \left(\frac{16-y}{4}\right)\right)^2 - \left(8 - \left(\sqrt{16-y}\right)\right)^2\right] dy$$

PROBLEM 7: Find the volume of a solid whose base is the region between the x-axis and the curve $y = 4 - x^2$, and whose cross-sections perpendicular to the x-axis are equilateral triangles with a side that lies on the base.

Answer: The curve $y = 4 - x^2$ intersects the x-axis at $x = -2$ and $x = 2$. The side of the triangle is $4 - x^2$, so all that we have to do is evaluate $\dfrac{\sqrt{3}}{4} \displaystyle\int_{-2}^{2} (4 - x^2)^2 \, dx$.

Expand the integrand to get

$$\frac{\sqrt{3}}{4} \int_{-2}^{2} (16 - 8x^2 + x^4) \, dx$$

Then integrate, which gives you

$$\frac{\sqrt{3}}{4} \left(16x - \frac{8x^3}{3} + \frac{x^5}{5} \right) \Bigg|_{-2}^{2} = \frac{128\sqrt{3}}{15} \approx 14.780$$

PRACTICE PROBLEM SET 23

Calculate the volumes below. The answers are in Chapter 19.

1. Find the volume of the solid that results when the region bounded by $y = \sqrt{9 - x^2}$ and the x-axis is revolved around the x-axis.

2. Find the volume of the solid that results when the region bounded by $y = \sec x$ and the x-axis from $x = -\dfrac{\pi}{4}$ to $x = \dfrac{\pi}{4}$ is revolved around the x-axis.

3. Find the volume of the solid that results when the region bounded by $x = 1 - y^2$ and the y-axis is revolved around the y-axis.

4. Find the volume of the solid that results when the region bounded by $x = \sqrt{5}y^2$ and the y-axis from $y = -1$ to $y = 1$ is revolved around the y-axis.

5. Find the volume of the solid that results when the region bounded by $y = x^3$, $x = 2$, and the x-axis is revolved around the line $x = 2$.

6. Use the method of cylindrical shells to find the volume of the solid that results when the region bounded by $y = x$, $x = 2$, and $y = -\dfrac{x}{2}$ is revolved around the y-axis.

7. Use the method of cylindrical shells to find the volume of the solid that results when the region bounded by $y = \sqrt{x}$, $y = 2x - 1$, and $x = 0$ is revolved around the y-axis.

8. Use the method of cylindrical shells to find the volume of the solid that results when the region bounded by $y = x^2$, $y = 4$, and $x = 0$ is revolved around the x-axis.

9. Use the method of cylindrical shells to find the volume of the solid that results when the region bounded by $y = 2\sqrt{x}$, $x = 4$, and $y = 0$ is revolved around the y-axis.

10. Use the method of cylindrical shells to find the volume of the solid that results when the region bounded by $y^2 = 8x$ and $x = 2$ is revolved around the line $x = 4$.

11. Find the volume of the solid whose base is the region between the semi-circle $y = \sqrt{16 - x^2}$ and the x-axis and whose cross-sections perpendicular to the x-axis are squares with a side on the base.

12. Find the volume of the solid whose base is the region between $y = x^2$ and $y = 4$ and whose perpendicular cross-sections are isosceles right triangles with the hypotenuse on the base.

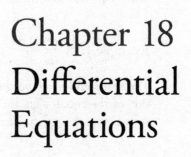

Chapter 18
Differential
Equations

There are many types of differential equations, but only a very small number of them appear on the AP exam. There are courses devoted to learning how to solve a wide variety of differential equations, but AP calculus provides only a very basic introduction to the topic.

SEPARATION OF VARIABLES

If you're given an equation in which the derivative of a function is equal to some other function, you can determine the original function by integrating both sides of the equation and then solving for the constant term.

Example 1: If $\dfrac{dy}{dx} = \dfrac{4x}{y}$ and $y(0) = 5$, find an equation for y in terms of x.

The first step in solving these is to put all of the terms that contain y on the left side of the equals sign and all of the terms that contain x on the right side. We then have $y\,dy = 4x\,dx$. The second step is to integrate both sides.

$$\int y\,dy = \int 4x\,dx$$

And then you integrate

$$\frac{y^2}{2} = 2x^2 + C$$

You're not done yet. The final step is to solve for the constant by plugging in $x = 0$ and $y = 5$.

$$\frac{5^2}{2} = 2(0^2) + C, \text{ so } C = \frac{25}{2}$$

The solution is $\dfrac{y^2}{2} = 2x^2 + \dfrac{25}{2}$.

That's all there is to it. Separate the variables, integrate both sides, and solve for the constant. Often, the equation will involve a logarithm. Let's do an example.

Example 2: If $\dfrac{dy}{dx} = 3x^2 y$ and $y(0) = 2$, find an equation for y in terms of x.

First, put the y terms on the left and the x terms on the right.

$$\frac{dy}{y} = 3x^2\,dx$$

Next, integrate both sides.

$$\int \frac{dy}{y} = \int 3x^2 \ dx$$

The result is: $\ln y = x^3 + C$. It's customary to solve this equation for y. You can do this by putting both sides into exponential form.

$$y = e^{x^3 + C}$$

This can be rewritten as $y = e^{x^3} \cdot e^C$ and, because e^C is a constant, the equation becomes

$$y = Ce^{x^3}$$

This is the preferred form of the equation. Now, solve for the constant. Plug in $x = 0$ and $y = 2$, and you get $2 = Ce^0$.

Because $e^0 = 1$, $C = 2$. The solution is $y = 2e^{x^3}$.

This is the typical differential equation that you'll see on the AP exam. Other common problem types involve position, velocity, and acceleration or exponential growths and decay (Problem 4). We did several problems of this type in Chapter 10, before you knew how to use integrals. In a sample problem, you're given the velocity and acceleration and told to find distance (the reverse of what we did before).

Example 3: If the acceleration of a particle is given by $a(t) = -32$ ft/sec^2, and the velocity of the particle is 64 ft/sec and the height of the particle is 32 ft at time $t = 0$, find: (a) the equation of the particle's velocity at time t; (b) the equation for the particle's height, h, at time t; and (c) the maximum height of the particle.

Part A: Because acceleration is the rate of change of velocity with respect to time, you can write that $\frac{dv}{dt} = -32$. Now separate the variables and integrate both sides.

$$\int dv = \int -32 \ dt$$

Integrating this expression, we get $v = -32t + C$. Now we can solve for the constant by plugging in $t = 0$ and $v = 64$. We get $64 = -32(0) + C$ and $C = 64$. Thus, velocity is $v = -32t + 64$.

Part B: Because velocity is the rate of change of displacement with respect to time, you know that

$$\frac{dh}{dt} = -32t + 64$$

Separate the variables and integrate both sides.

$$\int dh = \int \left(-32t + 64\right) dt$$

Integrate the expression: $h = -16t^2 + 64t + C$. Now solve for the constant by plugging in $t = 0$ and $h = 32$.

$$32 = -16\,(0^2) + 64(0) + C \text{ and } C = 32$$

Thus, the equation for height is $h = -16t^2 + 64t + 32$.

Part C: In order to find the maximum height, you need to take the derivative of the height with respect to time and set it equal to zero. Notice that the derivative of height with respect to time is the velocity; just set the velocity equal to zero and solve for t.

$$-32t + 64 = 0, \text{ so } t = 2$$

Thus, at time $t = 2$, the height of the particle is a maximum. Now, plug $t = 2$ into the equation for height.

$$h = -16(2)^2 + 64(2) + 32 = 96$$

Therefore, the maximum height of the particle is 96 feet.

Here are some solved problems. Do each problem, covering the answer first, then check your answer.

PROBLEM 1. If $\dfrac{dy}{dx} = \dfrac{3x}{2y}$ and $y(0) = 10$, find an equation for y in terms of x.

Answer: First, separate the variables.

$$2y\,dy = 3x\,dx$$

Then, we take the integral of both sides.

$$\int 2y\,dy = \int 3x\,dx$$

Next, integrate both sides.

$$y^2 = \frac{3x^2}{2} + C$$

Finally, solve for the constant.

$$10^2 = \frac{3(0)^2}{2} + C, \text{ so } C = 100$$

The solution is $y^2 = \frac{3x^2}{2} + 100$.

PROBLEM 2. If $\frac{dy}{dx} = 4xy^2$ and $y(0) = 1$, find an equation for y in terms of x.

Answer: First, separate the variables: $\frac{dy}{y^2} = 4x \, dx$. Then, take the integral of both sides.

$$\int \frac{dy}{y^2} = \int 4x \, dx$$

Next, integrate both sides: $-\frac{1}{y} = 2x^2 + C$. You can rewrite this as $y = -\frac{1}{2x^2 + C}$.

Finally, solve for the constant.

$$1 = -\frac{1}{2(0)^2 + C} = \frac{-1}{C}, \text{ so } C = -1$$

The solution is $y = -\frac{1}{2x^2 - 1}$.

PROBLEM 3. If $\frac{dy}{dx} = \frac{y^2}{x}$ and $y(1) = \frac{1}{3}$, find an equation for y in terms of x.

Answer: This time, separating the variables gives us this: $\frac{dy}{y^2} = \frac{dx}{x}$.

Then, take the integral of both sides: $\int \frac{dy}{y^2} = \int \frac{dx}{x}$.

Next, integrate both sides.

$$-\frac{1}{y} = \ln x + C$$

And rearrange the equation.

$$y = \frac{-1}{\ln x + C}$$

Finally, solve for the constant. $\frac{1}{3} = \frac{-1}{C}$, so $C = -3$. The solution is $y = \frac{-1}{\ln x - 3}$.

PROBLEM 4. A city had a population of 10,000 in 1980 and 13,000 in 1990. Assuming an exponential growth rate, estimate the city's population in 2000.

Answer: The phrase "exponential growth rate" means that $\frac{dy}{dt} = ky$, where k is a constant. Take the integral of both sides.

$$\int \frac{dy}{y} = \int k \, dt$$

Then, integrate both sides ($\ln y = kt + C$) and put them in exponential form.

$$y = e^{kt+c} = Ce^{kt}$$

Next, use the information about the population to solve for the constants. If you treat 1980 as $t = 0$ and 1990 as $t = 10$, then

$$10{,}000 = Ce^{k(0)} \text{ and } 13{,}000 = Ce^{k(10)}$$

So $C = 10{,}000$ and $k = \frac{1}{10}\ln 1.3 \approx 0.0262$.

The equation for population growth is approximately $y = 10{,}000e^{.0262t}$. We can estimate that the population in 2000 will be

$$y = 10{,}000e^{0.0262(20)} = 16{,}900$$

SLOPE FIELDS

The idea behind **slope fields**, also known as **direction fields**, is to make a graphical representation of the slope of a function at various points in the plane. We are given a differential equation, but not the equation itself. So how do we do this? Well, it's always easiest to start with an example.

Example 1: Given $\dfrac{dy}{dx} = x$, sketch the slope field of the function.

What does this mean? Look at the equation. It gives us the derivative of the function, which is the slope of the tangent line to the curve at any point x. In other words, the equation tells us that the slope of the curve at any point x is the x-value at that point.

For example, the slope of the curve at $x = 1$ is 1. The slope of the curve at $x = 2$ is 2. The slope of the curve at the origin is 0. The slope of the curve at $x = -1$ is -1. We will now represent these different slopes by drawing small segments of the tangent lines at those points. Let's make a sketch.

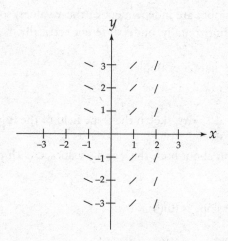

See how all of these slopes are independent of the y-values, so for each value of x, the slope is the same vertically, but is different horizontally. Compare this slope field to the next example.

Example 2: Given $\dfrac{dy}{dx} = y$, sketch the slope field of the function.

Here, the slope of the curve at $y = 1$ is 1. The slope of the curve at $y = 2$ is 2. The slope of the curve at the origin is 0. The slope of the curve at $y = -1$ is -1. Let's make a sketch.

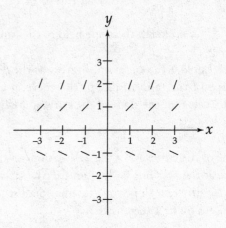

See how all of these slopes are independent of the x-values, so for each value of y, the slope is the same horizontally, but is different vertically.

Now let's do a slightly harder example.

Example 3: Given $\dfrac{dy}{dx} = xy$, sketch the slope field of the function.

Now, we have to think about both the x and y values at each point. Let's calculate a few slopes.

At $(0, 0)$, the slope is $(0)(0) = 0$.

At $(1, 0)$, the slope is $(1)(0) = 0$.

At $(2, 0)$, the slope is $(2)(0) = 0$.

At $(0, 1)$, the slope is $(0)(1) = 0$.

At $(0, 2)$, the slope is $(0)(2) = 0$.

So the slope will be zero at any point on the coordinate axes.

At $(1, 1)$, the slope is $(1)(1) = 1$.

At $(1, 2)$, the slope is $(1)(2) = 2$.

At (1, –1), the slope is (1)(–1) = –1.

At (1, –2), the slope is (1)(–2) = –2.

So the slope at any point where $x = 1$ will be the y-value. Similarly, you should see that the slope at any point where $y = 1$ will be the x-value. As we move out the coordinate axes, slopes will get steeper—whether positive or negative.

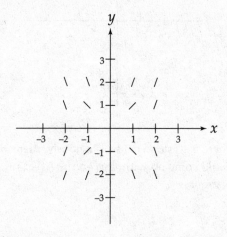

Let's do one more example.

Example 4: Given $\dfrac{dy}{dx} = y - x$, sketch the slope field of the function.

We have to think about both the x and y values at each point. This time, let's make a table of the values of the slope at different points.

	$y = -3$	$y = -2$	$y = -1$	$y = 0$	$y = 1$	$y = 2$	$y = 3$
$x = -3$	0	1	2	3	4	5	6
$x = -2$	–1	0	1	2	3	4	5
$x = -1$	–2	–1	0	1	2	3	4
$x = 0$	–3	–2	–1	0	1	2	3
$x = 1$	–4	–3	–2	–1	0	1	2
$x = 2$	–5	–4	–3	–2	–1	0	1
$x = 3$	–6	–5	–4	–3	–2	–1	0

Now let's make a sketch of the slope field. Notice that the slopes are zero along the line $y = x$ and that the slopes get steeper as we move away from the line in either direction.

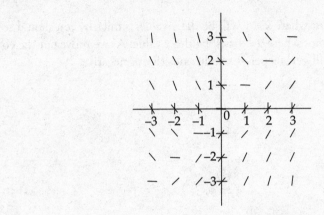

That's really all that there is to slope fields. Obviously, there are more complicated slope fields that one could come up with, but on the AP exam, they will ask you to sketch only the simplest ones.

PRACTICE PROBLEM SET 24

Now try these problems. The answers are in Chapter 19.

1. If $\dfrac{dy}{dx} = \dfrac{7x^2}{y^3}$ and $y(3) = 2$, find an equation for y in terms of x.

2. If $\dfrac{dy}{dx} = 5x^2 y$ and $y(0) = 6$, find an equation for y in terms of x.

3. If $\dfrac{dy}{dx} = \dfrac{1}{y + x^2 y}$ and $y(0) = 2$, find an equation for y in terms of x.

4. If $\dfrac{dy}{dx} = \dfrac{e^x}{y^2}$ and $y(0) = 1$, find an equation for y in terms of x.

5. If $\dfrac{dy}{dx} = \dfrac{y^2}{x^3}$ and $y(1) = 2$, find an equation for y in terms of x.

6. If $\dfrac{dy}{dx} = \dfrac{\sin x}{\cos y}$ and $y(0) = \dfrac{3\pi}{2}$, find an equation for y in terms of x.

7. A colony of bacteria grows exponentially and the colony's population is 4,000 at time $t = 0$ and 6,500 at time $t = 3$. How big is the population at time $t = 10$?

8. A rock is thrown upward with an initial velocity, $v(t)$, of 18 m/s from a height, $h(t)$, of 45 m. If the acceleration of the rock is a constant -9 m/s^2, find the height of the rock at time $t = 4$.

9. The rate of growth of the volume of a sphere is proportional to its volume. If the volume of the sphere is initially 36π ft^3, and expands to 90π ft^3 after 1 second, find the volume of the sphere after 3 seconds.

10. A radioactive element decays exponentially in proportion to its mass. One-half of its original amount remains after 5,750 years. If 10,000 grams of the element are present initially, how much will be left after 1,000 years?

11. Sketch the slope field for $\dfrac{dy}{dx} = 2x$.

12. Sketch the slope field for $\dfrac{dy}{dx} = -\dfrac{x}{y}$.

13. Sketch the slope field for $\dfrac{dy}{dx} = \dfrac{x}{y}$.

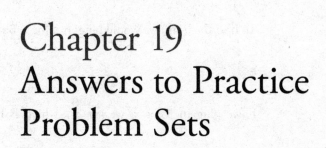

Chapter 19
Answers to Practice
Problem Sets

SOLUTIONS TO PRACTICE PROBLEM SET 1

1.   13

To find the limit, we simply plug in 8 for x: $\lim_{x \to 8} \left(x^2 - 5x - 11 \right) = \left(8^2 - (5)(8) - 11 \right) = 13$.

2. $\dfrac{4}{5}$

To find the limit, we simply plug in 5 for x: $\lim_{x \to 5} \left(\dfrac{x+3}{x^2-15} \right) = \dfrac{5+3}{5^2-15} = \dfrac{8}{10} = \dfrac{4}{5}$.

3. π^2

To find the limit, we would plug in π for x, but there is no x in the limit. So the limit is simply π^2.

4.   4

If we plug in 3 for x, we get $\dfrac{0}{0}$, which is indeterminate. When this happens, we try to factor the expression in order to get rid of the problem terms. Here we factor the top and get: $\lim_{x \to 3} \left(\dfrac{x^2 - 2x - 3}{x - 3} \right) = \lim_{x \to 3} \dfrac{(x-3)(x+1)}{x-3}$. Now we can cancel the term $x - 3$ to get $\lim_{x \to 3} (x+1)$.

Notice that we are allowed to cancel the terms because x is not 3 but very close to 3. Now we can plug in 3 for x: $\lim_{x \to 3} (x+1) = 3 + 1 = 4$.

5.   0

Here we are finding the limit as x goes to infinity. We divide the top and bottom by the highest power of x in the expression: $\lim_{x \to \infty} \left(\dfrac{10x^2 + 25x + 1}{x^4 - 8} \right) = \lim_{x \to \infty} \left(\dfrac{\dfrac{10x^2}{x^4} + \dfrac{25x}{x^4} + \dfrac{1}{x^4}}{\dfrac{x^4}{x^4} - \dfrac{8}{x^4}} \right)$. Next, simplify the top and bottom: $\lim_{x \to \infty} \left(\dfrac{\dfrac{10}{x^2} + \dfrac{25}{x^3} + \dfrac{1}{x^4}}{1 - \dfrac{8}{x^4}} \right)$. Now, if we take the limit as x goes to infinity, we get

$\lim_{x \to \infty} \left(\dfrac{\dfrac{10}{x^2} + \dfrac{25}{x^3} + \dfrac{1}{x^4}}{1 - \dfrac{8}{x^4}} \right) = \dfrac{0+0+0}{1+0} = 0$.

6. $+\infty$

Here we are finding the limit as x goes to infinity. We divide the top and bottom by the highest

power of x in the expression, which is x^4: $\lim\limits_{x\to\infty}\left(\dfrac{x^4-8}{10x^2+25x+1}\right) = \lim\limits_{x\to\infty}\left(\dfrac{\dfrac{x^4}{x^4}-\dfrac{8}{x^4}}{\dfrac{10x^2}{x^4}+\dfrac{25x}{x^4}+\dfrac{1}{x^4}}\right)$. Next,

simplify the top and bottom: $\lim\limits_{x\to\infty}\left(\dfrac{1-\dfrac{8}{x^4}}{\dfrac{10}{x^2}+\dfrac{25}{x^3}+\dfrac{1}{x^4}}\right)$. Now, if we take the limit as x goes to infinity,

we get $\lim\limits_{x\to\infty}\left(\dfrac{1-\dfrac{8}{x^4}}{\dfrac{10}{x^2}+\dfrac{25}{x^3}+\dfrac{1}{x^4}}\right) = \dfrac{1-0}{0+0+0} = \infty$.

7. $\dfrac{1}{10}$

Here we are finding the limit as x goes to infinity. We divide the top and bottom by the highest

power of x in the expression, which is x^4: $\lim\limits_{x\to\infty}\left(\dfrac{x^4-8}{10x^4+25x+1}\right) = \lim\limits_{x\to\infty}\left(\dfrac{\dfrac{x^4}{x^4}-\dfrac{8}{x^4}}{\dfrac{10x^4}{x^4}+\dfrac{25x}{x^4}+\dfrac{1}{x^4}}\right)$. Next,

simplify the top and bottom: $\lim\limits_{x\to\infty}\left(\dfrac{1-\dfrac{8}{x^4}}{10+\dfrac{25}{x^3}+\dfrac{1}{x^4}}\right)$. Now, if we take the limit as x goes to infinity,

we get $\lim\limits_{x\to\infty}\left(\dfrac{1-\dfrac{8}{x^4}}{10+\dfrac{25}{x^3}+\dfrac{1}{x^4}}\right) = \dfrac{1-0}{10+0+0} = \dfrac{1}{10}$.

8. $\sqrt{5}$

Here we are finding the limit as x goes to infinity. We divide the top and bottom by the highest power

of x in the expression, which is x^2. Notice that, under the radical, we divide by x^4 because $\sqrt{x^4}=x^2$:

$\lim\limits_{x\to\infty}\left(\dfrac{\sqrt{5x^4+2x}}{x^2}\right) = \lim\limits_{x\to\infty}\left(\dfrac{\sqrt{\dfrac{5x^4}{x^4}+\dfrac{2x}{x^4}}}{\dfrac{x^2}{x^2}}\right)$. Next, simplify the top and bottom: $\lim\limits_{x\to\infty}\left(\dfrac{\sqrt{5+\dfrac{2}{x^3}}}{1}\right)$.

Now, if we take the limit as x goes to infinity, we get $\lim\limits_{x\to\infty}\left(\dfrac{\sqrt{5+\dfrac{2}{x^3}}}{1}\right) = \dfrac{\sqrt{5+0}}{1} = \sqrt{5}$.

9. $+\infty$

Here we have to think about what happens when we plug in a value that is very close to 6, but a little bit more. The top expression will approach 8. The bottom expression will approach 0, but will be a little bit bigger. Thus, the limit will be $\dfrac{8}{0^+}$, which is $+\infty$.

10. $-\infty$

Here we have to think about what happens when we plug in a value that is very close to 6, but a little bit less. The top expression will approach 8. The bottom expression will approach 0 but will be a little bit less. Thus, the limit will be $\dfrac{8}{0^-}$, which is $-\infty$.

11. The limit *Does Not Exist.*

In order to evaluate the limit as x approaches 6, we find the limit as it approaches 6^+ (from the right) and the limit as it approaches 6^- (from the left). If the two limits approach the same value, or both approach positive infinity or both approach negative infinity, then the limit is that value, or the appropriately signed infinity. If the two limits do not agree, the limit "Does Not Exist." Here, if we look at the solutions to problems 9 and 10, we find that as x approaches 6^+, the limit is $+\infty$, but as x approaches 6^-, the limit is $-\infty$. Because the two limits are *not* the same, the limit *Does Not Exist.*

12. 1

Here we have to think about what happens when we plug in a value that is very close to 0, but a little bit more. The top and bottom expressions will both be positive and the same value, so we get $\lim\limits_{x \to 0^+} \dfrac{x}{|x|} = \dfrac{0^+}{0^+} = 1$.

13. -1

Here we have to think about what happens when we plug in a value that is very close to 0, but a little bit less. The top expression will be negative, and the bottom expression will be positive, so we get $\lim\limits_{x \to 0^-} \dfrac{x}{|x|} = \dfrac{0^-}{0^+} = -1$.

14.   +∞

Here we have to think about what happens when we plug in a value that is very close to 7, but a little bit more. The top expression will approach 7. The bottom expression will approach 0, but will be a little bit positive. Thus, the limit will be $\frac{7}{0^+}$, which is +∞.

15.   The limit *Does Not Exist*.

In order to evaluate the limit as x approaches 7, we find the limit as it approaches 7^+ (from the right) and the limit as it approaches 7^- (from the left). If the two limits approach the same value, or both approach positive infinity or both approach negative infinity, then the limit is that value, or the appropriately signed infinity. If the two limits do not agree, the limit "Does Not Exist." Here, if we look at the solutions to problem 14, we see that as x approaches 7^+, the limit is +∞. As x approaches 7^-, the top expression will approach 7. The bottom will approach 0, but will be a little bit negative. Thus, the limit will be $\frac{7}{0^-}$, which is −∞. Because the two limits are *not* the same, the limit *Does Not Exist*.

16.   +∞

In order to evaluate the limit as x approaches 7, we find the limit as it approaches 7^+ (from the right) and the limit as it approaches 7^- (from the left). If the two limits approach the same value, or both approach positive infinity or both approach negative infinity, then the limit is that value, or the appropriately-signed infinity. If the two limits do not agree, the limit "Does Not Exist." Here, we see that as x approaches 7^+, the top expression will approach 7. The bottom expression will approach 0, but will be a little bit positive. Thus, the limit will be $\frac{7}{0^+}$, which is +∞. As x approaches 7^-, the top expression will again approach 7. The bottom will approach 0 but will be a little bit positive. Thus, the limit will be $\frac{7}{0^+}$, which is +∞. Because the two limits are the same, the limit is +∞.

17.   (a) 4; (b) 5; (c) The limit *Does Not Exist*.

(a) Notice that $f(x)$ is a piecewise function, which means that we use the function $f(x) = x^2 - 5$ for all values of x less than or equal to 3. Thus, $\lim\limits_{x \to 3^-} f(x) = 3^2 - 5 = 4$.

(b) Here we use the function $f(x) = x + 2$ for all values of x greater than 3. Thus,

$\lim\limits_{x \to 3^+} f(x) = 3 + 2 = 5$.

(c) In order to evaluate the limit as x approaches 3, we find the limit as it approaches 3^+ (from the right) and the limit as it approaches 3^- (from the left). If the two limits approach the same value, or both approach positive infinity or both approach negative infinity, then the limit is that value, or the appropriately signed infinity. If the two limits do not agree, the limit "Does Not Exist."

Here, if we refer to the solutions in parts (a) and (b), we see that $\lim\limits_{x \to 3^-} f(x) = 4$ and $\lim\limits_{x \to 3^+} f(x) = 5$.

Because the two limits are *not* the same, the limit *Does Not Exist*.

18.   (a) 4; (b) 4; (c) 4

(a) Notice that $f(x)$ is a piecewise function, which means that we use the function $f(x) = x^2 - 5$ for all values of x less than or equal to 3. Thus, $\lim\limits_{x \to 3^-} f(x) = 3^2 - 5 = 4$.

(b) Here we use the function $f(x) = x + 1$ for all values of x greater than 3. Thus, $\lim\limits_{x \to 3^+} f(x) = 3 + 1 = 4$.

(c) In order to evaluate the limit as x approaches 3, we find the limit as it approaches 3^+ (from the right) and the limit as it approaches 3^- (from the left). If the two limits approach the same value, or both approach positive infinity or both approach negative infinity, then the limit is that value, or the appropriately signed infinity. If the two limits do not agree, the limit "Does Not Exist."

Here, if we refer to the solutions in parts (a) and (b), we see that $\lim\limits_{x \to 3^-} f(x) = 4$ and $\lim\limits_{x \to 3^+} f(x) = 4$.

Because the two limits are the same, the limit is 4.

19. $\dfrac{3}{\sqrt{2}}$

Here, if we plug in $\dfrac{\pi}{4}$ for x, we get $\lim\limits_{x \to \frac{\pi}{4}} 3 \cos x = 3 \cos \dfrac{\pi}{4} = \dfrac{3}{\sqrt{2}}$.

20.   0

Here, if we plug in 0 for x, we get $\lim\limits_{x \to 0} 3 \dfrac{x}{\cos x} = 3 \dfrac{0}{\cos 0} = 3 \dfrac{0}{1} = 0$.

21.   3

Remember Rule No. 1, which says that $\lim\limits_{x\to0}\dfrac{\sin x}{x}=1$. If we want to find the limit of its recip-

rocal, we can write this as $\lim\limits_{x\to0}\dfrac{1}{\dfrac{\sin x}{x}}=\dfrac{1}{\lim\limits_{x\to0}\dfrac{\sin x}{x}}=\dfrac{1}{1}=1$. Here, if we plug in 0 for x, we get

$\lim\limits_{x\to0}\left(3\dfrac{x}{\sin x}\right)=(3)(1)=3$.

22. $\dfrac{3}{8}$

Remember Rule No. 4, which says that $\lim\limits_{x\to0}\dfrac{\sin ax}{\sin bx}=\dfrac{a}{b}$. Here $\lim\limits_{x\to0}\dfrac{\sin 3x}{\sin 8x}=\dfrac{3}{8}$. If we want to eval-

uate the limit the long way, first we divide the numerator and the denominator of the expression by

x: $\lim\limits_{x\to0}\dfrac{\dfrac{\sin 3x}{x}}{\dfrac{\sin 8x}{x}}$. Next, we multiply the numerator and the denominator of the top expression by 3

and the numerator and the denominator of the bottom expression by 8. We get $\lim\limits_{x\to0}\dfrac{\dfrac{3\sin 3x}{3x}}{\dfrac{8\sin 8x}{8x}}$. Now,

we can evaluate the limit: $\lim\limits_{x\to0}\dfrac{\dfrac{3\sin 3x}{3x}}{\dfrac{8\sin 8x}{8x}}=\dfrac{(3)(1)}{(8)(1)}=\dfrac{3}{8}$.

23. $\dfrac{7}{5}$

Here, we can rewrite the expression as $\lim\limits_{x\to0}\dfrac{\tan 7x}{\sin 5x}=\lim\limits_{x\to0}\dfrac{\dfrac{\sin 7x}{\cos 7x}}{\sin 5x}=\lim\limits_{x\to0}\dfrac{\sin 7x}{(\cos 7x)(\sin 5x)}$. Remem-

ber Rule No. 4, which says that $\lim\limits_{x\to0}\dfrac{\sin ax}{\sin bx}=\dfrac{a}{b}$. Here, $\lim\limits_{x\to0}\dfrac{\sin 7x}{(\cos 7x)(\sin 5x)}=\dfrac{7}{(1)(5)}=\dfrac{7}{5}$. If we

want to evaluate the limit the long way, we first divide the numerator and the denominator of the

expression by x: $\lim\limits_{x\to0}\dfrac{\left(\dfrac{\sin 7x}{x}\right)}{(\cos 7x)\left(\dfrac{\sin 5x}{x}\right)}$. Next, we multiply the numerator and the denominator of the top expression by 7 and the numerator and the denominator of the bottom expression by 5. We

get $\displaystyle\lim_{x\to 0}\frac{\left(\dfrac{7\sin 7x}{7x}\right)}{(\cos 7x)\left(\dfrac{5\sin 5x}{5x}\right)}$. Now, we can evaluate the limit: $\displaystyle\lim_{x\to 0}\frac{(7)(1)}{(1)(5)(1)}=\frac{7}{5}$.

24.   The limit *Does Not Exist*.

The value of sin x oscillates between –1 and 1. Thus, as x approaches infinity, sin x does not approach a specific value. Therefore, the limit *Does Not Exist*.

25.   0

Here, as x approaches infinity, $\dfrac{1}{x}$ approaches 0. Thus, $\displaystyle\lim_{x\to\infty}\sin\frac{1}{x}=\sin 0=0$.

26.   0

Here, use the trigonometric identity $\sin^2 x=1-\cos^2 x$ to rewrite the bottom expression: $\displaystyle\lim_{x\to 0}\frac{x^2\sin x}{\sin^2 x}$.

Next, we can break up the limit into $\displaystyle\lim_{x\to 0}\left(\frac{x}{\sin x}\frac{x}{\sin x}\sin x\right)$. Remember that $\displaystyle\lim_{x\to 0}\frac{\sin x}{x}=1$ and

that $\displaystyle\lim_{x\to 0}\frac{x}{\sin x}=1$ as well. Now we can evaluate the limit: $\displaystyle\lim_{x\to 0}\left(\frac{x}{\sin x}\frac{x}{\sin x}\sin x\right)=(1)(1)(0)=0$.

27. $\dfrac{49}{121}$

We can break up the limit into $\displaystyle\lim_{x\to 0}\frac{\sin 7x}{\sin 11x}\frac{\sin 7x}{\sin 11x}$. Remember Rule No. 4, which says

that $\displaystyle\lim_{x\to 0}\frac{\sin ax}{\sin bx}=\frac{a}{b}$. Here, $\displaystyle\lim_{x\to 0}\frac{\sin 7x}{\sin 11x}\frac{\sin 7x}{\sin 11x}=\frac{7}{11}\cdot\frac{7}{11}=\frac{49}{121}$. If we want to evalu-

ate the limit the long way, we first divide the numerator and the denominator of the

expressions by x: $\displaystyle\lim_{x\to 0}\left(\frac{\dfrac{\sin 7x}{x}}{\dfrac{\sin 11x}{x}}\frac{\dfrac{\sin 7x}{x}}{\dfrac{\sin 11x}{x}}\right)$. Next, we multiply the numerator and the denomi-

nator of the top expression by 7 and the numerator and the denominator of the bottom expression by 11. We get $\lim\limits_{x\to 0}\left(\dfrac{\dfrac{7\sin 7x}{7x}}{\dfrac{11\sin 11x}{11x}}\dfrac{\dfrac{7\sin 7x}{7x}}{\dfrac{11\sin 11x}{11x}}\right)$. Now, we can evaluate the limit:

$$\lim\limits_{x\to 0}\left(\dfrac{\dfrac{7\sin 7x}{7x}}{\dfrac{11\sin 11x}{11x}}\dfrac{\dfrac{7\sin 7x}{7x}}{\dfrac{11\sin 11x}{11x}}\right)=\dfrac{(7)(1)}{(11)(1)}\dfrac{(7)(1)}{(11)(1)}=\left(\dfrac{7}{11}\right)\left(\dfrac{7}{11}\right)=\dfrac{49}{121}.$$

28. 6

Notice that if we plug in 0 for h, we get $\dfrac{0}{0}$, which is indeterminate. If we expand the expression in

the numerator, we get $\lim\limits_{h\to 0}\dfrac{9+6h+h^2-9}{h}$. This simplifies to $\lim\limits_{h\to 0}\dfrac{6h+h^2}{h}$. Next, factor h out

of the top expression: $\lim\limits_{h\to 0}\dfrac{h(6+h)}{h}$. Now, we can cancel the h and evaluate the limit to get:

$$\lim\limits_{h\to 0}\dfrac{h(6+h)}{h}=\lim\limits_{h\to 0}(6+h)=6+0=6\,.$$

29. cos x

Notice that if we plug in 0 for h, we get $\dfrac{0}{0}$, which is indeterminate. Recall that the trigono-

metric formula $\sin(A+B)=\sin A\cos B+\cos A\sin B$. Here we can rewrite the top expression

as: $\lim\limits_{h\to 0}\dfrac{\sin(x+h)-\sin x}{h}=\lim\limits_{h\to 0}\dfrac{\sin x\cos h+\cos x\sin h-\sin x}{h}$. We can break up

the limit into $\lim\limits_{h\to 0}\dfrac{\sin x\cos h-\sin x}{h}+\dfrac{\cos x\sin h}{h}$. Next, factor $\sin x$ out of the top

of the left-hand expression: $\lim\limits_{h\to 0}\dfrac{\sin x(\cos h-1)}{h}+\dfrac{\cos x\sin h}{h}$. Now, we can break

this into separate limits: $\lim\limits_{h\to 0}\dfrac{\sin x(\cos h-1)}{h}+\lim\limits_{h\to 0}\dfrac{\cos x\sin h}{h}$. The left-hand lim-

it is $\lim\limits_{h\to 0}\dfrac{\sin x(\cos h-1)}{h}=\sin x\lim\limits_{h\to 0}\dfrac{(\cos h-1)}{h}=\sin x\bullet 0=0$. The right-hand limit is

$\cos x\lim\limits_{h\to 0}\dfrac{\sin h}{h}=\cos x\bullet 1=\cos x$. Finally, combine the left-hand and right-hand limits:

$$\sin x\lim\nolimits_{h\to 0}\dfrac{\cos h-1}{h}+\cos x\lim\nolimits_{h\to 0}\dfrac{\sin h}{h}=0+\cos x=\cos x.$$

30. $-\dfrac{1}{x^2}$

Notice that if we plug in 0 for h, we get $\dfrac{0}{0}$, which is indeterminate. If we combine

the two expressions on top with a common denominator, we get

$$\lim_{h \to 0} \frac{\dfrac{1}{x+h} - \dfrac{1}{x}}{h} = \lim_{h \to 0} \frac{\dfrac{x}{x(x+h)} - \dfrac{x+h}{x(x+h)}}{h} = \lim_{h \to 0} \frac{\dfrac{x-(x+h)}{x(x+h)}}{h}.$$ We can simplify the top expression,

leaving us with: $\lim_{h \to 0} \dfrac{\dfrac{-h}{x(x+h)}}{h}$. Next, simplify the expression into $\lim_{h \to 0} \dfrac{\dfrac{-h}{x(x+h)}}{h} = \lim_{h \to 0} \dfrac{-h}{hx(x+h)}$.

We can cancel the h to get $\lim_{h \to 0} \dfrac{-1}{x(x+h)}$. Now, if we evaluate the limit we get

$$\lim_{h \to 0} \frac{-1}{x(x+h)} = \frac{-1}{x(x)} = -\frac{1}{x^2}.$$

SOLUTIONS TO PRACTICE PROBLEM SET 2

1.   Yes. It satisfies all three conditions.

In order for a function $f(x)$ to be continuous at a point $x = c$, it must fulfill *all three* of the following

conditions:

Condition 1: $f(c)$ exists.

Condition 2: $\lim\limits_{x \to c} f(x)$ exists.

Condition 3: $\lim\limits_{x \to c} f(x) = f(c)$

Let's test each condition.

$f(2) = 9$, which satisfies condition 1.

$\lim\limits_{x \to 2^-} f(x) = 9$ and $\lim\limits_{x \to 2^+} f(x) = 9$, so $\lim\limits_{x \to 2} f(x) = 9$, which satisfies condition 2.

$\lim\limits_{x \to 2} f(x) = 9 = f(2)$, which satisfies condition 3. Therefore, $f(x)$ is continuous at

$x = 2$.

2. No. It fails condition 3.

In order for a function $f(x)$ to be continuous at a point $x = c$, it must fulfill *all three* of the following conditions:

Condition 1: $f(c)$ exists.

Condition 2: $\lim\limits_{x \to c} f(x)$ exists.

Condition 3: $\lim\limits_{x \to c} f(x) = f(c)$

Let's test each condition.

$f(3) = 29$, which satisfies condition 1.

$\lim\limits_{x \to 3^-} f(x) = 30$ and $\lim\limits_{x \to 3^+} f(x) = 30$, so $\lim\limits_{x \to 3} f(x) = 30$, which satisfies condition 2.

But $\lim\limits_{x \to 3} f(x) \neq f(3)$. Therefore, $f(x)$ is not continuous at $x = 3$ because it fails condition 3.

3. No. It fails condition 1.

In order for a function $f(x)$ to be continuous at a point $x = c$, it must fulfill *all three* of the following conditions:

Condition 1: $f(c)$ exists.

Condition 2: $\lim\limits_{x \to c} f(x)$ exists.

Condition 3: $\lim\limits_{x \to c} f(x) = f(c)$

Notice that the function is not defined at $f(3)$. Therefore, $f(x)$ is not continuous at $x = 3$ because it fails condition 1.

4. No. It is discontinuous at any odd integral multiple of $\dfrac{\pi}{2}$.

Recall that $\sec x = \dfrac{1}{\cos x}$. This means that $\sec x$ is undefined at any value where $\cos x = 0$, which are the odd multiples of $\dfrac{\pi}{2}$. Therefore, $\sec x$ is not continuous everywhere.

5. No. It is discontinuous at the endpoints of the interval.

Recall that $\sec x = \dfrac{1}{\cos x}$. This means that sec x is undefined at any value where $\cos x = 0$. Also recall that $\cos \dfrac{\pi}{2} = 0$ and $\cos\left(-\dfrac{\pi}{2}\right) = 0$. Therefore, sec x is not continuous everywhere on the interval $\left[-\dfrac{\pi}{2}, \dfrac{\pi}{2}\right]$.

6. Yes.

Recall that $\sec x = \dfrac{1}{\cos x}$. This means that sec x is undefined at any value where $\cos x = 0$. Also recall that $\cos \dfrac{\pi}{2} = 0$ and $\cos\left(-\dfrac{\pi}{2}\right) = 0$. Therefore, sec x is continuous everywhere on the interval $\left(-\dfrac{\pi}{2}, \dfrac{\pi}{2}\right)$ because the interval does not include the endpoints.

7. The function is continuous for $k = \dfrac{9}{16}$.

In order for a function $f(x)$ to be continuous at a point $x = c$, it must fulfill *all three* of the following conditions:

Condition 1: $f(c)$ exists.

Condition 2: $\lim\limits_{x \to c} f(x)$ exists.

Condition 3: $\lim\limits_{x \to c} f(x) = f(c)$

We will need to find a value, or values, of k that enables $f(x)$ to satisfy each condition.

Condition 1: $f(4) = 0$

Condition 2: $\lim\limits_{x \to 4^-} f(x) = 0$ and $\lim\limits_{x \to 4^+} f(x) = 16k - 9$. In order for the limit to exist, the two limits must be the same. If we solve $16k - 9 = 0$, we get $k = \dfrac{9}{16}$.

Condition 3: If we now let $k = \dfrac{9}{16}$, $\lim\limits_{x \to 4} f(x) = 0 = f(4)$. Therefore, the solution is $k = \dfrac{9}{16}$.

8. The function is continuous for $k = 6$ or $k = -1$.

In order for a function $f(x)$ to be continuous at a point $x = c$, it must fulfill *all three* of the following conditions:

Condition 1: $f(c)$ exists.

Condition 2: $\lim\limits_{x \to c} f(x)$ exists.

Condition 3: $\lim\limits_{x \to c} f(x) = f(c)$

We will need to find a value, or values, of k that enables $f(x)$ to satisfy each condition.

Condition 1: $f(-3) = k^2 - 5k$

Condition 2: $\lim\limits_{x \to -3^-} f(x) = 6$ and $\lim\limits_{x \to -3^+} f(x) = 6$, so $\lim\limits_{x \to -3} f(x) = 6$.

Condition 3: Now we need to find a value, or values, of k such that $\lim\limits_{x \to -3} f(x) = 6 = f(3)$. If we set $k^2 - 5k = 6$, we obtain the solutions $k = 6$ and $k = -1$.

9. The removable discontinuity is at $\left(3, \dfrac{11}{5} \right)$.

A *removable discontinuity* occurs when you have a rational expression with common factors in the

numerator and denominator. Because these factors can be cancelled, the discontinuity is "remov-

able." In practical terms, this means that the discontinuity occurs where there is a "hole" in the

graph. If we factor $f(x) = \dfrac{x^2 + 5x - 24}{x^2 - x - 6}$, we get $f(x) = \dfrac{(x+8)(x-3)}{(x+2)(x-3)}$. If we cancel the common

factor, we get $f(x) = \dfrac{(x+8)}{(x+2)}$. Now, if we plug in $x = 3$, we get $f(x) = \dfrac{11}{5}$. Therefore, the remov-

able discontinuity is at $\left(3, \dfrac{11}{5} \right)$.

10.   (a) 0; (b) 0; (c) 1; (d) 1; (e) $f(3)$ *Does Not Exist*; (f) a jump discontinuity at $x = -3$; a removable discontinuity at $x = 3$ and an essential discontinuity at $x = 5$.

(a) If we look at the graph, we can see that $\lim\limits_{x \to -\infty} f(x) = 0$.

(b) If we look at the graph, we can see that $\lim\limits_{x \to \infty} f(x) = 0$.

(c) If we look at the graph, we can see that $\lim\limits_{x \to 3^-} f(x) = 1$.

(d) If we look at the graph, we can see that $\lim\limits_{x \to 3^+} f(x) = 1$.

(e) $f(3)$ Does Not Exist.

(f) There are three discontinuities: (1) a jump discontinuity at $x = -3$; (2) a removable discontinuity at $x = 3$; and (3) an essential discontinuity at $x = 5$.

SOLUTIONS TO PRACTICE PROBLEM SET 3

1.   5

We find the derivative of a function, $f(x)$, using the definition of the derivative, which is:

$$f'(x) = \lim\limits_{h \to 0} \frac{f(x+h) - f(x)}{h}.$$ Here $f(x) = 5x$ and $x = 3$. This means that $f(3) = 5(3) = 15$ and

$f(3+h) = 5(3+h) = 15 + 5h$. If we now plug these into the definition of the derivative, we get

$$f'(3) = \lim\limits_{h \to 0} \frac{f(3+h) - f(3)}{h} = \lim\limits_{h \to 0} \frac{15 + 5h - 15}{h}.$$ This simplifies to $f'(3) = \lim\limits_{h \to 0} \frac{5h}{h} = \lim\limits_{h \to 0} 5 = 5$.

If you noticed that the function is simply the equation of a line, then you would have seen that the derivative is simply the slope of the line, which is 5 everywhere.

2.   4

We find the derivative of a function, $f(x)$, using the definition of the derivative,

which is: $f'(x) = \lim\limits_{h \to 0} \frac{f(x+h) - f(x)}{h}$. Here $f(x) = 4x$ and $x = -8$. This means that

$f(-8) = 4(-8) = -32$ and $f(-8+h) = 4(-8+h) = -32 + 4h$. If we now plug these into the definition of the derivative, we get $f'(-8) = \lim\limits_{h \to 0} \dfrac{f(-8+h) - f(-8)}{h} = \lim\limits_{h \to 0} \dfrac{-32 + 4h + 32}{h}$. This sim-

plifies to $f'(-8) = \lim\limits_{h \to 0} \dfrac{4h}{h} = \lim\limits_{h \to 0} 4 = 4$.

If you noticed that the function is simply the equation of a line, then you would have seen that the derivative is simply the slope of the line, which is 4 everywhere.

3.   20

We find the derivative of a function, $f(x)$, using the definition of the derivative, which is:

$f'(x) = \lim\limits_{h \to 0} \dfrac{f(x+h) - f(x)}{h}$. Here $f(x) = 2x^2$ and $x = 5$. This means that $f(5) = 2(5)^2 = 50$

and $f(5+h) = 2(5+h)^2 = 2(25 + 10h + h^2) = 50 + 20h + 2h^2$. If we now plug these into the

definition of the derivative, we get $f'(5) = \lim\limits_{h \to 0} \dfrac{f(5+h) - f(5)}{h} = \lim\limits_{h \to 0} \dfrac{50 + 20h + 2h^2 - 50}{h}$. This

simplifies to $f'(5) = \lim\limits_{h \to 0} \dfrac{20h + 2h^2}{h}$. Now we can factor out the h from the numerator and cancel

it with the h in the denominator: $f'(5) = \lim\limits_{h \to 0} \dfrac{h(20 + 2h)}{h} = \lim\limits_{h \to 0} (20 + 2h)$. Now we take the limit

to get $f'(5) = \lim\limits_{h \to 0} (20 + 2h) = 20$.

4.   −10

We find the derivative of a function, $f(x)$, using the definition of the derivative, which is:

$f'(x) = \lim\limits_{h \to 0} \dfrac{f(x+h) - f(x)}{h}$. Here $f(x) = 5x^2$ and $x = -1$. This means that $f(-1) = 5(-1)^2 = 5$

and $f(-1+h) = 5(-1+h)^2 = 5(1 - 2h + h^2) = 5 - 10h + 5h^2$. If we now plug these into the defi-

nition of the derivative, we get $f'(-1) = \lim\limits_{h \to 0} \dfrac{f(-1+h) - f(-1)}{h} = \lim\limits_{h \to 0} \dfrac{5 - 10h + 5h^2 - 5}{h}$. This

simplifies to $f'(-1) = \lim\limits_{h \to 0} \dfrac{-10h + 5h^2}{h}$. Now we can factor out the h from the numerator and cancel it with the h in the denominator: $f'(-1) = \lim_{h \to 0} \dfrac{h(-10+5h)}{h} = \lim_{h \to 0}(-10+5h)$. Now we take

the limit to get $f'(-1) = \lim_{h \to 0}(-10+5h) = -10$.

5. $16x$

We find the derivative of a function, $f(x)$, using the definition of the

derivative, which is: $f'(x) = \lim_{h \to 0} \dfrac{f(x+h) - f(x)}{h}$.   Here $f(x) = 8x^2$ and

$f(x+h) = 8(x+h)^2 = 8(x^2 + 2xh + h^2) = 8x^2 + 16xh + 8h^2$. If we now plug these into the defi-

nition of the derivative, we get $f'(x) = \lim_{h \to 0} \dfrac{f(x+h) - f(x)}{h} = \lim_{h \to 0} \dfrac{8x^2 + 16xh + 8h^2 - 8x^2}{h}$. This

simplifies to $f'(x) = \lim_{h \to 0} \dfrac{16xh + 8h^2}{h}$. Now we can factor out the h from the numerator and cancel

it with the h in the denominator: $f'(x) = \lim_{h \to 0} \dfrac{h(16x + 8h)}{h} = \lim_{h \to 0}(16x + 8h)$. Now we take the

limit to get $f'(x) = \lim_{h \to 0}(16x + 8h) = 16x$.

6. $-20x$

We find the derivative of a function, $f(x)$, using the definition of the

derivative, which is $f'(x) = \lim_{h \to 0} \dfrac{f(x+h) - f(x)}{h}$.   Here $f(x) = -10x^2$ and

$f(x+h) = -10(x+h)^2 = -10(x^2 + 2xh + h^2) = -10x^2 - 20xh - 10h^2$.

If we now plug these into the definition of the derivative, we get

$f'(x) = \lim_{h \to 0} \dfrac{f(x+h) - f(x)}{h} = \lim_{h \to 0} \dfrac{-10x^2 - 20xh - 10h^2 + 10x^2}{h}$.   This   simplifies   to

$f'(x) = \lim_{h \to 0} \dfrac{-20xh - 10h^2}{h}$. Now we can factor out the h from the numerator and cancel it with the h in the denominator: $f'(x) = \lim_{h \to 0} \dfrac{h(-20x - 10h)}{h} = \lim_{h \to 0}(-20x - 10h)$. Now we take the limit

to get $f'(x) = \lim_{h \to 0}(-20x - 10h) = -20x$.

7.　　40a

We find the derivative of a function, $f(x)$, using the definition of the derivative, which is

$f'(x) = \lim_{h \to 0} \dfrac{f(x+h) - f(x)}{h}$. Here $f(x) = 20x^2$ and $x = a$. This means that $f(a) = 20a^2$ and

$f(a+h) = 20(a+h)^2 = 20(a^2 + 2ah + h^2) = 20a^2 + 40ah + 20h^2$. If we now plug these into the

definition of the derivative, we get $f'(a) = \lim_{h \to 0} \dfrac{f(a+h) - f(a)}{h} = \lim_{h \to 0} \dfrac{20a^2 + 40ah + 20h^2 - 20a^2}{h}$.

This simplifies to $f'(a) = \lim_{h \to 0} \dfrac{40ah + 20h^2}{h}$. Now we can factor out the h from the numerator and

cancel it with the h in the denominator: $f'(a) = \lim_{h \to 0} \dfrac{h(40a + 20h)}{h} = \lim_{h \to 0}(40a + 20h)$. Now we

take the limit to get $f'(a) = \lim_{h \to 0}(40a + 20h) = 40a$.

8.　　54

We find the derivative of a function, $f(x)$, using the definition of the

derivative, which is: $f'(x) = \lim_{h \to 0} \dfrac{f(x+h) - f(x)}{h}$. Here $f(x) = 2x^3$

and $x = -3$. This means that $f(-3) = 2(-3)^3 = -54$ and

$f(-3+h) = 2(-3+h)^3 = 2\left((-3)^3 + 3(-3)^2 h + 3(-3)h^2 + h^3\right) = -54 + 54h - 18h^2 + 2h^3$.

If we now plug these into the definition of the derivative, we get

$f'(-3) = \lim_{h \to 0} \dfrac{f(-3+h) - f(-3)}{h} = \lim_{h \to 0} \dfrac{-54 + 54h - 18h^2 + 2h^3 + 54}{h}$. This simplifies to

$f'(-3) = \lim\limits_{h \to 0} \dfrac{54h - 18h^2 + 2h^3}{h}$. Now we can factor out the h from the numerator and cancel it

with the h in the denominator: $f'(-3) = \lim\limits_{h \to 0} \dfrac{h\left(54 - 18h + 2h^2\right)}{h} = \lim\limits_{h \to 0}\left(54 - 18h + 2h^2\right)$. Now we

take the limit to get $f'(-3) = \lim\limits_{h \to 0}\left(54 - 18h + 2h^2\right) = 54$.

9.　　　$-9x^2$

We find the derivative of a function, $f(x)$, using the definition of the deriva-

tive, which is $f'(x) = \lim\limits_{h \to 0} \dfrac{f(x+h) - f(x)}{h}$. Here, $f(x) = -3x^3$. This means that

$f(x+h) = -3(x+h)^3 = -3\left(x^3 + 3x^2h + 3xh^2 + h^3\right) = -3x^3 - 9x^2h - 9xh^2 - 3h^3$.

If we now plug these into the definition of the derivative, we get

$f'(x) = \lim\limits_{h \to 0} \dfrac{f(x+h) - f(x)}{h} = \lim\limits_{h \to 0} \dfrac{-3x^3 - 9x^2h - 9xh^2 - 3h^3 + 3x^3}{h}$. This simplifies to

$f'(x) = \lim\limits_{h \to 0} \dfrac{-9x^2h - 9xh^2 - 3h^3}{h}$. Now we can factor out the h from the numerator and cancel it

with the h in the denominator: $f'(x) = \lim\limits_{h \to 0} \dfrac{h\left(-9x^2 - 9xh - 3h^2\right)}{h} = \lim\limits_{h \to 0}\left(-9x^2 - 9xh - 3h^2\right)$. Now

we take the limit to get $f'(x) = \lim\limits_{h \to 0}\left(-9x^2 - 9xh - 3h^2\right) = -9x^2$.

10.　　　$4x^3$

We find the derivative of a function, $f(x)$, using the definition of the deriva-

tive, which is $f'(x) = \lim\limits_{h \to 0} \dfrac{f(x+h) - f(x)}{h}$. Here $f(x) = x^4$. This means that

$f(x+h) = (x+h)^4 = \left(x^4 + 4x^3h + 6x^2h^2 + 4xh^3 + h^4\right)$. If we now plug these into the definition

of the derivative, we get $f'(x) = \lim\limits_{h \to 0} \dfrac{f(x+h) - f(x)}{h} = \lim\limits_{h \to 0} \dfrac{x^4 + 4x^3h + 6x^2h^2 + 4xh^3 + h^4 - x^4}{h}$.

This simplifies to $f'(x) = \lim_{h \to 0} \dfrac{4x^3 h + 6x^2 h^2 + 4xh^3 + h^4}{h}$. Now we can fac-

tor out the h from the numerator and cancel it with the h in the denominator:

$$f'(x) = \lim_{h \to 0} \frac{h\left(4x^3 + 6x^2 h + 4xh^2 + h^3\right)}{h} = \lim_{h \to 0}\left(4x^3 + 6x^2 h + 4xh^2 + h^3\right).$$ Now we take the limit

to get $f'(x) = \lim_{h \to 0}\left(4x^3 + 6x^2 h + 4xh^2 + h^3\right) = 4x^3$.

11. $5x^4$

We find the derivative of a function, $f(x)$, using the definition of the derivative, which is

$f'(x) = \lim_{h \to 0} \dfrac{f(x+h) - f(x)}{h}$. Here $f(x) = x^5$.

This means that $f(x+h) = (x+h)^5 = \left(x^5 + 5x^4 h + 10x^3 h^2 + 10x^2 h^3 + 5xh^4 + h^5\right)$.

If we now plug these into the definition of the derivative, we get

$$f'(x) = \lim_{h \to 0} \frac{f(x+h) - f(x)}{h} = \lim_{h \to 0} \frac{x^5 + 5x^4 h + 10x^3 h^2 + 10x^2 h^3 + 5xh^4 + h^5 - x^5}{h}.$$

This simplifies to $f'(x) = \lim_{h \to 0} \dfrac{5x^4 h + 10x^3 h^2 + 10x^2 h^3 + 5xh^4 + h^5}{h}$. Now we can fac-

tor out the h from the numerator and cancel it with the h in the denominator:

$$f'(x) = \lim_{h \to 0} \frac{h\left(5x^4 + 10x^3 h + 10x^2 h^2 + 5xh^3 + h^4\right)}{h} = \lim_{h \to 0}\left(5x^4 + 10x^3 h + 10x^2 h^2 + 5xh^3 + h^4\right).$$

Now we take the limit to get $f'(x) = \lim_{h \to 0}\left(5x^4 + 10x^3 h + 10x^2 h^2 + 5xh^3 + h^4\right) = 5x^4$.

12. $\dfrac{1}{3}$

We find the derivative of a function, $f(x)$, using the definition of the derivative, which is

$f'(x) = \lim_{h \to 0} \dfrac{f(x+h) - f(x)}{h}$. Here, $f(x) = 2\sqrt{x}$ and $x = 9$. This means that $f(9) = 2\sqrt{9} = 6$

and $f(9+h) = 2\sqrt{9+h}$. If we now plug these into the definition of the derivative, we get

$$f'(9) = \lim_{h \to 0} \frac{f(9+h) - f(9)}{h} = \lim_{h \to 0} \frac{2\sqrt{9+h} - 6}{h}.$$ Notice that if we now take the limit, we

get the indeterminate form $\dfrac{0}{0}$. With polynomials, we merely simplify the expression to elimi-

nate this problem. In any derivative of a square root, we first multiply the top and the bottom

of the expression by the conjugate of the numerator and then we can simplify. Here, the con-

jugate is $2\sqrt{9+h} + 6$. We get $f'(9) = \lim_{h \to 0} \dfrac{2\sqrt{9+h} - 6}{h} \times \dfrac{2\sqrt{9+h} + 6}{2\sqrt{9+h} + 6}$. This simplifies to

$$f'(9) = \lim_{h \to 0} \frac{4(9+h) - 36}{h\left(2\sqrt{9+h} + 6\right)} = \lim_{h \to 0} \frac{36 + 4h - 36}{h\left(2\sqrt{9+h} + 6\right)} = \lim_{h \to 0} \frac{4h}{h\left(2\sqrt{9+h} + 6\right)}.$$ Now we can cancel the

h in the numerator and the denominator to get $f'(9) = \lim_{h \to 0} \dfrac{4}{\left(2\sqrt{9+h} + 6\right)}$. Now we take the limit:

$$f'(9) = \lim_{h \to 0} \frac{4}{\left(2\sqrt{9+h} + 6\right)} = \frac{4}{\left(2\sqrt{9} + 6\right)} = \frac{1}{3}.$$

13. $\dfrac{5}{4}$

We find the derivative of a function, $f(x)$, using the definition of the derivative, which is

$$f'(x) = \lim_{h \to 0} \frac{f(x+h) - f(x)}{h}.$$ Here $f(x) = 5\sqrt{2x}$ and $x = 8$. This means that $f(8) = 5\sqrt{16} = 20$

and $f(8+h) = 5\sqrt{2(8+h)} = 5\sqrt{16 + 2h}$. If we now plug these into the definition of the deriva-

tive, we get $f'(8) = \lim_{h \to 0} \dfrac{f(8+h) - f(8)}{h} = \lim_{h \to 0} \dfrac{5\sqrt{16 + 2h} - 20}{h}$. Notice that if we now take the

limit, we get the indeterminate form $\dfrac{0}{0}$. With polynomials, we merely simplify the expression

to eliminate this problem. In any derivative of a square root, we first multiply the top and the

bottom of the expression by the conjugate of the numerator and then we simplify. Here the conjugate is $5\sqrt{16+2h}+20$. We get $f'(8)=\lim\limits_{h\to 0}\dfrac{5\sqrt{16+2h}-20}{h}\times\dfrac{5\sqrt{16+2h}+20}{5\sqrt{16+2h}+20}$. This simplifies

to $f'(8)=\lim\limits_{h\to 0}\dfrac{25(16+2h)-400}{h\left(5\sqrt{16+2h}+20\right)}=\lim\limits_{h\to 0}\dfrac{400+50h-400}{h\left(5\sqrt{16+2h}+20\right)}=\lim\limits_{h\to 0}\dfrac{50h}{h\left(5\sqrt{16+2h}+20\right)}$. Now we

can cancel the h in the numerator and the denominator to get $f'(8)=\lim\limits_{h\to 0}\dfrac{50}{\left(5\sqrt{16+2h}+20\right)}$.

Now, we take the limit: $f'(8)=\lim\limits_{h\to 0}\dfrac{50}{\left(5\sqrt{16+2h}+20\right)}=\dfrac{50}{(20+20)}=\dfrac{5}{4}$.

14. $\dfrac{1}{2}$

We find the derivative of a function, $f(x)$, using the definition of the derivative, which

is $f'(x)=\lim\limits_{h\to 0}\dfrac{f(x+h)-f(x)}{h}$. Here $f(x)=\sin x$ and $x=\dfrac{\pi}{3}$. This means that

$f\left(\dfrac{\pi}{3}\right)=\sin\dfrac{\pi}{3}=\dfrac{\sqrt{3}}{2}$ and $f\left(\dfrac{\pi}{3}+h\right)=\sin\left(\dfrac{\pi}{3}+h\right)$. If we now plug these into the definition of the

derivative, we get $f'\left(\dfrac{\pi}{3}\right)=\lim\limits_{h\to 0}\dfrac{f\left(\dfrac{\pi}{3}+h\right)-f\left(\dfrac{\pi}{3}\right)}{h}=\lim\limits_{h\to 0}\dfrac{\sin\left(\dfrac{\pi}{3}+h\right)-\dfrac{\sqrt{3}}{2}}{h}$. Notice that if we

now take the limit, we get the indeterminate form $\dfrac{0}{0}$. We cannot eliminate this problem

merely by simplifying the expression the way that we did with a polynomial. Recall that the

trigonometric formula $\sin(A+B)=\sin A\cos B+\cos A\sin B$. Here we can rewrite the top

expression as $f'\left(\dfrac{\pi}{3}\right)=\lim\limits_{h\to 0}\dfrac{\sin\left(\dfrac{\pi}{3}+h\right)-\dfrac{\sqrt{3}}{2}}{h}=\lim\limits_{h\to 0}\dfrac{\sin\dfrac{\pi}{3}\cos h+\cos\dfrac{\pi}{3}\sin h-\dfrac{\sqrt{3}}{2}}{h}$. We can

break up the limit into $\lim\limits_{h\to 0}\dfrac{\sin\dfrac{\pi}{3}\cos h-\dfrac{\sqrt{3}}{2}}{h}+\dfrac{\cos\dfrac{\pi}{3}\sin h}{h}=\lim\limits_{h\to 0}\dfrac{\dfrac{\sqrt{3}}{2}\cos h-\dfrac{\sqrt{3}}{2}}{h}+\dfrac{\left(\dfrac{1}{2}\right)\sin h}{h}$.

Next, factor $\dfrac{\sqrt{3}}{2}$ out of the top of the left-hand expression: $\lim\limits_{h\to 0}\dfrac{\dfrac{\sqrt{3}}{2}(\cos h-1)}{h}+\dfrac{\left(\dfrac{1}{2}\right)\sin h}{h}$.

Now, we can break this into separate limits: $\lim\limits_{h\to 0}\dfrac{\dfrac{\sqrt{3}}{2}(\cos h-1)}{h}+\lim\limits_{h\to 0}\dfrac{\left(\dfrac{1}{2}\right)\sin h}{h}$.

The left-hand limit is $\lim\limits_{h\to 0}\dfrac{\dfrac{\sqrt{3}}{2}(\cos h-1)}{h}=\dfrac{\sqrt{3}}{2}\lim\limits_{h\to 0}\dfrac{(\cos h-1)}{h}=\dfrac{\sqrt{3}}{2}\cdot 0=0$.

The right-hand limit is $\dfrac{1}{2}\lim\limits_{h\to 0}\dfrac{\sin h}{h}=\dfrac{1}{2}\cdot 1=\dfrac{1}{2}$. Therefore, the limit is $\dfrac{1}{2}$.

15. $-\sin x$

We find the derivative of a function, $f(x)$, using the definition of the derivative, which is $f'(x)=\lim\limits_{h\to 0}\dfrac{f(x+h)-f(x)}{h}$. Here $f(x)=\cos x$ and $f(x+h)=\cos(x+h)$. If we now plug these into the definition of the derivative, we get $f'(x)=\lim\limits_{h\to 0}\dfrac{f(x+h)-f(x)}{h}=\lim\limits_{h\to 0}\dfrac{\cos(x+h)-\cos x}{h}$. Notice that if we now take the limit, we get the indeterminate form $\dfrac{0}{0}$. We cannot eliminate this problem merely by simplifying the expression the way that we did with a polynomial. Recall that the trigonometric formula $\cos(A+B)=\cos A\cos B-\sin A\sin B$. Here, we can rewrite the top expression as $f'(x)=\lim\limits_{h\to 0}\dfrac{\cos(x+h)-\cos x}{h}=\lim\limits_{h\to 0}\dfrac{\cos x\cos h-\sin x\sin h-\cos x}{h}$. We can break up the limit into $\lim\limits_{h\to 0}\dfrac{\cos x\cos h-\cos x}{h}-\dfrac{\sin x\sin h}{h}$. Next, factor cos x out of the top of the left-hand expression: $\lim\limits_{h\to 0}\dfrac{\cos x(\cos h-1)}{h}-\dfrac{\sin x\sin h}{h}$. Now, we can break this into separate limits: $\lim\limits_{h\to 0}\dfrac{\cos x(\cos h-1)}{h}-\lim\limits_{h\to 0}\dfrac{\sin x\sin h}{h}$. The left-hand limit is $\displaystyle\lim_{h\to 0}\frac{\cos x(\cos h-1)}{h}=\cos x\lim_{h\to 0}\frac{(\cos h-1)}{h}=\cos x\cdot 0=0$. The right-hand limit is

$\sin x\displaystyle\lim_{h\to 0}\frac{\sin h}{h}=\sin x\cdot 1=\sin x$. Therefore, the limit is $-\sin x$.

16. $2x+1$

We find the derivative of a function, $f(x)$, using the definition of the

derivative, which is $f'(x)=\displaystyle\lim_{h\to 0}\frac{f(x+h)-f(x)}{h}$. Here $f(x)=x^2+x$ and

$f(x+h)=(x+h)^2+(x+h)=x^2+2xh+h^2+x+h$. If we now plug these into the definition of

the derivative, we get $f'(x)=\displaystyle\lim_{h\to 0}\frac{f(x+h)-f(x)}{h}=\lim_{h\to 0}\frac{x^2+2xh+h^2+x+h-(x^2+x)}{h}$. This

simplifies to $f'(x)=\displaystyle\lim_{h\to 0}\frac{2xh+h^2+h}{h}$. Now we can factor out the h from the numerator and cancel

it with the h in the denominator: $f'(x)=\displaystyle\lim_{h\to 0}\frac{2xh+h^2+h}{h}=\lim_{h\to 0}\frac{h(2x+h+1)}{h}=\lim_{h\to 0}(2x+h+1)$.

Now we take the limit to get $f'(x)=\displaystyle\lim_{h\to 0}(2x+h+1)=2x+1$.

17. $3x^2+3$

We find the derivative of a function, $f(x)$, using the definition of the

derivative, which is $f'(x)=\displaystyle\lim_{h\to 0}\frac{f(x+h)-f(x)}{h}$. Here $f(x)=x^3+3x+2$ and

$f(x+h)=(x+h)^3+3(x+h)+2=x^3+3x^2h+3xh^2+h^3+3x+3h+2$.

If we now plug these into the definition of the derivative, we get

$f'(x)=\displaystyle\lim_{h\to 0}\frac{f(x+h)-f(x)}{h}=\lim_{h\to 0}\frac{x^3+3x^2h+3xh^2+h^3+3x+3h+2-(x^3+3x+2)}{h}$.

This simplifies to $f'(x)=\displaystyle\lim_{h\to 0}\frac{3x^2h+3xh^2+h^3+3h}{h}$. Now we can fac-

tor out the h from the numerator and cancel it with the h in the denominator:

$$f'(x) = \lim_{h \to 0} \frac{3x^2h + 3xh^2 + h^3 + 3h}{h} \lim_{h \to 0} \frac{h\left(3x^2 + 3xh + h^2 + 3\right)}{h} = \lim_{h \to 0}\left(3x^2 + 3xh + h^2 + 3\right). \quad \text{Now}$$

we take the limit to get $f'(x) = \lim_{h \to 0}\left(3x^2 + 3xh + h^2 + 3\right) = 3x^2 + 3$.

18. $-\dfrac{1}{x^2}$

We find the derivative of a function, $f(x)$, using the definition of the derivative, which is

$f'(x) = \lim_{h \to 0} \dfrac{f(x+h) - f(x)}{h}$. Here $f(x) = \dfrac{1}{x}$ and $f(x+h) = \dfrac{1}{x+h}$. If we now plug these

into the definition of the derivative, we get $f'(x) = \lim_{h \to 0} \dfrac{f(x+h) - f(x)}{h} = \lim_{h \to 0} \dfrac{\dfrac{1}{x+h} - \dfrac{1}{x}}{h}$.

Notice that if we now take the limit, we get the indeterminate form $\dfrac{0}{0}$. We cannot elimi-

nate this problem merely by simplifying the expression the way that we did with a poly-

nomial. Here, we combine the two terms in the numerator of the expression to get

$$f'(x) = \lim_{h \to 0} \frac{\dfrac{1}{x+h} - \dfrac{1}{x}}{h} = \lim_{h \to 0} \frac{\dfrac{x}{x(x+h)} - \dfrac{x+h}{x(x+h)}}{h} = \lim_{h \to 0} \frac{\dfrac{-h}{x(x+h)}}{h}. \quad \text{This} \quad \text{simplifies} \quad \text{to}$$

$f'(x) = \lim_{h \to 0} \dfrac{\dfrac{-h}{x(x+h)}}{h} = \lim_{h \to 0} \dfrac{-h}{x(x+h)h}$. Now we can cancel the factor h in the numerator

and the denominator to get $f'(x) = \lim_{h \to 0} \dfrac{-h}{x(x+h)h} = \lim_{h \to 0} \dfrac{-1}{x(x+h)}$. Now we take the limit:

$f'(x) = \lim_{h \to 0} \dfrac{-1}{x(x+h)} = -\dfrac{1}{x^2}$.

19. $2ax+b$

We find the derivative of a function, $f(x)$, using the definition of the

derivative, which is $f'(x) = \lim\limits_{h \to 0} \dfrac{f(x+h)-f(x)}{h}$. Here $f(x)$ = ax^2 + bx + c and

$f(x+h) = a(x+h)^2 + b(x+h) + c = ax^2 + 2axh + ah^2 + bx + bh + c.$

If we now plug these into the definition of the derivative, we get

$f'(x) = \lim\limits_{h \to 0} \dfrac{f(x+h)-f(x)}{h} = \lim\limits_{h \to 0} \dfrac{ax^2 + 2axh + ah^2 + bx + bh + c - \left(ax^2 + bx + c\right)}{h}$. This simplifies

to $f'(x) = \lim\limits_{h \to 0} \dfrac{2axh + ah^2 + bh}{h}$. Now we can factor out the h from the numerator and cancel it with

the h in the denominator: $f'(x) = \lim\limits_{h \to 0} \dfrac{2axh + ah^2 + bh}{h} = \lim\limits_{h \to 0} \dfrac{h(2ax + ah + b)}{h} = \lim\limits_{h \to 0} (2ax + ah + b)$.

Now we take the limit to get $f'(x) = \lim\limits_{h \to 0} (2ax + ah + b) = 2ax + b$.

20. $-\dfrac{2}{x^3}$

We find the derivative of a function, $f(x)$, using the definition of the derivative, which is

$f'(x) = \lim\limits_{h \to 0} \dfrac{f(x+h)-f(x)}{h}$. Here $f(x) = \dfrac{1}{x^2}$ and that $f(x+h) = \dfrac{1}{(x+h)^2}$. If we now plug

these into the definition of the derivative, we get $f'(x) = \lim\limits_{h \to 0} \dfrac{f(x+h)-f(x)}{h} = \lim\limits_{h \to 0} \dfrac{\dfrac{1}{(x+h)^2} - \dfrac{1}{x^2}}{h}$.

Notice that if we now take the limit, we get the indeterminate form $\dfrac{0}{0}$. We cannot elimi-

nate this problem merely by simplifying the expression the way that we did with a polynomial. Here we combine the two terms in the numerator of the expression to get

$$f'(x) = \lim_{h\to 0} \frac{\dfrac{1}{(x+h)^2} - \dfrac{1}{x^2}}{h} = \lim_{h\to 0} \frac{\dfrac{x^2}{x^2(x+h)^2} - \dfrac{(x+h)^2}{x^2(x+h)^2}}{h} = \lim_{h\to 0} \frac{\dfrac{x^2 - (x+h)^2}{x^2(x+h)^2}}{h}$$. This simplifies to

$$f'(x) = \lim_{h\to 0} \frac{\dfrac{x^2 - (x^2 + 2xh + h^2)}{x^2(x+h)^2}}{h} = \lim_{h\to 0} \frac{\dfrac{-2xh - h^2}{x^2(x+h)^2}}{h} = \lim_{h\to 0} \frac{-2xh - h^2}{x^2(x+h)^2\,h}$$. Now we can cancel the

h in the numerator and the denominator to get $f'(x) = \lim_{h\to 0} \dfrac{h(-2x-h)}{x^2(x+h)^2\,h} = \lim_{h\to 0} \dfrac{(-2x-h)}{x^2(x+h)^2}$. Now

we take the limit: $f'(x) = \lim_{h\to 0} \dfrac{(-2x-h)}{x^2(x^*+h)^2} = \dfrac{-2x}{x^2(x^2)} = -\dfrac{2}{x^3}$.

SOLUTIONS TO PRACTICE PROBLEM SET 4

1. $64x^3 + 16x$

First, expand $(4x^2 + 1)^2$ to get $16x^4 + 8x^2 + 1$. Now, use the Power Rule to take the derivative

of each term. The derivative of $16x^4 = 16(4x^3) = 64x^3$. The derivative of $8x^2 = 8(2x) = 16x$.

The derivative of 1 = 0 (because the derivative of a constant is zero). Therefore, the derivative is

$64x^3 + 16x$.

2. $10x^9 + 36x^5 + 18x$

First, expand $(x^5 + 3x)^2$ to get $x^{10} + 6x^6 + 9x^2$. Now, use the Power Rule to take the derivative of

each term. The derivative of $x^{10} = 10x^9$. The derivative of $6x^6 = 6(6x^5) = 36x^5$. The derivative of

$9x^2 = 9(2x) = 18x$. Therefore, the derivative is $10x^9 + 36x^5 + 18x$.

3. $77x^6$

Simply use the Power Rule. The derivative is $11x^7 = 11(7x^6) = 77x^6$.

4. $80x^9$

Simply use the Power Rule. The derivative is $8x^{10} = 8(10x^9) = 80x^9$.

5. $54x^2 + 12$

Use the Power Rule to take the derivative of each term. The derivative of $18x^3 = 18(3x^2) = 54x^2$.
The derivative of $12x = 12$. The derivative of $11 = 0$ (because the derivative of a constant is zero).
Therefore, the derivative is $54x^2 + 12$.

6. $6x^{11}$

Use the Power Rule to take the derivative of each term. The derivative of $x^{12} = 12x^{11}$. The derivative of
$17 = 0$ (because the derivative of a constant is zero). Therefore, the derivative is $\frac{1}{2}(12x^{11}) = 6x^{11}$.

7. $-3x^8 - 2x^2$

Use the Power Rule to take the derivative of each term. The derivative of $x^9 = 9x^8$. The derivative of $2x^3 = 2(3x^2) = 6x^2$. The derivative of $9 = 0$ (because the derivative of a constant is zero).
Therefore, the derivative is $-\frac{1}{3}(9x^8 + 6x^2) = -3x^8 - 2x^2$.

8.   0

Don't be fooled by the power. π^5 is a constant so the derivative is zero.

9. $\frac{2}{ab}x - \frac{2}{a^2} + \frac{d}{ax^2}$

Use the Power Rule to take the derivative of each term. The derivative of $\frac{1}{b}x^2 = \frac{1}{b}(2x) = \frac{2}{b}x$. The

derivative of $\frac{2}{a}x = \frac{2}{a}$. The derivative of $\frac{d}{x} = -\frac{d}{x^2}$ (remember the shortcut that we showed you on

page 69). Therefore, the derivative is $\frac{1}{a}\left(\frac{2}{b}x - \frac{2}{a} + \frac{d}{x^2}\right) = \frac{2}{ab}x - \frac{2}{a^2} + \frac{d}{ax^2}$.

10. $64x^{-9} + \dfrac{6}{\sqrt{x}}$

Use the Power Rule to take the derivative of each term. The derivative of $-8x^{-8} = -8(-8x^{-9}) = 64x^{-9}$.

The derivative of $12\sqrt{x} = \dfrac{12}{2\sqrt{x}} = \dfrac{6}{\sqrt{x}}$ (remember the shortcut that we showed you on page 69).

Therefore, the derivative is $64x^{-9} + \dfrac{6}{\sqrt{x}}$.

11. $-42x^{-8} - \dfrac{2}{\sqrt{x}}$

Use the Power Rule to take the derivative of each term. The derivative of $6x^{-7} = 6(-7x^{-8}) = -42x^{-8}$.

The derivative of $4\sqrt{x} = \dfrac{4}{2\sqrt{x}} = \dfrac{2}{\sqrt{x}}$ (remember the shortcut that we showed you on page 69).

Therefore, the derivative is $-42x^{-8} - \dfrac{2}{\sqrt{x}}$.

12. $\dfrac{-5}{x^6} - \dfrac{8}{x^9}$

Use the Power Rule to take the derivative of each term. The derivative of $x^{-5} = -5x^{-6}$. To find the

derivative of $\dfrac{1}{x^8}$, we first rewrite it as x^{-8}. The derivative of $x^{-8} = -8x^{-9}$. Therefore, the derivative

is $-5x^{-6} - 8x^{-9} = \dfrac{-5}{x^6} - \dfrac{8}{x^9}$.

13. $\dfrac{1}{2\sqrt{x}} - \dfrac{3}{x^4}$

Use the Power Rule to take the derivative of each term. The derivative of $\sqrt{x} = \dfrac{1}{2\sqrt{x}}$ (remember

the shortcut that we showed you on page 69). To find the derivative of $\dfrac{1}{x^3}$, we first rewrite it as x^{-3}.

The derivative of $x^{-3} = -3x^{-4}$. Therefore, the derivative is $\dfrac{1}{2\sqrt{x}} - 3x^{-4} = \dfrac{1}{2\sqrt{x}} - \dfrac{3}{x^4}$.

14. $216x^2 - 48x + 36$

First, expand $(6x^2 + 3)(12x - 4)$ to get $72x^3 - 24x^2 + 36x - 12$. Now, use the Power Rule to take the derivative of each term. The derivative of $72x^3 = 72(3x^2) = 216x^2$. The derivative of $24x^2 = 24(2x) = 48x$. The derivative of $36x = 36$. The derivative of $12 = 0$ (because the derivative of a constant is zero). Therefore, the derivative is $216x^2 - 48x + 36$.

15. $-6 - 36x^2 + 12x^3 - 5x^4 - 14x^6$

First, expand $(3 - x - 2x^3)(6 + x^4)$ to get $18 - 6x - 12x^3 + 3x^4 - x^5 - 2x^7$. Now, use the Power Rule to take the derivative of each term. The derivative of $18 = 0$ (because the derivative of a constant is zero). The derivative of $6x = 6$. The derivative of $12x^3 = 12(3x^2) = 36x^2$. The derivative of $3x^4 = 3(4x^3) = 12x^3$. The derivative of $x^5 = 5x^4$. The derivative of $2x^7 = 2(7x^6) = 14x^6$. Therefore, the derivative is $-6 - 36x^2 + 12x^3 - 5x^4 - 14x^6$.

16. 0

Don't be fooled by the powers. Each term is a constant so the derivative is zero.

17. $-\dfrac{16}{x^5} + \dfrac{10}{x^6} + \dfrac{36}{x^7}$

First, expand $\left(\dfrac{1}{x} + \dfrac{1}{x^2}\right)\left(\dfrac{4}{x^3} - \dfrac{6}{x^4}\right)$ to get $\dfrac{4}{x^4} - \dfrac{6}{x^5} + \dfrac{4}{x^5} - \dfrac{6}{x^6} = \dfrac{4}{x^4} - \dfrac{2}{x^5} - \dfrac{6}{x^6}$. Next, rewrite the terms as: $4x^{-4} - 2x^{-5} - 6x^{-6}$. Now, use the Power Rule to take the derivative of each term. The derivative of $4x^{-4} = 4(-4x^{-5}) = -16x^{-5}$. The derivative of $2x^{-5} = 2(-5x^{-6}) = -10x^{-6}$. The derivative of $6x^{-6} = -36x^{-7}$. Therefore, the derivative is $-16x^{-5} + 10x^{-6} + 36x^{-7} = -\dfrac{16}{x^5} + \dfrac{10}{x^6} + \dfrac{36}{x^7}$.

18. $\dfrac{1}{2\sqrt{x}}$

Use the Power Rule to take the derivative of each term. The derivative of $\sqrt{x} = \dfrac{1}{2\sqrt{x}}$ (remember

the shortcut that we showed you on page 69). The derivative of $\dfrac{1}{\sqrt{3}} = 0$ (because the derivative of

a constant is zero). Therefore, the derivative is $\dfrac{1}{2\sqrt{x}}$.

19. $-\dfrac{16}{x^2} + \dfrac{14}{x^3} - \dfrac{24}{x^4} + \dfrac{16}{x^5}$

First, expand $\left(x^2 + 8x - 4\right)\left(2x^{-2} + x^{-4}\right)$ to get $2 + 16x^{-1} - 7x^{-2} + 8x^{-3} - 4x^{-4}$. Now, use the Power

Rule to take the derivative of each term. The derivative of $2 = 0$ (because the derivative of a constant is

zero). The derivative of $16x^{-1} = 16\left(-1x^{-2}\right) = -16x^{-2}$. The derivative of $7x^{-2} = 7\left(-2x^{-3}\right) = -14x^{-3}$.

The derivative of $8x^{-3} = 8\left(-3x^{-4}\right) = -24x^{-4}$. The derivative of $4x^{-4} = 4\left(-4x^{-5}\right) = -16x^{-5}$. There-

fore, the derivative is $-16x^{-2} + 14x^{-3} - 24x^{-4} + 16x^{-5} = -\dfrac{16}{x^2} + \dfrac{14}{x^3} - \dfrac{24}{x^4} + \dfrac{16}{x^5}$.

20. 0

The derivative of a constant is zero.

21. $3x^2 + 6x + 3$

First, expand $\left(x + 1\right)^3$ to get $x^3 + 3x^2 + 3x + 1$. Now, use the Power Rule to take the derivative

of each term. The derivative of $x^3 = 3x^2$. The derivative of $3x^2 = 3\left(2x\right) = 6x$. The derivative of

$3x = 3$. The derivative of $1 = 0$ (because the derivative of a constant is zero). Therefore, the deriva-

tive is $3x^2 + 6x + 3$.

22. $\dfrac{1}{2\sqrt{x}} + \dfrac{1}{3\sqrt[3]{x^2}} + \dfrac{2}{3\sqrt[3]{x}}$

Use the Power Rule to take the derivative of each term. The derivative of $\sqrt{x} = \dfrac{1}{2\sqrt{x}}$

(remember the shortcut that we showed you on page 69). Rewrite $\sqrt[3]{x}$ as $x^{\frac{1}{3}}$ and $\sqrt[3]{x^2}$ as $x^{\frac{2}{3}}$.

The derivative of $x^{\frac{1}{3}} = \frac{1}{3}x^{-\frac{2}{3}}$. The derivative of $x^{\frac{2}{3}} = \frac{2}{3}x^{-\frac{1}{3}}$. Therefore, the derivative is

$$\frac{1}{2\sqrt{x}} + \frac{1}{3}x^{-\frac{2}{3}} + \frac{2}{3}x^{-\frac{1}{3}} = \frac{1}{2\sqrt{x}} + \frac{1}{3\sqrt[3]{x^2}} + \frac{2}{3\sqrt[3]{x}}.$$

23. $6x^2 + 6x - 14$

First, expand $x(2x+7)(x-2)$ to get $x(2x^2 + 3x - 14) = 2x^3 + 3x^2 - 14x$. Now, use the Power Rule to take the derivative of each term. The derivative of $2x^3 = 2(3x^2) = 6x^2$. The derivative of $3x^2 = 3(2x) = 6x$. The derivative of $14x = 14$. Therefore, the derivative is $6x^2 + 6x - 14$.

24. $\dfrac{5}{6\sqrt[6]{x}} + \dfrac{7}{10\sqrt[10]{x^3}}$

First, rewrite the terms as $x^{\frac{1}{2}}\left(x^{\frac{1}{3}} + x^{\frac{1}{5}}\right)$. Next, distribute to get: $x^{\frac{1}{2}}\left(x^{\frac{1}{3}} + x^{\frac{1}{5}}\right) = x^{\frac{5}{6}} + x^{\frac{7}{10}}$. Now,

use the Power Rule to take the derivative of each term. The derivative of $x^{\frac{5}{6}} = \frac{5}{6}x^{-\frac{1}{6}}$. The derivative of $x^{\frac{7}{10}} = \frac{7}{10}x^{-\frac{3}{10}}$. Therefore, the derivative is $\frac{5}{6}x^{-\frac{1}{6}} + \frac{7}{10}x^{-\frac{3}{10}} = \frac{5}{6\sqrt[6]{x}} + \frac{7}{10\sqrt[10]{x^3}}$.

25. $5ax^4 + 4bx^3 + 3cx^2 + 2dx + e$

Use the Power Rule to take the derivative of each term. The derivative of $ax^5 = a(5x^4) = 5ax^4$.

The derivative of $bx^4 = b(4x^3) = 4bx^3$. The derivative of $cx^3 = c(3x^2) = 3cx^2$. The derivative of

$dx^2 = d(2x) = 2dx$. The derivative of $ex = e$. The derivative of $f = 0$ (because the derivative of a

constant is zero). Therefore, the derivative is $5ax^4 + 4bx^3 + 3cx^2 + 2dx + e$.

SOLUTIONS TO PRACTICE PROBLEM SET 5

1. $$\dfrac{-80x^9 + 75x^8 + 12x^2 - 6x}{\left(5x^7 + 1\right)^2}$$

    We find the derivative using the Quotient Rule, which says that if $f(x) = \dfrac{u}{v}$, then

    $f'(x) = \dfrac{v\dfrac{du}{dx} - u\dfrac{dv}{dx}}{v^2}$. Here $f(x) = \dfrac{4x^3 - 3x^2}{5x^7 + 1}$, so $u = 4x^3 - 3x^2$ and $v = 5x^7 + 1$. Using the Quo-

    tient Rule, we get $f'(x) = \dfrac{\left(5x^7 + 1\right)\left(12x^2 - 6x\right) - \left(4x^3 - 3x^2\right)\left(35x^6\right)}{\left(5x^7 + 1\right)^2}$. This can be simplified to

    $f'(x) = \dfrac{-80x^9 + 75x^8 + 12x^2 - 6x}{\left(5x^7 + 1\right)^2}$.

2. $3x^2 - 6x - 1$

    We find the derivative using the Product Rule, which says that if $f(x) = uv$, then

    $f'(x) = u\dfrac{dv}{dx} + v\dfrac{du}{dx}$. Here $f(x) = \left(x^2 - 4x + 3\right)(x + 1)$, so $u = x^2 - 4x + 3$ and $v = x + 1$. Us-

    ing the Product Rule, we get $f'(x) = \left(x^2 - 4x + 3\right)(1) + (x + 1)(2x - 4)$. This can be simplified to

    $f'(x) = 3x^2 - 6x - 1$.

3. $10(x + 1)^9$

    We find the derivative using the Chain Rule, which says that if $y = f\big(g(x)\big)$,

    then $y' = \left(\dfrac{df\big(g(x)\big)}{dg}\right)\left(\dfrac{dg}{dx}\right)$. Here $f(x) = (x + 1)^{10}$. Using the Chain Rule, we get

    $f'(x) = 10(x + 1)^9 (1) = 10(x + 1)^9$.

4. $\dfrac{16x^2 - 32}{\sqrt{(x^2 - 4)}}$

We find the derivative using the Chain Rule, which says that if $y = f\big(g(x)\big)$, then $y' = \left(\dfrac{df\big(g(x)\big)}{dg}\right)\left(\dfrac{dg}{dx}\right)$. Here $f(x) = 8\sqrt{(x^4 - 4x^2)}$, which can be written as $f(x) = 8(x^4 - 4x^2)^{\frac{1}{2}}$. Using the Chain Rule, we get $f'(x) = 8\left(\dfrac{1}{2}\right)(x^4 - 4x^2)^{-\frac{1}{2}}(4x^3 - 8x)$.

This can be simplified to $f'(x) = \dfrac{16x^2 - 32}{\sqrt{(x^2 - 4)}}$.

5. $\dfrac{3x^2 - 3x^4}{(x^2 + 1)^4}$

Here we will find the derivative using the Chain Rule. We will also need the Quotient Rule to take the derivative of the expression inside the parentheses. The Chain Rule says that if $y = f\big(g(x)\big)$, then $y' = \left(\dfrac{df\big(g(x)\big)}{dg}\right)\left(\dfrac{dg}{dx}\right)$, and the Quotient Rule says that if $f(x) = \dfrac{u}{v}$, then $f'(x) = \dfrac{v\dfrac{du}{dx} - u\dfrac{dv}{dx}}{v^2}$. We get $f'(x) = 3\left(\dfrac{x}{x^2 + 1}\right)^2\left(\dfrac{(x^2 + 1)(1) - x(2x)}{(x^2 + 1)^2}\right)$. This can be simplified to $f'(x) = \dfrac{3x^2 - 3x^4}{(x^2 + 1)^4}$.

6. $\dfrac{1}{4}\left(\dfrac{2x-5}{5x+2}\right)^{-\frac{3}{4}}\left(\dfrac{29}{\left(5x+2\right)^{2}}\right)$

Here we will find the derivative using the Chain Rule. We will also need the Quotient Rule to take the derivative of the expression inside the parentheses. The Chain Rule says that if $y=f\left(g\left(x\right)\right)$, then $y'=\left(\dfrac{df\left(g\left(x\right)\right)}{dg}\right)\left(\dfrac{dg}{dx}\right)$, and the Quotient Rule says that if $f\left(x\right)=\dfrac{u}{v}$, then $f'\left(x\right)=\dfrac{v\dfrac{du}{dx}-u\dfrac{dv}{dx}}{v^{2}}$. We get $f'\left(x\right)=\dfrac{1}{4}\left(\dfrac{2x-5}{5x+2}\right)^{-\frac{3}{4}}\left(\dfrac{\left(5x+2\right)\left(2\right)-\left(2x-5\right)\left(5\right)}{\left(5x+2\right)^{2}}\right)$. This can be simplified to $f'\left(x\right)=\dfrac{1}{4}\left(\dfrac{2x-5}{5x+2}\right)^{-\frac{3}{4}}\left(\dfrac{29}{\left(5x+2\right)^{2}}\right)$.

7. $\dfrac{32x^{11}+7x^{\frac{7}{2}}}{16x^{8}}$

We find the derivative using the Quotient Rule, which says that if $f\left(x\right)=\dfrac{u}{v}$, then $f'\left(x\right)=\dfrac{v\dfrac{du}{dx}-u\dfrac{dv}{dx}}{v^{2}}$. Here $f\left(x\right)=\dfrac{4x^{8}-\sqrt{x}}{8x^{4}}$, so $u=4x^{8}-\sqrt{x}$ and $v=8x^{4}$. Using the Quotient Rule, we get $f'\left(x\right)=\dfrac{\left(8x^{4}\right)\left(32x^{7}-\dfrac{1}{2\sqrt{x}}\right)-\left(4x^{8}-\sqrt{x}\right)\left(32x^{3}\right)}{\left(8x^{4}\right)^{2}}$. This can be simplified to $f'\left(x\right)=\dfrac{32x^{11}+7x^{\frac{7}{2}}}{16x^{8}}$.

8.　　$3x^2 + 1 + \dfrac{1}{x^2} + \dfrac{3}{x^4}$

We have two ways that we could solve this. We could expand the expression first and then take the derivative of each term, or we could find the derivative using the Product Rule. Let's do both methods just to see that they both give us the same answer. First, let's use the Product Rule, which says that if $f(x) = uv$, then $f'(x) = u\dfrac{dv}{dx} + v\dfrac{du}{dx}$. Here

$f(x) = \left(x + \dfrac{1}{x} \right)\left(x^2 - \dfrac{1}{x^2} \right)$, so $u = \left(x + \dfrac{1}{x} \right)$ and $v = \left(x^2 - \dfrac{1}{x^2} \right)$. Using the Product Rule, we get

$f'(x) = \left(x + \dfrac{1}{x} \right)\left(2x - (-2)x^{-3} \right) + \left(x^2 - \dfrac{1}{x^2} \right)\left(1 - \dfrac{1}{x^2} \right) = \left(x + \dfrac{1}{x} \right)\left(2x + \dfrac{2}{x^3} \right) + \left(x^2 - \dfrac{1}{x^2} \right)\left(1 - \dfrac{1}{x^2} \right)$

This can be simplified to $f'(x) = \left(2x^2 + \dfrac{2}{x^2} + 2 + \dfrac{2}{x^4} \right) + \left(x^2 - 1 - \dfrac{1}{x^2} + \dfrac{1}{x^4} \right) = 3x^2 + 1 + \dfrac{1}{x^2} + \dfrac{3}{x^4}$.

The other way we could find the derivative is to expand the expression first and then take the derivative. We get $f(x) = \left(x + \dfrac{1}{x} \right)\left(x^2 - \dfrac{1}{x^2} \right) = x^3 - \dfrac{1}{x} + x - \dfrac{1}{x^3}$. Now we can take the derivative of each term. We get $f'(x) = 3x^2 + \dfrac{1}{x^2} + 1 - (-3)x^{-4} = 3x^2 + \dfrac{1}{x^2} + 1 + \dfrac{3}{x^4}$. As we can see, the second method is a little quicker, and they both give the same result.

9. $$\frac{4x^3}{(x+1)^5}$$

Here we will find the derivative using the Chain Rule. We will also need the Quotient Rule

to take the derivative of the expression inside the parentheses. The Chain Rule says that if

$y = f\big(g(x)\big)$, then $y' = \left(\dfrac{df\big(g(x)\big)}{dg}\right)\left(\dfrac{dg}{dx}\right)$, and the Quotient Rule says that if $f(x) = \dfrac{u}{v}$, then

$f'(x) = \dfrac{v\dfrac{du}{dx} - u\dfrac{dv}{dx}}{v^2}$. We get $f'(x) = 4\left(\dfrac{x}{x+1}\right)^3\left(\dfrac{(x+1)(1) - (x)(1)}{(x+1)^2}\right)$. This can be simplified

to $f'(x) = 4\left(\dfrac{x}{x+1}\right)^3\left(\dfrac{1}{(x+1)^2}\right) = \dfrac{4x^3}{(x+1)^5}$.

10. $$100\big(x^2 + x\big)^{99}(2x+1)$$

Here we will find the derivative using the Chain Rule. The Chain Rule says that if $y = f\big(g(x)\big)$,

then $y' = \left(\dfrac{df\big(g(x)\big)}{dg}\right)\left(\dfrac{dg}{dx}\right)$. We get $f'(x) = 100\big(x^2 + x\big)^{99}(2x+1)$.

11. $$\frac{-2x}{\big(x^2 + 1\big)^{\frac{1}{2}}\big(x^2 - 1\big)^{\frac{3}{2}}}$$

Here we will find the derivative using the Chain Rule. We will also need the Quotient Rule

to take the derivative of the expression inside the parentheses. The Chain Rule says that if

$y = f\big(g(x)\big)$, then $y' = \left(\dfrac{df\big(g(x)\big)}{dg}\right)\left(\dfrac{dg}{dx}\right)$, and the Quotient Rule says that if $f(x) = \dfrac{u}{v}$, then

$f'(x)=\dfrac{v\dfrac{du}{dx}-u\dfrac{dv}{dx}}{v^2}$. We get $f'(x)=\dfrac{1}{2}\left(\dfrac{x^2+1}{x^2-1}\right)^{-\frac{1}{2}}\left(\dfrac{(x^2-1)(2x)-(x^2+1)(2x)}{(x^2-1)^2}\right)$. This can be

simplified to $f'(x)=\dfrac{1}{2}\left(\dfrac{x^2+1}{x^2-1}\right)^{-\frac{1}{2}}\left(\dfrac{-4x}{(x-1)^2}\right)=\dfrac{-2x}{(x^2+1)^{\frac{1}{2}}(x^2-1)^{\frac{3}{2}}}$.

12. $\dfrac{9}{64}$

We find the derivative using the Quotient Rule, which says that if $f(x)=\dfrac{u}{v}$, then

$f'(x)=\dfrac{v\dfrac{du}{dx}-u\dfrac{dv}{dx}}{v^2}$. Here $f(x)=\dfrac{(x+4)(x-8)}{(x+6)(x-6)}$, so $u=(x+4)(x-8)$ and $v=(x+6)(x-6)$.

Before we take the derivative, we can simplify the numerator and denominator of the

expression: $f(x)=\dfrac{(x+4)(x-8)}{(x+6)(x-6)}=\dfrac{x^2-4x-32}{x^2-36}$. Now using the Quotient Rule, we get

$f(x)=\dfrac{(x^2-36)(2x-4)-(x^2-4x-32)(2x)}{(x^2-36)^2}$. Next, we don't simplify. We simply plug in $x=2$

to get $f(x)=\dfrac{((2)^2-36)(2(2)-4)-((2)^2-4(2)-32)(2(2))}{((2)^2-36)^2}=\dfrac{(-32)(0)-(-36)(4)}{(-32)^2}=\dfrac{9}{64}$.

13.   106

We find the derivative using the Quotient Rule, which says that if $f(x)=\dfrac{u}{v}$,

then $f'(x)=\dfrac{v\dfrac{du}{dx}-u\dfrac{dv}{dx}}{v^2}$. We will also need the Chain Rule to take the derivative

of the expression in the denominator. The Chain Rule says that if $y=f(g(x))$, then

$y' = \left(\dfrac{df\left(g(x)\right)}{dg} \right)\left(\dfrac{dg}{dx} \right)$, here $f(x) = \dfrac{x^6 + 4x^3 + 6}{\left(x^4 - 2\right)^2}$, so $u = x^6 + 4x^3 + 6$ and $v = \left(x^4 - 2\right)^2$. We

get $f(x) = \dfrac{\left(x^4 - 2\right)^2 \left(6x^5 + 12x^2\right) - \left(x^6 + 4x^3 + 6\right)2\left(x^4 - 2\right)\left(4x^3\right)}{\left(x^4 - 2\right)^4}$. Now we don't simplify.

We simply plug in $x = 1$ to get

$$f(x) = \dfrac{\left((1)^4 - 2\right)^2 \left(6(1)^5 + 12(1)^2\right) - \left((1)^6 + 4(1)^3 + 6\right)2\left((1)^4 - 2\right)\left(4(1)^3\right)}{\left((1)^4 - 2\right)^4} =$$

$$\dfrac{(1)(18) - (11)2(-1)(4)}{(-1)^4} = 106$$

14.   0

Here we will find the derivative using the Chain Rule. We will also need the Quotient Rule to

take the derivative of the expression inside the parentheses. The Chain Rule

says that if $y = f\left(g(x)\right)$, then $y' = \left(\dfrac{df\left(g(x)\right)}{dg} \right)\left(\dfrac{dg}{dx} \right)$, and the Quo-

tient Rule says that if $f(x) = \dfrac{u}{v}$, then $f'(x) = \dfrac{v\dfrac{du}{dx} - u\dfrac{dv}{dx}}{v^2}$. We get

$f'(x) = 2\left(\dfrac{x - \sqrt{x}}{x + \sqrt{x}} \right)\left(\dfrac{\left(x + \sqrt{x}\right)\left(1 - \dfrac{1}{2\sqrt{x}}\right) - \left(x - \sqrt{x}\right)\left(1 + \dfrac{1}{2\sqrt{x}}\right)}{\left(x + \sqrt{x}\right)^2} \right)$. Now we don't simplify. We

simply plug in $x = 1$ to get $f'(x) = 2\left(\dfrac{1 - \sqrt{1}}{1 + \sqrt{1}} \right)\left(\dfrac{\left(1 + \sqrt{1}\right)\left(1 - \dfrac{1}{2\sqrt{1}}\right) - \left(1 - \sqrt{1}\right)\left(1 + \dfrac{1}{2\sqrt{1}}\right)}{\left(1 + \sqrt{1}\right)^2} \right) = 0.$

15. $\dfrac{x^2-6x+3}{(x-3)^2}$

We find the derivative using the Quotient Rule, which says that if $f(x)=\dfrac{u}{v}$, then

$f'(x)=\dfrac{v\dfrac{du}{dx}-u\dfrac{dv}{dx}}{v^2}$. Here, $f(x)=\dfrac{x^2-3}{x-3}$, so $u=x^2-3$ and $v=x-3$. Using the Quotient

Rule, we get $f'(x)=\dfrac{(x-3)(2x)-(x^2-3)(1)}{(x-3)^2}$. This can be simplified to $f'(x)=\dfrac{x^2-6x+3}{(x-3)^2}$.

16. 6

We find the derivative using the Product Rule, which says that if $f(x)=uv$, then

$f'(x)=u\dfrac{dv}{dx}+v\dfrac{du}{dx}$. Here $f(x)=(x^4-x^2)(2x^3+x)$, so $u=x^4-x^2$ and $v=2x^3+x$. Using the

Product Rule, we get $f'(x)=(x^4-x^2)(6x^2+1)+(2x^3+x)(4x^3-2x)$. Now we don't simplify.

We simply plug in $x=1$ to get $f'(x)=\left((1)^4-(1)^2\right)\left(6(1)^2+1\right)+\left(2(1)^3+(1)\right)\left(4(1)^3-2(1)\right)=6$.

17. $-\dfrac{7}{4}$

We find the derivative using the Quotient Rule, which says that if $f(x)=\dfrac{u}{v}$,

then $f'(x)=\dfrac{v\dfrac{du}{dx}-u\dfrac{dv}{dx}}{v^2}$. Here $f(x)=\dfrac{x^2+2x}{x^4-x^3}$, so $u=x^2+2x$ and $v=x^4-x^3$.

Using the Quotient Rule, we get $f'(x)=\dfrac{(x^4-x^3)(2x+2)-(x^2+2x)(4x^3-3x^2)}{(x^4-x^3)^2}$.

Now, we don't simplify. We simply plug in $x=2$ to get

$f'(x)=\dfrac{\left((2)^4-(2)^3\right)(2(2)+2)-\left((2)^2+2(2)\right)\left(4(2)^3-3(2)^2\right)}{\left((2)^4-(2)^3\right)^2}=-\dfrac{7}{4}$.

18. $\dfrac{2x^2+1}{\sqrt{x^2+1}}$

We find the derivative using the Chain Rule, which says that if $y = f\big(g(x)\big)$, then

$y' = \left(\dfrac{df\big(g(x)\big)}{dg}\right)\left(\dfrac{dg}{dx}\right)$. Here $f(x) = \sqrt{x^4+x^2} = \left(x^4+x^2\right)^{\frac{1}{2}}$. Using the Chain Rule, we get

$f'(x) = \dfrac{1}{2}\left(x^4+x^2\right)^{-\frac{1}{2}}\left(4x^3+2x\right)$. This can be simplified to $f'(x) = \dfrac{2x^2+1}{\sqrt{x^2+1}}$.

19. $-\dfrac{1}{4}$

We find the derivative using the Quotient Rule, which says that if $f(x) = \dfrac{u}{v}$,

then $f'(x) = \dfrac{v\dfrac{du}{dx} - u\dfrac{dv}{dx}}{v^2}$. We will also need the Chain Rule to take the deriva-

tive of the expression in the denominator. The Chain Rule says that if $y = f\big(g(x)\big)$,

then $y' = \left(\dfrac{df\big(g(x)\big)}{dg}\right)\left(\dfrac{dg}{dx}\right)$. Here $f(x) = \dfrac{x}{\left(1+x^2\right)^2}$, so $u = x$ and $v = \left(1+x^2\right)^2$. We get

$f(x) = \dfrac{\left(1+x^2\right)^2(1) - (x)2\left(1+x^2\right)(2x)}{\left(1+x^2\right)^4}$. Now we don't simplify. We simply plug in $x = 1$ to get

$f(x) = \dfrac{\left(1+(1)^2\right)^2(1) - (1)2\left(1+(1)^2\right)(2(1))}{\left(1+(1)^2\right)^4} = -\dfrac{1}{4}$.

20. $-\dfrac{2}{(x-1)^3}$

We find the derivative using the Chain Rule, which says that if $y = y(v)$ and $v = v(x)$, then

$\dfrac{dy}{dx} = \dfrac{dy}{dv}\dfrac{dv}{dx}$. Here $\dfrac{dy}{du} = 2u$ and $\dfrac{du}{dx} = -1(x-1)^{-2} = -\dfrac{1}{(x-1)^2}$. Thus, $\dfrac{dy}{dx} = (2u)\dfrac{-1}{(x-1)^2}$ and

because $u = \dfrac{1}{x-1}$, $\dfrac{dy}{dx} = \left(\dfrac{2}{x-1}\right)\left(\dfrac{-1}{(x-1)^2}\right) = -\dfrac{2}{(x-1)^3}$.

21.   –24

We find the derivative using the Chain Rule, which says that if $y = y(v)$ and

$v = v(x)$, then $\dfrac{dy}{dx} = \dfrac{dy}{dv}\dfrac{dv}{dx}$. Here $\dfrac{dy}{dt} = \dfrac{(t^2 - 2)(2t) - (t^2 + 2)(2t)}{(t^2 - 2)^2}$ and $\dfrac{dt}{dx} = 3x^2$.

Now we plug $x = 1$ into the derivative. Note that where $x = 1$, $t = (1)^3 = 1$.

We get $\dfrac{dy}{dt} = \dfrac{((1)^2 - 2)(2(1)) - ((1)^2 + 2)(2(1))}{((1)^2 - 2)^2} = -8$ and $\dfrac{dt}{dx} = 3(1)^2 = 3$. Thus,

$\dfrac{dy}{dx} = \dfrac{dy}{dt}\dfrac{dt}{dx} = (-8)(3) = -24$.

22. $\left(t^3 - 6t^{\frac{5}{2}}\right)\left(5 + \dfrac{1}{2\sqrt{t}}\right) + \left(5t + \sqrt{t}\right)\left(3t^2 - 15t^{\frac{3}{2}}\right)$

Here the solution will be much simpler if we first substitute $x = \sqrt{t}$ into the expression for y.

We get $y = \left(t^3 - 6t^{\frac{5}{2}}\right)\left(5t + \sqrt{t}\right)$. Now we find the derivative using the Product Rule, which says

that if $f(x) = uv$, then $f'(x) = u\dfrac{dv}{dx} + v\dfrac{du}{dx}$. Here $y = \left(t^3 - 6t^{\frac{5}{2}}\right)\left(5t + \sqrt{t}\right)$, so $u = t^3 - 6t^{\frac{5}{2}}$ and

$v = 5t + \sqrt{t}$. Using the Product Rule, we get $\dfrac{dy}{dt} = \left(t^3 - 6t^{\frac{5}{2}}\right)\left(5 + \dfrac{1}{2\sqrt{t}}\right) + \left(5t + \sqrt{t}\right)\left(3t^2 - 15t^{\frac{3}{2}}\right)$.

23. $-\dfrac{7\sqrt{2}}{16\sqrt{3}}$

We find the derivative using the Chain Rule, which says that if $y = y(v)$ and

$v = v(x)$, then $\dfrac{dy}{dx} = \dfrac{dy}{dv}\dfrac{dv}{dx}$. Here $\dfrac{du}{dx} = \dfrac{1}{2}(x^3 + x^2)^{-\frac{1}{2}}(3x^2 + 2x)$ and $\dfrac{dx}{dv} = -\dfrac{1}{v^2}$.

Now, we plug $v = 2$ into the derivative. Note that, where $v = 2$, $x = \dfrac{1}{2}$.

We get $\dfrac{du}{dx} = \dfrac{1}{2}\left(\left(\dfrac{1}{2}\right)^3 + \left(\dfrac{1}{2}\right)^2\right)^{-\frac{1}{2}}\left(3\left(\dfrac{1}{2}\right)^2 + 2\left(\dfrac{1}{2}\right)\right) = \dfrac{7}{8}\left(\dfrac{3}{8}\right)^{-\frac{1}{2}}$ and $\dfrac{dx}{dv} = -\dfrac{1}{4}$, so

$$\dfrac{du}{dv} = \dfrac{7}{8}\left(\dfrac{3}{8}\right)^{-\frac{1}{2}}\left(-\dfrac{1}{4}\right) = -\dfrac{7\sqrt{2}}{16\sqrt{3}}.$$

24.   2

We find the derivative using the Chain Rule, which says that if $y = y(v)$ and $v = v(x)$, then

$\dfrac{dy}{dx} = \dfrac{dy}{dv}\dfrac{dv}{dx}$. Here $\dfrac{dy}{du} = \dfrac{\left(1 + u^2\right)(1) - (1 + u)(2u)}{\left(1 + u^2\right)^2}$ and $\dfrac{du}{dx} = 2x$. Now, we plug $x = 1$ into the

derivative. Note that, where $x = 1$, $u = 0$. We get $\dfrac{dy}{du} = \dfrac{(1 + 0)(1) - (1 + 0)(0)}{(1 + 0)^2} = 1$ and $\dfrac{du}{dx} = 2$, so

$$\dfrac{dy}{dx} = (1)(2) = 2.$$

25. $\dfrac{48v^5}{\left(v^2 + 8\right)^4}$

We find the derivative using the Chain Rule, which says that if $y = y(v)$ and

$v = v(x)$, then $\dfrac{dy}{dx} = \dfrac{dy}{dv}\dfrac{dv}{dx}$, although, in this case, we have $u(y)$, $y(x)$, and $x(v)$,

so we will find $\dfrac{du}{dv}$ by $\dfrac{du}{dv} = \dfrac{du}{dy}\dfrac{dy}{dx}\dfrac{dx}{dv}$. Here $\dfrac{du}{dy} = 3y^2$, $\dfrac{dy}{dx} = \dfrac{(x + 8)(1) - (x)(1)}{(x + 8)^2} = \dfrac{8}{(x + 8)^2}$,

and $\dfrac{dx}{dv} = 2v$. Next, $\dfrac{du}{dv} = \dfrac{du}{dy}\dfrac{dy}{dx}\dfrac{dx}{dv} = \left(3y^2\right)\left(\dfrac{8}{(x + 8)^2}\right)(2v)$. Now because $x = v^2$ and

$y = \dfrac{x}{x + 8} = \dfrac{v^2}{v^2 + 8}$, we get $\dfrac{du}{dv} = \left(3\dfrac{\left(v^2\right)^2}{\left(v^2 + 8\right)^2}\right)\left(\dfrac{8}{\left(v^2 + 8\right)^2}\right)(2v) = \dfrac{48v^5}{\left(v^2 + 8\right)^4}.$

SOLUTIONS TO PRACTICE PROBLEM SET 6

1. $2\sin x \cos x$ or $\sin 2x$

    Recall that $\dfrac{d}{dx}(\sin x) = \cos x$. Here we use the Chain Rule to find the derivative:

    $\dfrac{dy}{dx} = 2(\sin x)(\cos x)$. If you recall your trigonometric identities, $2\sin x \cos x = \sin 2x$. Either

    answer is acceptable.

2. $-2x\sin(x^2)$

    Recall that $\dfrac{d}{dx}(\cos x) = -\sin x$. Here, we use the Chain Rule to find the derivative:

    $\dfrac{dy}{dx} = \left(-\sin(x^2)\right)(2x) = -2x\sin(x^2)$.

3. $2\sec^3 x - \sec x$

    Recall that $\dfrac{d}{dx}(\tan x) = \sec^2 x$ and that $\dfrac{d}{dx}(\sec x) = \sec x \tan x$. Using the Product Rule, we get $\dfrac{dy}{dx} = (\tan x)(\sec x \tan x) + (\sec x)(\sec^2 x)$. This can be simplified to

    $\sec^3 x + \sec x \tan^2 x = 2\sec^3 x - \sec x$.

4. $-4\csc^2(4x)$

    Recall that $\dfrac{d}{dx}(\cot x) = -\csc^2 x$. Here we use the Chain Rule to find the derivative:

    $\dfrac{dy}{dx} = \left(-\csc^2 4x\right)(4) = -4\csc^2(4x)$.

5. $\dfrac{3\cos 3x}{2\sqrt{\sin 3x}}$

    Recall that $\dfrac{d}{dx}(\sin x) = \cos x$. Here we use the Chain Rule to find the derivative:

    $\dfrac{dy}{dx} = \dfrac{1}{2}(\sin 3x)^{-\frac{1}{2}}(\cos 3x)(3)$. This can be simplified to $\dfrac{dy}{dx} = \dfrac{3\cos 3x}{2\sqrt{\sin 3x}}$.

6. $\dfrac{2\cos x}{\left(1-\sin x\right)^{2}}$

Recall that $\dfrac{d}{dx}(\sin x) = \cos x$. Here we use the Quotient Rule to find the derivative:

$\dfrac{dy}{dx} = \dfrac{(1-\sin x)(\cos x)-(1+\sin x)(-\cos x)}{(1-\sin x)^{2}}$. This can be simplified to $\dfrac{2\cos x}{(1-\sin x)^{2}}$.

7. $-4x\csc^{2}\left(x^{2}\right)\cot\left(x^{2}\right)$

Recall that $\dfrac{d}{dx}(\csc x) = -\csc x\cot x$. Here we use the Chain Rule to find the derivative:

$\dfrac{dy}{dx} = \left(2\csc x^{2}\right)\left(-\csc x^{2}\cot x^{2}\right)\left(2x\right) = -4x\csc^{2}\left(x^{2}\right)\cot\left(x^{2}\right)$.

8. $6\cos 3x\cos 4x - 8\sin 3x\sin 4x$

Recall that $\dfrac{d}{dx}(\sin x) = \cos x$ and that $\dfrac{d}{dx}(\cos x) = -\sin x$. Here we use the Product Rule to find the derivative: $\dfrac{dy}{dx} = 2\left[(\sin 3x)(-\sin 4x)(4)+(\cos 4x)(\cos 3x)(3)\right]$. This can be simplified to $\dfrac{dy}{dx} = 6\cos 3x\cos 4x - 8\sin 3x\sin 4x$.

9. $16\sin 2x$

Recall that $\dfrac{d}{dx}(\sin x) = \cos x$ and that $\dfrac{d}{dx}(\cos x) = -\sin x$. Here we will use the Chain Rule four times to find the fourth derivative.

The first derivative is: $\dfrac{dy}{dx} = (\cos 2x)(2) = 2\cos 2x$.

The second derivative is: $\dfrac{d^{2}y}{dx^{2}} = 2(-\sin 2x)(2) = -4\sin 2x$.

The third derivative is: $\dfrac{d^{3}y}{dx^{3}} = -4(\cos 2x)(2) = -8\cos 2x$.

And the fourth derivative is: $\dfrac{d^{4}y}{dx^{4}} = -8(-\sin 2x)(2) = 16\sin 2x$.

10. $$\left[\cos\left(1+\cos^2 x\right)+\sin\left(1+\cos^2 x\right)\right]\left(-2\sin x\cos x\right)$$

Recall that $\dfrac{d}{dx}(\sin x)=\cos x$ and that $\dfrac{d}{dx}(\cos x)=-\sin x$. Here we will use the Chain Rule to find the derivative: $\dfrac{dy}{dt}=\cos t-(-\sin t)=\cos t+\sin t$ and

$\dfrac{dt}{dx}=2(\cos x)(-\sin x)=-2\sin x\cos x$. Next, because $\dfrac{dy}{dx}=\dfrac{dy}{dt}\dfrac{dt}{dx}$ and $t=1+\cos^2 x$, we get

$$\dfrac{dy}{dx}=(\cos t+\sin t)(-2\sin x\cos x)=\left[\cos\left(1+\cos^2 x\right)+\sin\left(1+\cos^2 x\right)\right]\left(-2\sin x\cos x\right).$$

11. $$\dfrac{2\tan x\sec^2 x}{\left(1-\tan x\right)^3}$$

Recall that $\dfrac{d}{dx}(\tan x)=\sec^2 x$. Using the Quotient Rule and the Chain Rule, we get

$$\dfrac{dy}{dx}=2\left(\dfrac{\tan x}{1-\tan x}\right)\dfrac{\left(1-\tan x\right)\left(\sec^2 x\right)-\left(\tan x\right)\left(-\sec^2 x\right)}{\left(1-\tan x\right)^2}.$$ This simplifies to $\dfrac{2\tan x\sec^2 x}{\left(1-\tan x\right)^3}$.

12. $$\left(\sec\theta\right)\left(\sec^2\left(2\theta\right)\right)(2)+\left(\tan 2\theta\right)\left(\sec\theta\tan\theta\right)$$

Recall that $\dfrac{d}{dx}(\tan x)=\sec^2 x$ and that $\dfrac{d}{dx}(\sec x)=\sec x\tan x$. Using the Product Rule and the

Chain Rule, we get $\dfrac{dr}{d\theta}=\left(\sec\theta\right)\left(\sec^2\left(2\theta\right)\right)(2)+\left(\tan 2\theta\right)\left(\sec\theta\tan\theta\right).$

13. $$-\left[\sin\left(1+\sin\theta\right)\right]\left(\cos\theta\right)$$

Recall that $\dfrac{d}{dx}(\sin x)=\cos x$ and that $\dfrac{d}{dx}(\cos x)=-\sin x$. Using the Chain Rule, we get

$$\dfrac{dr}{d\theta}=-\left[\sin\left(1+\sin\theta\right)\right]\left(\cos\theta\right).$$

14. $$\frac{\sec\theta(\tan\theta-1)}{(1+\tan\theta)^2}$$

Recall that $\frac{d}{dx}(\tan x)=\sec^2 x$ and that $\frac{d}{dx}(\sec x)=\sec x\tan x$. Using the Quotient Rule, we

get $\frac{dr}{d\theta}=\dfrac{(1+\tan\theta)(\sec\theta\tan\theta)-(\sec\theta)(\sec^2\theta)}{(1+\tan\theta)^2}$. This can be simplified (using trigonometric

identities) to $\dfrac{\sec\theta(\tan\theta-1)}{(1+\tan\theta)^2}$.

15. $$-\frac{\left(\dfrac{4}{x^2}\right)\left(\csc^2\left(\dfrac{2}{x}\right)\right)}{\left(1+\cot\left(\dfrac{2}{x}\right)\right)^3}$$

Recall that $\frac{d}{dx}(\cot x)=-\csc^2 x$. Here we use the Chain Rule to find the derivative:

$$\frac{dy}{dx}=-2\left(1+\cot\left(\frac{2}{x}\right)\right)^{-3}\left(-\csc^2\left(\frac{2}{x}\right)\right)\left(\frac{-2}{x^2}\right)=-\frac{\left(\dfrac{4}{x^2}\right)\left(\csc^2\left(\dfrac{2}{x}\right)\right)}{\left(1+\cot\left(\dfrac{2}{x}\right)\right)^3}.$$

16. $$\cos\left(\cos\left(\sqrt{x}\right)\right)\left(-\sin\sqrt{x}\right)\left(\frac{1}{2\sqrt{x}}\right)$$

Recall that $\frac{d}{dx}(\sin x)=\cos x$ and that $\frac{d}{dx}(\cos x)=-\sin x$. Using the Chain Rule, we get

$$\frac{dy}{dx}=\cos\left(\cos\left(\sqrt{x}\right)\right)\left(-\sin\sqrt{x}\right)\left(\frac{1}{2\sqrt{x}}\right).$$

SOLUTIONS TO PRACTICE PROBLEM SET 7

1. $\dfrac{3x^2}{1+3y^2}$

   We take the derivative of each term with respect to x: $\left(3x^2\right)\left(\dfrac{dx}{dx}\right)-\left(3y^2\right)\left(\dfrac{dy}{dx}\right)=(1)\left(\dfrac{dy}{dx}\right)$.

   Next, because $\dfrac{dx}{dx}=1$, we can eliminate that term and get $\left(3x^2\right)-\left(3y^2\right)\left(\dfrac{dy}{dx}\right)=\left(\dfrac{dy}{dx}\right)$.

   Next, group the terms containing $\dfrac{dy}{dx}$: $\left(3x^2\right)=\left(\dfrac{dy}{dx}\right)+\left(3y^2\right)\left(\dfrac{dy}{dx}\right)$.

   Factor out the term $\dfrac{dy}{dx}$: $\left(3x^2\right)=\left(\dfrac{dy}{dx}\right)\left(1+3y^2\right)$. Now, we can isolate $\dfrac{dy}{dx}$: $\dfrac{dy}{dx}=\dfrac{3x^2}{1+3y^2}$.

2. $\dfrac{8y-x}{y-8x}$

   We take the derivative of each term with respect to x:

   $(2x)\left(\dfrac{dx}{dx}\right)-16\left[(x)\left(\dfrac{dy}{dx}\right)+(y)\left(\dfrac{dx}{dx}\right)\right]+(2y)\left(\dfrac{dy}{dx}\right)=0$.

   Next, because $\dfrac{dx}{dx}=1$ we can eliminate that term and we can distribute the -16 to get

   $2x-16x\left(\dfrac{dy}{dx}\right)-16y+2y\left(\dfrac{dy}{dx}\right)=0$.

   Next, group the terms containing $\dfrac{dy}{dx}$ on one side of the equal sign and the other terms on the

   other side: $-16x\left(\dfrac{dy}{dx}\right)+2y\left(\dfrac{dy}{dx}\right)=16y-2x$.

   Factor out the term $\dfrac{dy}{dx}$: $\left(\dfrac{dy}{dx}\right)(2y-16x)=16y-2x$. Now we can isolate $\dfrac{dy}{dx}$: $\dfrac{dy}{dx}=\dfrac{16y-2x}{2y-16x}$,

   which can be reduced to $\dfrac{dy}{dx}=\dfrac{8y-x}{y-8x}$.

3. $\dfrac{1}{2}$

First, cross-multiply so that we don't have to use the Quotient Rule: $x + y = 3x - 3y$. Next, simplify $4y = 2x$, which reduces to $y = \dfrac{1}{2}x$. Now, we can take the derivative: $\dfrac{dy}{dx} = \dfrac{1}{2}$. Note that just because a problem has the x's and y's mixed together doesn't mean that we need to use implicit differentiation to solve it!

4. $-\dfrac{\sin x + \cos x}{\sin y + \cos y}$

We take the derivative of each term with respect to x:

$$(-\sin y)\left(\dfrac{dy}{dx}\right) - (\cos x)\left(\dfrac{dx}{dx}\right) = (\cos y)\left(\dfrac{dy}{dx}\right) - (-\sin x)\left(\dfrac{dx}{dx}\right).$$

Next, because $\dfrac{dx}{dx} = 1$ we can eliminate that term to get

$$(-\sin y)\left(\dfrac{dy}{dx}\right) - \cos x = (\cos y)\left(\dfrac{dy}{dx}\right) + \sin x.$$

Next, group the terms containing $\dfrac{dy}{dx}$ on one side of the equal sign and the other terms on the other side: $(-\sin y)\left(\dfrac{dy}{dx}\right) - (\cos y)\left(\dfrac{dy}{dx}\right) = \sin x + \cos x$.

Factor out the term $\dfrac{dy}{dx}: (-\sin y - \cos y)\left(\dfrac{dy}{dx}\right) = \sin x + \cos x$. Now, we can isolate $\dfrac{dy}{dx}$:

$\dfrac{dy}{dx} = \dfrac{\sin x + \cos x}{-\sin y - \cos y}$, which can be simplified to $\dfrac{dy}{dx} = -\dfrac{\sin x + \cos x}{\sin y + \cos y}$.

5. $\dfrac{8}{7}$

We take the derivative of each term with respect to x:

$$(32x)\left(\dfrac{dx}{dx}\right) - 16\left[(x)\left(\dfrac{dy}{dx}\right) + (y)\left(\dfrac{dx}{dx}\right)\right] + (2y)\left(\dfrac{dy}{dx}\right) = 0.$$

Next, because $\dfrac{dx}{dx} = 1$ we can eliminate that term to get $32x - 16x\left(\dfrac{dy}{dx}\right) - 16y + 2y\left(\dfrac{dy}{dx}\right) = 0$.

Next, don't simplify. Plug in (1, 1) for x and y: $32(1) - 16(1)\left(\dfrac{dy}{dx}\right) - 16(1) + 2(1)\left(\dfrac{dy}{dx}\right) = 0$,

which simplifies to $16 - 14\left(\dfrac{dy}{dx}\right) = 0$.

Finally, we can solve for $\dfrac{dy}{dx}$: $\dfrac{dy}{dx} = \dfrac{8}{7}$.

6. $\dfrac{1}{7}$

We take the derivative of each term with respect to x: $\left(\dfrac{1}{2}x^{-\frac{1}{2}}\right)\left(\dfrac{dx}{dx}\right) + \left(\dfrac{1}{2}y^{-\frac{1}{2}}\right)\left(\dfrac{dy}{dx}\right) = (4y)\left(\dfrac{dy}{dx}\right)$.

Next, because $\dfrac{dx}{dx} = 1$ we can eliminate that term to get $\dfrac{1}{2}x^{-\frac{1}{2}} + \left(\dfrac{1}{2}y^{-\frac{1}{2}}\right)\left(\dfrac{dy}{dx}\right) = (4y)\left(\dfrac{dy}{dx}\right)$.

Next, don't simplify. Plug in (1, 1) for x and y, $\dfrac{1}{2}(1)^{-\frac{1}{2}} + \left(\dfrac{1}{2}(1)^{-\frac{1}{2}}\right)\left(\dfrac{dy}{dx}\right) = 4(1)\left(\dfrac{dy}{dx}\right)$, which

simplifies to $\dfrac{1}{2} + \left(\dfrac{1}{2}\right)\left(\dfrac{dy}{dx}\right) = 4\left(\dfrac{dy}{dx}\right)$.

Finally, we can solve for $\dfrac{dy}{dx}$: $\dfrac{dy}{dx} = \dfrac{1}{7}$.

7. -1

We take the derivative of each term with respect to x:

$(x\cos y)\left(\dfrac{dy}{dx}\right) + (\sin y)\left(\dfrac{dx}{dx}\right) + (y\cos x)\left(\dfrac{dx}{dx}\right) + (\sin x)\left(\dfrac{dy}{dx}\right) = 0$.

Next, because $\dfrac{dx}{dx} = 1$ we can eliminate that term to get

$(x\cos y)\left(\dfrac{dy}{dx}\right) + \sin y + y\cos x + (\sin x)\left(\dfrac{dy}{dx}\right) = 0$. Next, don't simplify. Plug in

$\left(\dfrac{\pi}{4}, \dfrac{\pi}{4}\right)$ for x and y: $\left(\dfrac{\pi}{4}\cos\dfrac{\pi}{4}\right)\left(\dfrac{dy}{dx}\right) + \sin\dfrac{\pi}{4} + \dfrac{\pi}{4}\cos\dfrac{\pi}{4} + \left(\sin\dfrac{\pi}{4}\right)\left(\dfrac{dy}{dx}\right) = 0$, which simplifies to $\left(\dfrac{\pi}{4}\dfrac{1}{\sqrt{2}}\right)\left(\dfrac{dy}{dx}\right)+\dfrac{1}{\sqrt{2}}+\dfrac{\pi}{4}\dfrac{1}{\sqrt{2}}+\dfrac{1}{\sqrt{2}}\left(\dfrac{dy}{dx}\right)=0$. If we multiply through by $\sqrt{2}$ we get

$\dfrac{\pi}{4}\dfrac{dy}{dx}+1+\dfrac{\pi}{4}+\dfrac{dy}{dx}=0$.

Finally, we can solve for $\dfrac{dy}{dx}$: $\dfrac{dy}{dx}=-1$.

8. $-\dfrac{1}{16y^3}$

We take the derivative of each term with respect to x: $\left(2x\right)\left(\dfrac{dx}{dx}\right)+\left(8y\right)\left(\dfrac{dy}{dx}\right)=0$.

Next, because $\dfrac{dx}{dx}=1$ we can eliminate that term to get $2x+\left(8y\right)\left(\dfrac{dy}{dx}\right)=0$. Next, we can iso-

late $\dfrac{dy}{dx}$: $\dfrac{dy}{dx}=-\dfrac{x}{4y}$. Now, we take the derivative again: $\dfrac{d^2y}{dx^2}=-\dfrac{\left(4y\right)\left(\dfrac{dx}{dx}\right)-\left(x\right)\left(4\dfrac{dy}{dx}\right)}{16y^2}$. Next,

because $\dfrac{dx}{dx}=1$ and $\dfrac{dy}{dx}=-\dfrac{x}{4y}$ we get $\dfrac{d^2y}{dx^2}=-\dfrac{4y-4x\left(-\dfrac{x}{4y}\right)}{16y^2}$. This can be simplified to

$\dfrac{d^2y}{dx^2}=-\dfrac{4y+\dfrac{x^2}{y}}{16y^2}=-\dfrac{4y^2+x^2}{16y^3}=-\dfrac{1}{16y^3}$.

9. $\dfrac{\sin x \sin^2 y-\cos y\cos^2 x}{\sin^3 y}$

We take the derivative of each term with respect to x: $\left(\cos x\right)\left(\dfrac{dx}{dx}\right)=\left(-\sin y\right)\left(\dfrac{dy}{dx}\right)$.

Next, because $\dfrac{dx}{dx}=1$ we can eliminate that term to get

$\cos x=\left(-\sin y\right)\left(\dfrac{dy}{dx}\right)$. Next, we can isolate $\dfrac{dy}{dx}$: $\dfrac{dy}{dx}=-\dfrac{\cos x}{\sin y}$. Now, we take the deriva-

tive again: $\dfrac{d^2y}{dx^2}=\dfrac{\left(\sin y\right)\left(\sin x\right)\left(\dfrac{dx}{dx}\right)+\left(\cos x\right)\left(\cos y\right)\left(\dfrac{dy}{dx}\right)}{\sin^2 y}$. Next, because $\dfrac{dx}{dx}=1$ and

$$\frac{dy}{dx} = -\frac{\cos x}{\sin y} \quad \text{we get} \quad \frac{d^2y}{dx^2} = \frac{(\sin y)(\sin x) + (\cos x)(\cos y)\left(-\dfrac{\cos x}{\sin y}\right)}{\sin^2 y}. \text{ This can be simplified to}$$

$$\frac{d^2y}{dx^2} = \frac{\sin x \sin^2 y - \cos y \cos^2 x}{\sin^3 y}.$$

10.   1

We can easily isolate y in this equation: $y = \dfrac{1}{2}x^2 - 2x + 1$. We take the derivative: $\dfrac{dy}{dx} = x - 2$. And

we take the derivative again: $\dfrac{d^2y}{dx^2} = 1$. Note that just because a problem has the x's and y's mixed

together doesn't mean that we need to use implicit differentiation to solve it!

SOLUTIONS TO PRACTICE PROBLEM SET 8

1. $y - 2 = 5(x - 1)$

Remember that the equation of a line through a point (x_1, y_1) with slope m is $y - y_1 = m(x - x_1)$.

We find the y-coordinate by plugging $x = 1$ into the equation $y = 3x^2 - x$, and we find the slope

by plugging $x = 1$ into the derivative of the equation.

First, we find the y-coordinate, y_1: $y = 3(1)^2 - 1 = 2$. This means that the line passes through the

point $(1, 2)$.

Next, we take the derivative: $\dfrac{dy}{dx} = 6x - 1$. Now, we can find the slope, m: $\dfrac{dy}{dx}\bigg|_{x=1} = 6(1) - 1 = 5$.

Finally, we plug in the point $(1, 2)$ and the slope $m = 5$ to get the equation of the tangent line:

$y - 2 = 5(x - 1)$.

2. $y - 18 = 24(x - 3)$

Remember that the equation of a line through a point (x_1, y_1) with slope m is $y - y_1 = m(x - x_1)$.

We find the y-coordinate by plugging $x = 3$ into the equation $y = x^3 - 3x$, and we find the slope by

plugging $x = 3$ into the derivative of the equation.

First, we find the y-coordinate, y_1: $y = (3)^3 - 3(3) = 18$. This means that the line passes through

the point $(3, 18)$.

Next, we take the derivative: $\dfrac{dy}{dx} = 3x^2 - 3$. Now, we can find the slope, m: $\dfrac{dy}{dx}\Big|_{x=3} = 3(3)^2 - 3 = 24$.

Finally, we plug in the point $(3, 18)$ and the slope $m = 24$ to get the equation of the tangent line:

$y - 18 = 24(x - 3)$.

3. $y - 4 = -(x - 2)$

Remember that the equation of a line through a point (x_1, y_1) with slope m is $y - y_1 = m(x - x_1)$.

We find the y-coordinate by plugging $x = 2$ into the equation $y = \sqrt{8x}$, and we find the slope by

plugging $x = 2$ into the derivative of the equation.

First, we find the y-coordinate, y_1: $y = \sqrt{8(2)} = 4$. This means that the line passes through the

point $(2, 4)$.

Next, we take the derivative: $\dfrac{dy}{dx} = \dfrac{4}{\sqrt{8x}}$. Now, we can find the slope, m: $\dfrac{dy}{dx}\Big|_{x=2} = \dfrac{4}{\sqrt{8(2)}} = 1$.

However, this is the slope of the *tangent* line. The *normal* line is perpendicular to the tangent line,

so its slope will be the negative reciprocal of the tangent line's slope. In this case, the slope of the

normal line is $\dfrac{-1}{1} = -1$. Finally, we plug in the point $(2, 4)$ and the slope $m = -1$ to get the equa-

tion of the normal line: $y - 4 = -(x - 2)$.

4. $$y - \frac{1}{4} = -\frac{3}{64}(x - 3)$$

Remember that the equation of a line through a point (x_1, y_1) with slope m is $y - y_1 = m(x - x_1)$.

We find the y-coordinate by plugging $x = 3$ into the equation $y = \dfrac{1}{\sqrt{x^2 + 7}}$, and we find the slope

by plugging $x = 3$ into the derivative of the equation.

First, we find the y-coordinate, y_1: $y = \dfrac{1}{\sqrt{(3)^2 + 7}} = \dfrac{1}{4}$. This means that the line passes through the

point $\left(3, \dfrac{1}{4}\right)$.

Next, we take the derivative: $\dfrac{dy}{dx} = -\dfrac{1}{2}(x^2 + 7)^{-\frac{3}{2}}(2x) = -\dfrac{x}{\left(\sqrt{x^2 + 7}\right)^3}$. Now, we can find the

slope, m: $\dfrac{dy}{dx}\bigg|_{x=3} = -\dfrac{3}{\left(\sqrt{(3)^2 + 7}\right)^3} = -\dfrac{3}{64}$. Finally, we plug in the point $\left(3, \dfrac{1}{4}\right)$ and the slope

$m = -\dfrac{3}{64}$ to get the equation of the tangent line: $y - \dfrac{1}{4} = -\dfrac{3}{64}(x - 3)$.

5. $$y - 7 = \frac{1}{6}(x - 4)$$

Remember that the equation of a line through a point (x_1, y_1) with slope m is $y - y_1 = m(x - x_1)$.

We find the y-coordinate by plugging $x = 4$ into the equation $y = \dfrac{x + 3}{x - 3}$, and we find the slope by

plugging $x = 4$ into the derivative of the equation.

First, we find the y-coordinate, y_1: $y = \dfrac{4 + 3}{4 - 3} = 7$. This means that the line passes through the

point $(4, 7)$.

Next, we take the derivative: $\dfrac{dy}{dx} = \dfrac{(x - 3)(1) - (x + 3)(1)}{(x - 3)^2} = -\dfrac{6}{(x - 3)^2}$. Now, we can find the

slope, m: $\dfrac{dy}{dx}\bigg|_{x=4} = -\dfrac{6}{(4 - 3)^2} = -6$. However, this is the slope of the *tangent* line. The *normal* line is perpendicular to the tangent line, so its slope will be the negative reciprocal of the tangent line's slope. In this case, the slope of the normal line is $\dfrac{-1}{-6} = \dfrac{1}{6}$. Finally, we plug in the point $(4, 7)$ and the slope $m = \dfrac{1}{6}$ to get the equation of the normal line: $y - 7 = \dfrac{1}{6}(x - 4)$.

6. $y = -3x + 4$

Remember that the equation of a line through a point (x_1, y_1) with slope m is $y - y_1 = m(x - x_1)$.

We find the slope by plugging $x = 0$ into the derivative of the equation $y = 4 - 3x - x^2$. First, we take the derivative: $\dfrac{dy}{dx} = -3 - 2x$. Now, we can find the slope, m: $\left.\dfrac{dy}{dx}\right|_{x=0} = -3 - 3(0) = -3$.

Finally, we plug in the point $(0, 4)$ and the slope $m = -3$ to get the equation of the tangent line:

$y - 4 = -3(x - 0)$ or $y = -3x + 4$.

7. $y = 0$

Remember that the equation of a line through a point (x_1, y_1) with slope m is $y - y_1 = m(x - x_1)$.

We find the y-coordinate by plugging $x = 2$ into the equation $y = 2x^3 - 3x^2 - 12x + 20$, and we find the slope by plugging $x = 2$ into the derivative of the equation.

First, we find the y-coordinate, y_1: $y = 2(2)^3 - 3(2)^2 - 12(2) + 20 = 0$. This means that the line passes through the point $(2, 0)$.

Next, we take the derivative: $\dfrac{dy}{dx} = 6x^2 - 6x - 12$. Now, we can find the slope, m: $\left.\dfrac{dy}{dx}\right|_{x=2} = 6(2)^2 - 6(2) - 12 = 0$. Finally, we plug in the point $(2, 0)$ and the slope $m = 0$ to get the equation of the tangent line: $y - 0 = 0(x - 2)$ or $y = 0$.

8. $y + 29 = -39(x - 5)$

Remember that the equation of a line through a point (x_1, y_1) with slope m is $y - y_1 = m(x - x_1)$.

We find the y-coordinate by plugging $x = 5$ into the equation $y = \dfrac{x^2 + 4}{x - 6}$, and we find the slope by plugging $x = 5$ into the derivative of the equation.

First, we find the y-coordinate, y_1: $y = \dfrac{(5)^2 + 4}{(5) - 6} = -29$. This means that the line passes through the point $(5, -29)$.

Next, we take the derivative: $\dfrac{dy}{dx} = \dfrac{(x - 6)(2x) - (x^2 + 4)(1)}{(x - 6)^2} = \dfrac{x^2 - 12x - 4}{(x - 6)^2}$. Now, we can find the slope, m: $\dfrac{dy}{dx}\bigg|_{x=5} = \dfrac{(5)^2 - 12(5) - 4}{(5 - 6)^2} = -39$. Finally, we plug in the point $(5, -29)$ and the slope $m = -39$ to get the equation of the tangent line: $y + 29 = -39(x - 5)$.

9. $y - 7 = \dfrac{24}{7}(x - 4)$

Remember that the equation of a line through a point (x_1, y_1) with slope m is $y - y_1 = m(x - x_1)$.

We find the slope by plugging $x = 4$ into the derivative of the equation $y = \sqrt{x^3 - 15}$. First, we take the derivative: $\dfrac{dy}{dx} = \dfrac{1}{2}(x^3 - 15)^{-\frac{1}{2}}(3x^2) = \dfrac{3x^2}{2\sqrt{x^3 - 15}}$. Now, we can find the slope, m: $\dfrac{dy}{dx}\bigg|_{x=4} = \dfrac{3(4)^2}{2\sqrt{(4)^3 - 15}} = \dfrac{24}{7}$. Finally, we plug in the point $(4, 7)$ and the slope $m = \dfrac{24}{7}$ to get the equation of the tangent line: $y - 7 = \dfrac{24}{7}(x - 4)$.

10. $y = 0$

Remember that the equation of a line through a point (x_1, y_1) with slope m is $y - y_1 = m(x - x_1)$.

We find the y-coordinate by plugging $x = -2$ into the equation $y = (x^2 + 4x + 4)^2$, and we find the

slope by plugging $x = -2$ into the derivative of the equation.

First, we find the y-coordinate, y_1: $y = ((-2)^2 + 4(-2) + 4)^2 = 0$. This means that the line passes

through the point $(-2, 0)$.

Next, we take the derivative: $\dfrac{dy}{dx} = 2(x^2 + 4x + 4)(2x + 4)$. Now, we can find the slope,

m: $\dfrac{dy}{dx}\Big|_{x=-2} = 2((-2)^2 + 4(-2) + 4)(2(-2) + 4) = 0$. Finally, we plug in the point $(-2, 0)$ and the

slope $m = 0$ to get the equation of the tangent line: $y - 0 = 0(x + 2)$ or $y = 0$.

11. $x = \pm\sqrt{\dfrac{3}{2}}$

The slope of the line $y = x$ is 1, so we want to know where the slope of the tangent line is equal

to 1. We find the slope of the tangent line by taking the derivative: $\dfrac{dy}{dx} = 6x^2 - 8$. Now we set the

derivative equal to 1: $6x^2 - 8 = 1$. If we solve for x, we get $x = \pm\sqrt{\dfrac{3}{2}}$.

12. $y - 7 = \dfrac{1}{2}(x - 3)$

Remember that the equation of a line through a point (x_1, y_1) with slope m is $y - y_1 = m(x - x_1)$.

We find the y-coordinate by plugging $x = 3$ into the equation $y = \dfrac{3x + 5}{x - 1}$, and we find the slope by

plugging $x = 3$ into the derivative of the equation.

First, we find the y-coordinate, y_1: $y = \dfrac{3(3) + 5}{(3) - 1} = 7$. This means that the line passes through the

point $(3, 7)$.

Next, we take the derivative: $\dfrac{dy}{dx} = \dfrac{(x-1)(3)-(3x+5)(1)}{(x-1)^2} = \dfrac{-8}{(x-1)^2}$. Now, we can find the

slope, m: $\dfrac{dy}{dx}\Big|_{x=3} = \dfrac{-8}{(3-1)^2} = -2$. However, this is the slope of the *tangent* line. The *normal* line is

perpendicular to the tangent line, so its slope will be the negative reciprocal of the tangent line's

slope. In this case, the slope of the normal line is $\dfrac{-1}{-2} = \dfrac{1}{2}$. Finally, we plug in the point $(3,7)$ and

the slope $m = \dfrac{1}{2}$ to get the equation of the normal line: $y - 7 = \dfrac{1}{2}(x-3)$.

13. $x = 9$

A line that is parallel to the y-axis has an infinite (or undefined) slope. In order to find where the

normal line has an infinite slope, we first take the derivative to find the slope of the tangent line:

$\dfrac{dy}{dx} = 2(x-9)(1) = 2x - 18$. Next, because the normal line is perpendicular to the tangent line, the

slope of the normal line is the negative reciprocal of the slope of the tangent line: $m = \dfrac{-1}{2x-18}$. Now,

we need to find where the slope is infinite. This is simply where the denominator of the slope is zero:

$x = 9$.

14. $\left(-\dfrac{3}{2}, \dfrac{41}{4}\right)$

A line that is parallel to the x-axis has a zero slope. In order to find where the tangent line has a

zero slope, we first take the derivative: $\dfrac{dy}{dx} = -3 - 2x$. Now we need to find where the slope is zero.

The derivative $-3 - 2x = 0$ at $x = -\dfrac{3}{2}$. Now, we need to find the y-coordinate, which we get by

plugging $x = -\dfrac{3}{2}$ into the equation for y: $8 - 3\left(-\dfrac{3}{2}\right) - \left(-\dfrac{3}{2}\right)^2 = \dfrac{41}{4}$. Therefore, the answer is

$\left(-\dfrac{3}{2}, \dfrac{41}{4}\right)$.

15. $a = 1$, $b = 0$, and $c = 1$.

The two equations will have a common tangent line where they have the same slope, which we find by taking the derivative of each equation. The derivative of the first equation is: $\dfrac{dy}{dx} = 2x + a$. The derivative of the second equation is $\dfrac{dy}{dx} = c + 2x$. Setting the two derivatives equal to each other, we get $a = c$. Each equation will pass through the point $(-1, 0)$. If we plug $(-1, 0)$ into the first equation, we get $0 = (-1)^2 + a(-1) + b$, which simplifies to: $a - b = 1$. If we plug $(-1, 0)$ into the second equation, we get $0 = c(-1) + (-1)^2$, which simplifies to $c = 1$. Now we can find the values for a, b, and c. We get $a = 1$, $b = 0$, and $c = 1$.

SOLUTIONS TO PRACTICE PROBLEM SET 9

1. $c = 0$

The Mean Value Theorem says that: If $f(x)$ is continuous on the interval $[a, b]$ and is differentiable everywhere on the interval (a, b), then there exists at least one number c on the interval (a, b) such that $f'(c) = \dfrac{f(b) - f(a)}{b - a}$. Here, the function is $f(x) = 3x^2 + 5x - 2$ and the interval is $[-1, 1]$. Thus, the Mean Value Theorem says that $f'(c) = \dfrac{\left(3(1)^2 + 5(1) - 2\right) - \left(3(-1)^2 + 5(-1) - 2\right)}{(1 + 1)}$. This simplifies to $f'(c) = 5$. Next, we need to find $f'(c)$. The derivative of $f(x)$ is $f'(x) = 6x + 5$, so $f'(c) = 6c + 5$.

Now, we can solve for c: $6c + 5 = 5$ and $c = 0$. Note that 0 is in the interval $(-1, 1)$, just as we expected.

2. $c = \dfrac{4}{\sqrt{3}}$

The Mean Value Theorem says that: If $f(x)$ is continuous on the interval $[a, b]$ and is differentiable everywhere on the interval (a, b), then there exists at least one number c on the interval (a, b) such that

$f'(c) = \dfrac{f(b) - f(a)}{b - a}$. Here the function is $f(x) = x^3 + 24x - 16$ and the interval is [0, 4]. Thus, the

Mean Value Theorem says that $f'(c) = \dfrac{\left((4)^3 + 24(4) - 16\right) - \left((0)^3 + 24(0) - 16\right)}{(4 - 0)}$. This simplifies to

$f'(c) = 40$. Next, we need to find $f'(c)$ from the equation. The derivative of $f(x)$ is $f'(x) = 3x^2 + 24$, so

$f'(c) = 3c^2 + 24$. Now, we can solve for c: $3c^2 + 24 = 40$ and $c = \pm\dfrac{4}{\sqrt{3}}$. Note that $\dfrac{4}{\sqrt{3}}$ is in the

interval (0, 4), but $-\dfrac{4}{\sqrt{3}}$ is *not* in the interval. Thus, the answer is only $c = \dfrac{4}{\sqrt{3}}$. It's *very* im-

portant to check that the answers you get for c fall in the given interval when doing Mean Value

Theorem problems.

3. $c = \dfrac{-12 + 8\sqrt{3}}{3} \approx 0.62$

The Mean Value Theorem says that: If $f(x)$ is continuous on the interval $[a, b]$ and

is differentiable everywhere on the interval (a, b), then there exists at least one num-

ber c on the interval (a, b) such that $f'(c) = \dfrac{f(b) - f(a)}{b - a}$. Here, the function is

$f(x) = x^3 + 12x^2 + 7x$ and the interval is [−4, 4]. Thus, the Mean Value Theorem says

that $f'(c) = \dfrac{\left((4)^3 + 12(4)^2 + 7(4)\right) - \left((-4)^3 + 12(-4)^2 + 7(-4)\right)}{(4 + 4)}$.   This   simplifies   to

$f'(c) = 23$. Next, we need to find $f'(c)$. The derivative of $f(x)$ is $f'(x) = 3x^2 + 24x + 7$, so

$f'(c) = 3c^2 + 24c + 7$. Now, we can solve for c: $3c^2 + 24c + 7 = 23$ and $c = \dfrac{-12 \pm 8\sqrt{3}}{3}$. Note that

$c = \dfrac{-12 + 8\sqrt{3}}{3}$ is in the interval (−4, 4), but $\dfrac{-12 - 8\sqrt{3}}{3}$ is *not* in the interval. Thus, the answer

is only $c = \dfrac{-12 + 8\sqrt{3}}{3} \approx 0.62$. It's *very* important to check that the answers you get for c fall in the

given interval when doing Mean Value Theorem problems.

4. $c = \sqrt{2}$

The Mean Value Theorem says that: If $f(x)$ is continuous on the interval $[a, b]$ and is differentiable everywhere on the interval (a, b), then there exists at least one number c on the interval (a, b) such that $f'(c) = \dfrac{f(b) - f(a)}{b - a}$. Here the function is $f(x) = \dfrac{6}{x} - 3$ and the interval is [1, 2]. Thus, the Mean Value Theorem says that $f'(c) = \dfrac{\left(\dfrac{6}{2} - 3\right) - \left(\dfrac{6}{1} - 3\right)}{(2-1)}$. This simplifies to $f'(c) = -3$. Next, we need to find $f'(c)$ from the equation. The derivative of $f(x)$ is $f'(x) = -\dfrac{6}{x^2}$, so $f'(c) = -\dfrac{6}{c^2}$. Now, we can solve for c: $-\dfrac{6}{c^2} = -3$ and $c = \pm\sqrt{2}$. Note that $c = \sqrt{2}$ is in the interval (1, 2), but $-\sqrt{2}$ is *not* in the interval. Thus, the answer is only $c = \sqrt{2}$. It's *very* important to check that the answers you get for c fall in the given interval when doing Mean Value Theorem problems.

5.   *No Solution.*

The Mean Value Theorem says that: If $f(x)$ is continuous on the interval $[a, b]$ and is differentiable everywhere on the interval (a, b), then there exists at least one number c on the interval (a, b) such that $f'(c) = \dfrac{f(b) - f(a)}{b - a}$. Here the function is $f(x) = \dfrac{6}{x} - 3$ and the interval is [−1, 2]. Note that the function is *not* continuous on the interval. It has an essential discontinuity (vertical asymptote) at $x = 0$. Thus, the Mean Value Theorem does not apply on the interval, and there is no solution.

Suppose that we were to apply the theorem anyway. We would get $f'(c) = \dfrac{\left(\dfrac{6}{2} - 3\right) - \left(\dfrac{6}{-1} - 3\right)}{(2 + 1)}$.

This simplifies to $f'(c) = 3$. Next, we need to find $f'(c)$ from the equation. The derivative of $f(x)$ is $f'(x) = -\dfrac{6}{x^2}$, so $f'(c) = -\dfrac{6}{c^2}$. Now, we can solve for c: $-\dfrac{6}{c^2} = 3$. This has no real solution.

Therefore, remember that it's *very* important to check that the function is continuous and differentiable everywhere on the given interval (it does not have to be differentiable at the endpoints) when doing Mean Value Theorem problems. If it is not, then the theorem does not apply.

6. $c = 4$

Rolle's theorem says that if $f(x)$ is continuous on the interval $[a, b]$ and is differentiable everywhere on the interval (a, b), and if $f(a) = f(b) = 0$, then there exists at least one number c on the interval (a, b) such that $f'(c) = 0$. Here the function is $f(x) = x^2 - 8x + 12$ and the interval is $[2, 6]$. First, we check if the function is equal to zero at both of the endpoints: $f(6) = (6)^2 - 8(6) + 12 = 0$ and $f(2) = (2)^2 - 8(2) + 12 = 0$. Next, we take the derivative to find $f'(c)$: $f'(x) = 2x - 8$, so $f'(c) = 2c - 8$. Now, we can solve for c: $2c - 8 = 0$ and $c = 4$. Note that 4 is in the interval $(2, 6)$, just as we expected.

7. $c = \pm\dfrac{1}{\sqrt{3}}$

Rolle's theorem says that if $f(x)$ is continuous on the interval $[a, b]$ and is differentiable everywhere on the interval (a, b), and if $f(a) = f(b) = 0$, then there exists at least one number c on the interval (a, b) such that $f'(c) = 0$. Here the function is $f(x) = x^3 - x$ and the interval is $[-1, 1]$. First, we check if the function is equal to zero at both of the endpoints: $f(-1) = (-1)^3 - (-1) = 0$ and $f(1) = (1)^3 - (1) = 0$. Next, we take the derivative to find $f'(c)$: $f'(x) = 3x^2 - 1$, so $f'(c) = 3c^2 - 1$. Now, we can solve for c: $3c^2 - 1 = 0$ and $c = \pm\dfrac{1}{\sqrt{3}}$. Note that $\pm\dfrac{1}{\sqrt{3}}$ are both in the interval $(-1, 1)$, just as we expected.

8. $c = \dfrac{1}{2}$

Rolle's theorem says that if $f(x)$ is continuous on the interval $[a, b]$ and is differentiable everywhere on the interval (a, b), and if $f(a) = f(b) = 0$, then there exists at least one number c on the interval (a, b) such that $f'(c) = 0$. Here the function is $f(x) = x(1 - x)$ and the interval is $[0, 1]$. First, we check if the function is equal to zero at both of the endpoints: $f(0) = (0)(1 - 0) = 0$ and $f(1) = (1)(1 - 1) = 0$. Next, we take the derivative to find $f'(c)$: $f'(x) = 1 - 2x$, so $f'(c) = 1 - 2c$. Now, we can solve for c: $1 - 2c = 0$ and $c = \dfrac{1}{2}$. Note that $\dfrac{1}{2}$ is in the interval $(0, 1)$, just as we expected.

9.   *No Solution.*

Rolle's theorem says that if $f(x)$ is continuous on the interval $[a, b]$ and is differentiable everywhere on the interval (a, b), and if $f(a) = f(b) = 0$, then there exists at least one number c on the interval (a, b) such that $f'(c) = 0$. Here the function is $f(x) = 1 - \dfrac{1}{x^2}$, and the interval is $[-1, 1]$. Note that the function is *not* continuous on the interval. It has an essential discontinuity (vertical asymptote) at $x = 0$. Thus, Rolle's theorem does not apply on the interval, and there is no solution.

Suppose we were to apply the theorem anyway. First, we check if the function is equal to zero at both of the endpoints: $f(1) = 1 - \dfrac{1}{(1)^2} = 0$ and $f(-1) = 1 - \dfrac{1}{(-1)^2} = 0$. Next, we take the derivative to find $f'(c)$: $f'(x) = \dfrac{2}{x^3}$, so $f'(c) = \dfrac{2}{c^3}$. This has no solution.

Therefore, remember that it's *very* important to check that the function is continuous and differentiable everywhere on the given interval (it does not have to be differentiable at the endpoints) when doing Rolle's theorem problems. If it is not, then the theorem does not apply.

10. $c = \dfrac{1}{8}$

Rolle's theorem says that if $f(x)$ is continuous on the interval $[a, b]$ and is differentiable everywhere on the interval (a, b), and if $f(a) = f(b) = 0$, then there exists at least one number c on the interval (a, b) such that $f'(c) = 0$. Here the function is $f(x) = x^{\frac{2}{3}} - x^{\frac{1}{3}}$ and the interval is $[0, 1]$. First, we check if the function is equal to zero at both of the endpoints: $f(0) = (0)^{\frac{2}{3}} - (0)^{\frac{1}{3}} = 0$ and $f(1) = (1)^{\frac{2}{3}} - (1)^{\frac{1}{3}} = 0$. Next, we

take the derivative to find $f'(c)$ $f(x) = \frac{2}{3}x^{-\frac{1}{3}} - \frac{1}{3}x^{-\frac{2}{3}} = \frac{2}{3\sqrt[3]{x}} - \frac{1}{3\sqrt[3]{x^2}}$, so $f(c) = \frac{2}{3\sqrt[3]{c}} - \frac{1}{3\sqrt[3]{c^2}}$. Now,

we can solve for c: $\frac{2}{3\sqrt[3]{c}} - \frac{1}{3\sqrt[3]{c^2}} = 0$ and $c = \frac{1}{8}$. Note that $\frac{1}{8}$ is in the interval (0, 1), just as we

expected.

SOLUTIONS TO PRACTICE PROBLEM SET 10

1.   The area is 32.

First, let's draw a picture.

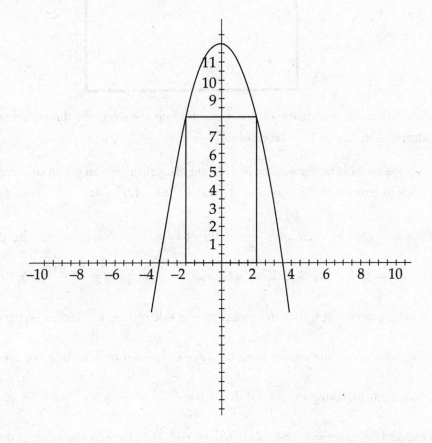

The rectangle can be expressed as a function of x, where the height is $12 - x^2$ and the base is $2x$.

The area is: $A = 2x(12 - x^2) = 24x - 2x^3$. Now, we take the derivative: $\frac{dA}{dx} = 24 - 6x^2$. Next, we set the derivative equal to zero: $24 - 6x^2 = 0$. If we solve this for x, we get $x = \pm 2$. A negative answer doesn't make any sense in the case, so we use the solution $x = 2$. We can then find the area by plugging in $x = 2$ to get $A = 24(2) - 2(2)^3 = 32$. We can verify that this is a maximum by taking the second derivative: $\dfrac{d^2A}{dx^2} = -12x$. Next, we plug in $x = 2$: $\dfrac{d^2A}{dx^2} = -12(2) = -24$. Because the value of the second derivative is negative, according to the second derivative test (see page 123), the area is a maximum at $x = 2$.

2. $\quad x = \dfrac{7 - \sqrt{13}}{2} \approx 1.697$ inches

First, let's draw a picture.

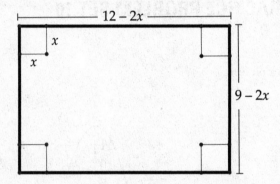

After we cut out the squares of side x and fold up the sides, the dimensions of the box will be: width: $9 - 2x$; length: $12 - 2x$; depth: x.

Using the formula for the volume of a rectangular prism, we can get an equation for the volume of the box in terms of x: $V = x(9 - 2x)(12 - 2x) = 108x - 42x^2 + 4x^3$.

Now, we take the derivative: $\dfrac{dV}{dx} = 108 - 84x + 12x^2$. Next, we set the derivative equal to zero: $108 - 84x + 12x^2 = 0$. If we solve this for x, we get $x = \dfrac{7 \pm \sqrt{13}}{2} \approx 5.303, 1.697$. We can't cut two squares of length 5.303 inches from a side of length 9 inches, so we can get rid of that answer. Therefore, the answer must be $x = \dfrac{7 - \sqrt{13}}{2} \approx 1.697$ inches. We can verify that this is a maximum by taking the second derivative: $\dfrac{d^2V}{dx^2} = -84 + 24x$. Next, we plug in $x = 1.697$ to get approximately $\dfrac{d^2V}{dx^2} = -84 + 24(1.697) = -43.272$. Because the value of the second derivative is negative, according to the second derivative test (see page 123), the volume is a maximum at $x = \dfrac{7 - \sqrt{13}}{2} \approx 1.697$ inches.

3. 16 meters by 24 meters

First, let's draw a picture.

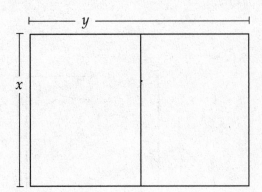

If we call the length of the plot y and the width x, the area of the plot is $A = xy = 384$. The perimeter is $P = 3x + 2y$. So, if we want to minimize the length of the fence, we need to minimize the perimeter of the plot. If we solve the area equation for y, we get $y = \dfrac{384}{x}$. Now we can substitute this for y in the perimeter equation: $P = 3x + 2\left(\dfrac{384}{x}\right) = 3x + \dfrac{768}{x}$. Now we take the derivative of P: $\dfrac{dP}{dx} = 3 - \dfrac{768}{x^2}$. If we solve this for x, we get $x = \pm 16$. A negative answer doesn't make any sense in the case, so we use the solution $x = 16$ meters. Now we can solve for y: $y = \dfrac{384}{16} = 24$ meters.

We can verify that this is a minimum by taking the second derivative: $\dfrac{d^2P}{dx^2} = \dfrac{1536}{x^3}$. Next, we plug in $x = 16$ to get $\dfrac{d^2P}{dx^2} = \dfrac{1536}{16^3} = \dfrac{3}{8}$. Because the value of the second derivative is positive, according to the second derivative test (see page 123), the perimeter is a minimum at $x = 16$.

4.   Radius is $\sqrt[3]{\dfrac{256}{\pi}}$ inches.

First, let's draw a picture.

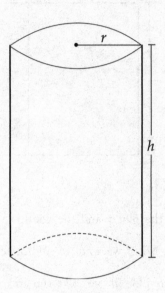

The volume of a cylinder is $V = \pi r^2 h = 512$. The material for the can is the surface area of the cylinder (don't forget the ends!) $S = 2\pi rh + 2\pi r^2$. If we solve the volume equation for h, we get $h = \dfrac{512}{\pi r^2}$.

Now we can substitute this for h in the surface area equation: $S = 2\pi r\left(\dfrac{512}{\pi r^2}\right) + 2\pi r^2 = \dfrac{1024}{r} + 2\pi r^2$.

Now we take the derivative of S: $\dfrac{dS}{dr} = -\dfrac{1024}{r^2} + 4\pi r$. If we solve this for r, we get $r = \sqrt[3]{\dfrac{256}{\pi}}$

inches. We can verify that this is a minimum by taking the second derivative: $\dfrac{d^2S}{dx^2} = \dfrac{2048}{r^3} + 4\pi$.

Next, we plug in $\sqrt[3]{\dfrac{256}{\pi}}$ and we can see that the value of the second derivative is positive. Therefore,

according to the second derivative test (see page 123), the perimeter is a minimum at $r = \sqrt[3]{\dfrac{256}{\pi}}$.

5.   1,352.786 meters

Let's think about the situation. If the swimmer swims the whole distance to the cottage, she will be traveling the entire time at her slowest speed. If she swims straight to shore first, minimizing her swimming distance, she will be maximizing her running distance. Therefore, there should be a point, somewhere between the cottage and the point on the shore directly opposite her, where the swimmer should come on land to switch from swimming to running to get to the cottage in the shortest time.

Let's draw a picture.

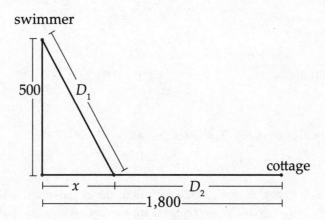

Let x be the distance from the point on the shore directly opposite the swimmer to the point where she comes on land. We have two distances to consider. The first is the diagonal distance that the swimmer swims. This distance, which we'll call D_1 is $D_1 = \sqrt{500^2 + x^2} = \sqrt{250,000 + x^2}$.

The second distance, which we'll call D_2, is simply $D_2 = 1,800 - x$. Remember that *rate × time = distance*? We'll use this formula to find the total time that the swimmer needs.

The time for the swimmer to travel D_1 is $T_1 = \dfrac{D_1}{4} = \dfrac{\sqrt{250,000 + x^2}}{4}$ (because she swims at 4 m/s) and the time for the swimmer to travel D_2 is $T_2 = \dfrac{D_2}{6} = \dfrac{1,800 - x}{6} = 300 - \dfrac{x}{6}$.

Therefore, the total time is $T = \dfrac{\sqrt{250,000 + x^2}}{4} + 300 - \dfrac{x}{6}$. Now we simply take the derivative:

$\dfrac{dT}{dx} = \dfrac{1}{4}\left(\dfrac{1}{2}\right)\left(250,000 + x^2\right)^{-\frac{1}{2}}(2x) - \dfrac{1}{6} = \dfrac{x}{4\sqrt{250,000 + x^2}} - \dfrac{1}{6}$. Next, we set this equal to zero and

solve: $\dfrac{x}{4\sqrt{250,000 + x^2}} - \dfrac{1}{6} = 0$. The best way to solve this is to move the $\dfrac{1}{6}$ to the other side of

the equals sign and cross-multiply:

$$\dfrac{x}{4\sqrt{250,000 + x^2}} = \dfrac{1}{6}$$

$$6x = 4\sqrt{250,000 + x^2}$$

Next, we square both sides: $36x^2 = 16(250,000 + x^2)$.

Simplify: $20x^2 = 4,000,000$.

Solve for x: $x = \pm 447.214$ m. (We can ignore the negative answer.) Therefore, she should land $1,800 - 447.214 = 1,352.786$ meters from the cottage. We could verify that this is a minimum by taking the second derivative but that would be messy. It is simpler to use the calculator to check the sign of the derivative at a point on either side of the answer, or to graph the equation for the time.

6. $\left(\dfrac{2}{\sqrt{5}}, \dfrac{1}{\sqrt{5}}\right)$

First, let's draw a picture.

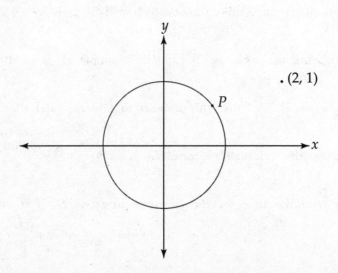

We need to find an expression for the distance from the point P to the point $(2, 1)$ and then minimize the distance. If we call the coordinates of P (x, y) then we can find the distance to $(2, 1)$ using the distance formula: $D^2 = (x - 2)^2 + (y - 1)^2$. Next, just as we did in sample problem 4 (page 132), we can let $L = D^2$ and minimize L: $L = (x - 2)^2 + (y - 1)^2 = x^2 - 4x + 4 + y^2 - 2y + 1$.

Because $x^2 + y^2 = 1$, we can substitute for y to get: $L = x^2 - 4x + 4 + \left(1 - x^2\right) - 2\sqrt{1 - x^2} + 1$, which simplifies to: $L = -4x + 6 - 2\sqrt{1 - x^2}$. Next, we take the derivative: $\dfrac{dL}{dx} = -4 - 2\left(\dfrac{1}{2}\right)\left(1 - x^2\right)^{-\frac{1}{2}}\left(-2x\right) = -4 + \dfrac{2x}{\sqrt{1 - x^2}}$. Next, we set the derivative equal to zero: $-4 + \dfrac{2x}{\sqrt{1 - x^2}} = 0$. The best way to solve this is to move the 4 to the other side of the equals sign and cross-multiply.

$$\frac{2x}{\sqrt{1 - x^2}} = 4$$

$$2x = 4\sqrt{1 - x^2}$$

Next, we can simplify and square both sides: $x^2 = 4(1 - x^2)$. Now, we can solve this easily. We get $x = \pm\dfrac{2}{\sqrt{5}}$. Next, we find the y-coordinate: $y = \pm\dfrac{1}{\sqrt{5}}$. There are thus four possible answers but, if we look at the picture, the answer is obviously the point $\left(\dfrac{2}{\sqrt{5}}, \dfrac{1}{\sqrt{5}}\right)$. We could verify that this is a minimum by taking the second derivative, but that will be messy. It is simpler to use the calculator to check the sign of the derivative at a point on either side of the answer, or to graph the equation for the distance.

7. $r = \dfrac{288}{4 + \pi}$ inches

First, let's draw a picture.

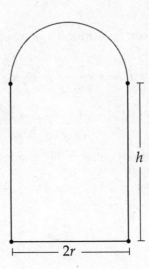

Call the width of the window $2r$. Notice that this is the diameter of the semicircle. Call the height of the window h. The area of the rectangular portion of the window is $2rh$ and the perimeter is $2r + 2h$. The area of the semicircular portion of the window is $\dfrac{\pi r^2}{2}$ and the perimeter is $\dfrac{2\pi r}{2} = \pi r$. Therefore, the area of the window is $A = 2rh + \dfrac{\pi r^2}{2}$, and the perimeter is $2r + 2h + \pi r = 288$. We can use the equation for the perimeter to eliminate a variable from the equation for the area. Let's isolate h: $h = 144 - r - \dfrac{\pi r}{2}$. Now we can substitute for h in the equation for the area: $A = 2r\left(144 - r - \dfrac{\pi r}{2}\right) + \dfrac{\pi r^2}{2} = 288r - 2r^2 - \dfrac{\pi r^2}{2}$. Next, we can take the derivative: $\dfrac{dA}{dr} = 288 - 4r - \pi r$. If we set this equal to zero and solve for r, we get $r = \dfrac{288}{4 + \pi}$ inches. We can verify that this is a maximum by taking the second derivative: $\dfrac{d^2 A}{dr^2} = -4 - \pi$. We can see that the value of the second derivative is negative. Therefore, according to the second derivative test (see page 123), the area is a maximum at $r = \dfrac{288}{4 + \pi}$ inches.

8. $\theta = \dfrac{\pi}{4}$ radians (or 45 degrees)

We simply take the derivative: $\dfrac{dR}{d\theta} = \dfrac{{v_0}^2}{g}\left(2\cos 2\theta\right)$. Note that v_0 and g are constants, so the only variable we need to take the derivative with respect to is θ. Now we set the derivative equal to zero: $\dfrac{{v_0}^2}{g}\left(2\cos 2\theta\right) = 0$. Although this has an infinite number of solutions, we are interested only in values of θ between 0 and $\dfrac{\pi}{2}$ radians (why?). The value of θ that makes the derivative zero is $\theta = \dfrac{\pi}{4}$ radians (or 45 degrees). We can verify that this is a maximum by taking the second derivative: $\dfrac{d^2 R}{d\theta^2} = \dfrac{{v_0}^2}{g}\left(-4\sin 2\theta\right)$. We can see that the value of the second derivative is negative at $\theta = \dfrac{\pi}{4}$. Therefore, according to the second derivative test (see page 123), the range is a maximum at $\theta = \dfrac{\pi}{4}$ radians.

9. $15 billion

We simply take the derivative: $\dfrac{dP}{dx} = 3x^2 - 96x + 720$. Now we set the derivative equal to zero: $3x^2 - 96x + 720 = 0$. The solutions to this are $x = 12$ and $x = 20$. Note, however, that the function is bounded by $x = 0$ and $x = 40$.

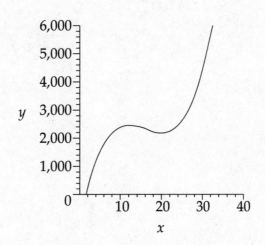

Thus, in order to find the maximum profit, we need to plug all four values of x into the profit equation to find which gives the greatest value. We get

$P(0) = -1,000$

$P(12) = (12)^3 - 48(12)^2 + 720(12) - 1,000 = 2,456$

$P(20) = (20)^3 - 48(20)^2 + 720(20) - 1,000 = 2,200$

$P(40) = (40)^3 - 48(40)^2 + 720(40) - 1,000 = 15,000$

Thus, even though we have a relative maximum at $x = 12$, the absolute maximum profit occurs when $x = 40$. The profit at that value is 15,000, or \$15 billion dollars.

If we look at a graph of the profit function, we can see that the critical points gave us relative maximum and minimum, but not the absolute maximum or minimum. This is why it is always important to check the endpoints of a function in any max/min problem.

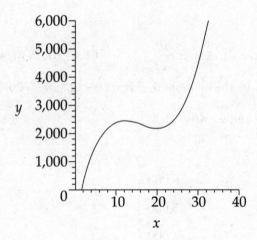

SOLUTIONS TO PRACTICE PROBLEM SET 11

1. Minimum at $\left(\sqrt{3}, -6\sqrt{3} - 6\right)$; Maximum at $\left(-\sqrt{3}, 6\sqrt{3} - 6\right)$; Point of inflection at $(0, 6)$.

First, let's find the y-intercept. We set $x = 0$ to get: $y = (0)^3 - 9(0) - 6 = -6$. Therefore, the y-intercept is $(0, -6)$. Next, we find any critical points using the first derivative. The derivative is $\dfrac{dy}{dx} = 3x^2 - 9$. If we set this equal to zero and solve for x, we get $x = \pm\sqrt{3}$. Plug $x = \sqrt{3}$ and $x = -\sqrt{3}$ into the original equation to find the y-coordinates of the critical points: When $x = \sqrt{3}$, $y = \left(\sqrt{3}\right)^3 - 9\left(\sqrt{3}\right) - 6 = -6\sqrt{3} - 6$. When $x = -\sqrt{3}$, $y = \left(-\sqrt{3}\right)^3 - 9\left(-\sqrt{3}\right) - 6 = 6\sqrt{3} - 6$. Thus, we have critical points at $\left(\sqrt{3}, -6\sqrt{3} - 6\right)$ and $\left(-\sqrt{3}, 6\sqrt{3} - 6\right)$. Next, we take the second derivative to find any points of inflection. The second derivative is: $\dfrac{d^2 y}{dx^2} = 6x$, which is equal to zero at $x = 0$. Note that this is the y-intercept $(0, -6)$, which we already found, so there is a point of inflection at $(0, -6)$. Next, we need to determine if each critical point is maximum, minimum, or something else. If we plug $x = \sqrt{3}$ into the second derivative, the value is obviously positive, so $\left(\sqrt{3}, -6\sqrt{3} - 6\right)$ is a minimum. If we plug $x = -\sqrt{3}$ into the second derivative, the value is obviously negative, so $\left(-\sqrt{3}, 6\sqrt{3} - 6\right)$ is a maximum. Now, we can draw the curve. It looks like the following:

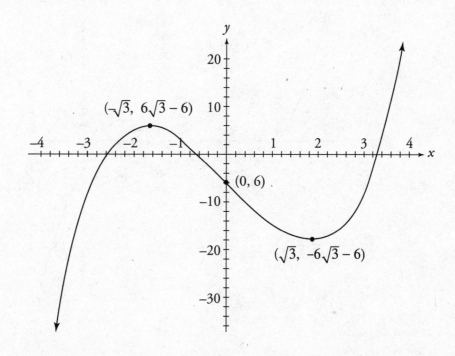

2.   Minimum at (–3, –4); Maximum at (–1, 0); Point of inflection at (–2, –2).

First, let's find the y-intercept. We set $x = 0$ to get: $y = -(0)^3 - 6(0)^2 - 9(0) - 4 = -4$. Therefore, the y-intercept is $(0, -4)$. Next, we find any critical points using the first derivative. The derivative is $\dfrac{dy}{dx} = -3x^2 - 12x - 9$. If we set this equal to zero and solve for x, we get $x = -1$ and $x = -3$. Plug $x = -1$ and $x = -3$ into the original equation to find the y-coordinates of the critical points: When $x = -1$, $y = -(-1)^3 - 6(-1)^2 - 9(-1) - 4 = 0$. When $x = -3$, $y = -(-3)^3 - 6(-3)^2 - 9(-3) - 4 = -4$. Thus, we have critical points at $(-1, 0)$ and $(-3, -4)$. Next, we take the second derivative to find any points of inflection. The second derivative is $\dfrac{d^2 y}{dx^2} = -6x - 12$, which is equal to zero at $x = -2$. Plug $x = -2$ into the original equation to find the y-coordinate: $y = -(-2)^3 - 6(-2)^2 - 9(-2) - 4 = -2$, so there is a point of inflection at $(-2, -2)$. Next, we need to determine if each critical point is maximum, minimum, or something else. If we plug $x = -1$ into the second derivative, the value is negative, so $(-1, 0)$ is a maximum. If we plug $x = -3$ into the second derivative, the value is positive, so $(-3, -4)$ is a minimum. Now, we can draw the curve. It looks like the following:

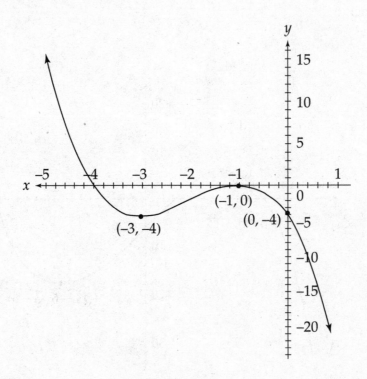

3.   Minimum at (0, −36); Maxima at $\left(\sqrt{\dfrac{13}{2}}, \dfrac{25}{4}\right)$ and $\left(-\sqrt{\dfrac{13}{2}}, \dfrac{25}{4}\right)$; points of inflection at $\left(\sqrt{\dfrac{13}{6}}, -\dfrac{451}{36}\right)$ and $\left(-\sqrt{\dfrac{13}{6}}, -\dfrac{451}{36}\right)$.

First, we can easily see that the graph has x-intercepts at $x = \pm 2$ and $x = \pm 3$. Next, before we take the derivative, let's multiply the two terms so we don't have to use the Product Rule. We get

$y = -x^4 + 13x^2 - 36$. Now we can take the derivative: $\dfrac{dy}{dx} = -4x^3 + 26x$. Next, we set the derivative

equal to zero to find the critical points. There are three solutions: $x = 0$, $x = \sqrt{\dfrac{13}{2}}$, and $x = -\sqrt{\dfrac{13}{2}}$.

We plug these values into the original equation to find the y-coordinates of the critical points: When

$x = 0$, $y = -(0)^4 + 13(0)^2 - 36 = -36$. When $x = \sqrt{\dfrac{13}{2}}$, $y = -\left(\sqrt{\dfrac{13}{2}}\right)^4 + 13\left(\sqrt{\dfrac{13}{2}}\right)^2 - 36 = \dfrac{25}{4}$.

When $x = -\sqrt{\dfrac{13}{2}}$, $y = -\left(-\sqrt{\dfrac{13}{2}}\right)^4 + 13\left(-\sqrt{\dfrac{13}{2}}\right)^2 - 36 = \dfrac{25}{4}$.

Thus, we have critical points at $(0, -36)$, $\left(\sqrt{\dfrac{13}{2}}, \dfrac{25}{4}\right)$ and $\left(-\sqrt{\dfrac{13}{2}}, \dfrac{25}{4}\right)$.

Next, we take the second derivative to find any points of inflection. The second derivative is

$\dfrac{d^2y}{dx^2} = -12x^2 + 26$, which is equal to zero at $x = \sqrt{\dfrac{13}{6}}$ and $x = -\sqrt{\dfrac{13}{6}}$. We plug these values into

the original equation to find the y-coordinates.

When $x = \sqrt{\dfrac{13}{6}}$, then $y = -\left(\sqrt{\dfrac{13}{6}}\right)^4 + 13\left(\sqrt{\dfrac{13}{6}}\right)^2 - 36 = -\dfrac{451}{36}$.

When $x = -\sqrt{\dfrac{13}{6}}$, then $y = -\left(-\sqrt{\dfrac{13}{6}}\right)^4 + 13\left(-\sqrt{\dfrac{13}{6}}\right)^2 - 36 = -\dfrac{451}{36}$. So there are points of inflection at $\left(\sqrt{\dfrac{13}{6}}, -\dfrac{451}{36} \right)$ and $\left(-\sqrt{\dfrac{13}{6}}, -\dfrac{451}{36} \right)$. Next, we need to determine if each critical

point is maximum, minimum, or something else. If we plug $x = 0$ into the second derivative, the

value is positive, so $(0, -36)$ is a minimum. If we plug $x = \sqrt{\dfrac{13}{2}}$ into the second derivative, the

value is negative, so $\left(\sqrt{\dfrac{13}{2}}, \dfrac{25}{4} \right)$ is a maximum. If we plug $x = -\sqrt{\dfrac{13}{2}}$ into the second derivative,

the value is negative, so $\left(-\sqrt{\dfrac{13}{2}}, \dfrac{25}{4} \right)$ is a maximum. Now, we can draw the curve. It looks like

the following:

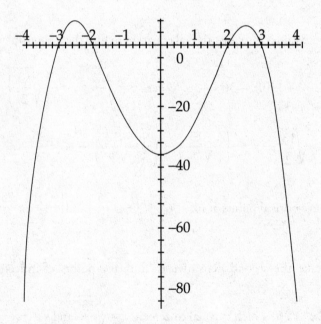

4.   Maximum at $(0, 0)$; Minima at $(2, -4)$ and $(-2, -4)$; Points of inflection at $\left(\dfrac{2}{\sqrt{3}}, -\dfrac{20}{9} \right)$ and
$\left(-\dfrac{2}{\sqrt{3}}, -\dfrac{20}{9} \right)$.

First, we can easily see that the graph has x-intercepts at $x = 0$ and $x = \pm\sqrt{8}$. Next, we take the

derivative: $\dfrac{dy}{dx} = x^3 - 4x$. Next, we set the derivative equal to zero to find the critical points. There

are three solutions: $x = 0$, $x = 2$, and $x = -2$.

We plug these values into the original equation to find the y-coordinates of the criti-

cal points: When $x = 0$, $y = \dfrac{(0)^4}{4} - 2(0)^2 = 0$. When $x = 2$, $y = \dfrac{(2)^4}{4} - 2(2)^2 = -4$. When

$x = -2$, $y = \dfrac{(-2)^4}{4} - 2(-2)^2 = -4$. Thus, we have critical points at $(0, 0)$, $(2, -4)$, and

$(-2, -4)$. Next, we take the second derivative to find any points of inflection. The second deriva-

tive is $\dfrac{d^2 y}{dx^2} = 3x^2 - 4$, which is equal to zero at $x = \dfrac{2}{\sqrt{3}}$ and $x = -\dfrac{2}{\sqrt{3}}$. We plug these values into

the original equation to find the y-coordinates. When $x = \dfrac{2}{\sqrt{3}}$: $y = \dfrac{\left(\dfrac{2}{\sqrt{3}}\right)^4}{4} - 2\left(\dfrac{2}{\sqrt{3}}\right)^2 = -\dfrac{20}{9}$.

When $x = -\dfrac{2}{\sqrt{3}}$, then $y = \dfrac{\left(-\dfrac{2}{\sqrt{3}}\right)^4}{4} - 2\left(-\dfrac{2}{\sqrt{3}}\right)^2 = -\dfrac{20}{9}$. So there are points of inflection at

$\left(\dfrac{2}{\sqrt{3}}, -\dfrac{20}{9}\right)$ and $\left(-\dfrac{2}{\sqrt{3}}, -\dfrac{20}{9}\right)$. Next, we need to determine if each critical point is maximum,

minimum, or something else. If we plug $x = 0$ into the second derivative, the value is negative, so

$(0, 0)$ is a maximum. If we plug $x = 2$ into the second derivative, the value is positive, so $(2, -4)$

is a minimum. If we plug $x = -2$ into the second derivative, the value is negative, so $(-2, -4)$ is a

minimum. Now, we can draw the curve. It looks like the following:

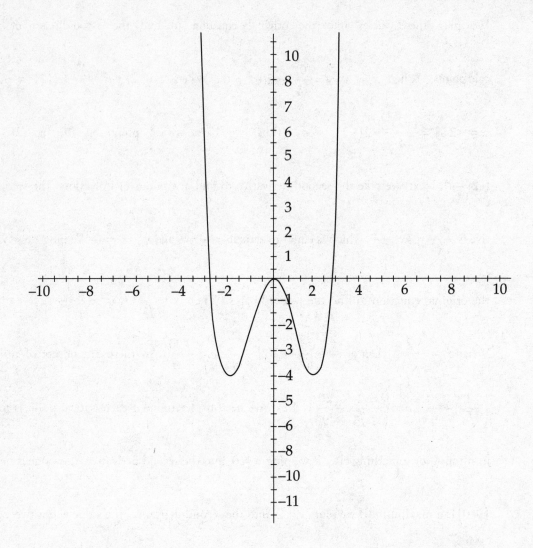

5. Vertical asymptote at $x = -8$; Horizontal asymptote at $y = 1$; No maxima, minima, or points of inflection.

First, notice that there is an x-intercept at $x = 3$ and that the y-intercept is $\left(0, -\dfrac{3}{8}\right)$. Next, we take the derivative: $\dfrac{dy}{dx} = \dfrac{(x+8)(1) - (x-3)(1)}{(x+8)^2} = \dfrac{11}{(x+8)^2}$. Next, we set the derivative equal to zero to find the critical points. There is no solution. Next, we take the second derivative: $\dfrac{d^2y}{dx^2} = -\dfrac{22}{(x+8)^3}$. If we set this equal to zero, there is also no solution. Therefore, there are no maxima, minima, or points of inflection. Note that the first derivative is always positive. This means that the curve is always increasing. Also notice that the second derivative changes sign from positive to negative at $x = -8$. This means that the curve is concave up for values of x less than $x = -8$ and concave down for values of x greater than $x = -8$.

Now, we can draw the curve. It looks like the following:

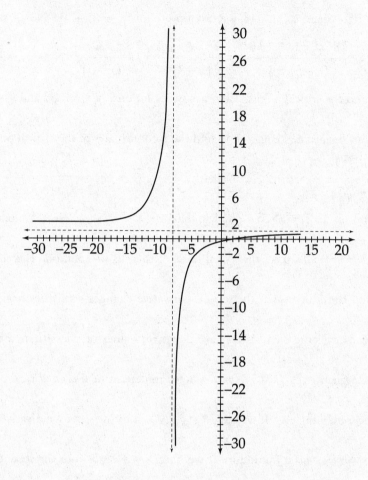

6. Vertical asymptote at $x = 3$; Oblique asymptote of $y = x + 3$; Maximum at $\left(3 + \sqrt{5}, 6 + 2\sqrt{5}\right)$; Minimum at $\left(3 - \sqrt{5}, 6 - 2\sqrt{5}\right)$; No points of inflection.

First, notice that there are x-intercepts at $x = \pm 2$ and that the y-intercept is $\left(0, \dfrac{4}{3}\right)$. There is a vertical asymptote at $x = 3$. There is no horizontal asymptote, but notice that the degree of the numerator of the function is 1 greater than the denominator. This means that there is an oblique (slant) asymptote. We find this by dividing the denominator into the numerator and looking at the quotient. We get

$$x - 3 \overline{\smash{\big)}\ x^2 + 0x - 4} \quad \Rightarrow \quad x + 3 + \dfrac{5}{x - 3}$$

This means that as $x \to \pm\infty$, the function will behave like the function $y = x + 3$. This means that there is an oblique asymptote of $y = x + 3$. Next, we take the derivative:

$\dfrac{dy}{dx} = \dfrac{(x-3)(2x) - (x^2-4)(1)}{(x-3)^2} = \dfrac{x^2-6x+4}{(x-3)^2} = 1 - \dfrac{5}{(x-3)^2}$. Next, we set the derivative equal to

zero to find the critical points. There are two solutions: $x = 3 + \sqrt{5}$ and $x = 3 - \sqrt{5}$. We plug these

values into the original equation to find the y-coordinates of the critical points: When $x = 3 + \sqrt{5}$,

$y = \dfrac{\left(3+\sqrt{5}\right)^2 - 4}{\left(3+\sqrt{5}\right) - 3} = 6 + 2\sqrt{5}$. When $x = 3 - \sqrt{5}$, $y = \dfrac{\left(3-\sqrt{5}\right)^2 - 4}{\left(3-\sqrt{5}\right) - 3} = 6 - 2\sqrt{5}$. Thus, we have

critical points at $\left(3+\sqrt{5}, 6+2\sqrt{5}\right)$ and $\left(3-\sqrt{5}, 6-2\sqrt{5}\right)$. Next, we take the second derivative:

$\dfrac{d^2 y}{dx^2} = \dfrac{10}{(x-3)^3}$. If we set this equal to zero, there is no solution. Therefore, there is no point of

inflection. But, notice that the second derivative changes sign from negative to positive at $x = 3$.

This means that the curve is concave down for values of x less than $x = 3$ and concave up for

values of x greater than $x = 3$. Next, we need to determine if each critical point is maximum, mini-

mum, or something else. If we plug $x = 3 - \sqrt{5}$ into the second derivative, the value is positive, so

$\left(3+\sqrt{5}, 6+2\sqrt{5}\right)$ is a minimum. If we plug $x = 3 - \sqrt{5}$ into the second derivative, the value is

positive, so $\left(3-\sqrt{5}, 6-2\sqrt{5}\right)$ is a maximum.

Now, we can draw the curve. It looks like the following:

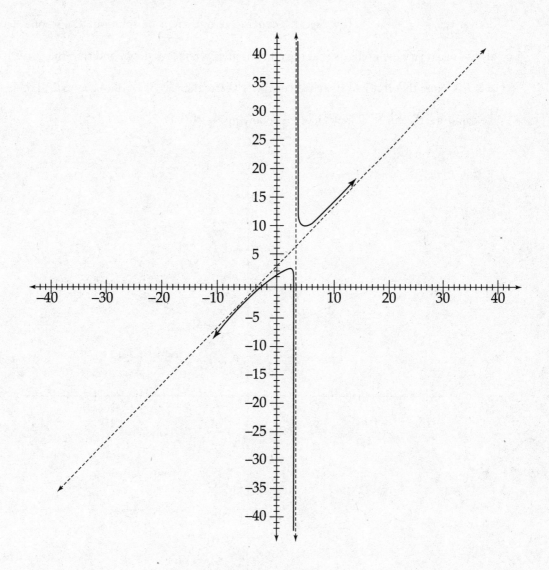

7.   No maxima, minima, or points of inflection; Cusp at (0, 3).

First, notice that the function is always positive. There is a y-intercept at (0, 3). There is no x-intercept. Next, we take the derivative: $\dfrac{dy}{dx} = \dfrac{2}{3}x^{-\frac{1}{3}}$. If we set the derivative equal to zero, there is no solution. But, notice that the derivative is not defined at $x = 0$. This means that the function has either a vertical tangent or a cusp at $x = 0$. We'll be able to determine which after we take the second derivative. Notice also that the derivative is negative for $x < 0$ and positive for $x > 0$.

Therefore, the curve is decreasing for $x < 0$ and increasing for $x > 0$. Next, we take the second derivative: $\dfrac{dy}{dx} = -\dfrac{2}{9}x^{-\frac{4}{3}}$. If we set this equal to zero, there is no solution. The second derivative is always negative which tells us that the curve is always concave down and that the curve has a cusp at $x = 0$. Note that if it had switched concavity there, then $x = 0$ would be a vertical tangent. Now, we can draw the curve. It looks like the following:

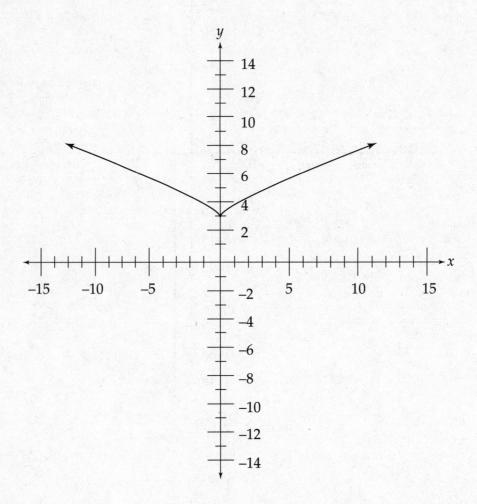

8.   Maximum at $(1, 1)$; No point of inflection; Cusp at $(0, 0)$.

First, we notice that the curve has x-intercepts at $x = 0$ and $x = \dfrac{27}{8}$, and that there is a y-intercept at $(0, 0)$. Before we take the derivative, let's expand the expression. That way, we won't have to use the Product Rule. We get $y = 3x^{\frac{2}{3}} - 2x$. Next, we take the derivative: $\dfrac{dy}{dx} = 2x^{-\frac{1}{3}} - 2$. Next, we set the derivative equal to zero to find the critical points. There is one solution: $x = 1$. We plug this value into the original equation to find the y-coordinate of the critical point: When $x = 1$, $y = 3(1)^{\frac{2}{3}} - 2(1) = 1$. Thus, we have a critical point at $(1, 1)$. But, notice that the derivative is not defined at $x = 0$. This means that the function has either a vertical tangent or a cusp at $x = 0$. We'll be able to determine which after we take the second derivative. Notice also that the derivative is negative for $x < 0$ and for $x > 1$ and positive for $0 < x < 1$. Therefore, the curve is decreasing for $x < 0$ and for $x > 1$, and increasing for $0 < x < 1$. Next, we take the second derivative: $\dfrac{d^2 y}{dx^2} = -\dfrac{2}{3} x^{-\frac{4}{3}}$. If we set this equal to zero, there is no solution. Therefore, there is no point of inflection. The second derivative is always negative which tells us that the curve is always concave down and that the curve has a cusp at $x = 0$. Note that if it had switched concavity there, then $x = 0$ would be a vertical tangent. Next, we need to determine if each critical point is maximum, minimum, or something else. If we plug $x = 1$ into the second derivative, the value is negative, so $(1, 1)$ is a maximum. Now, we can draw the curve. It looks like the following:

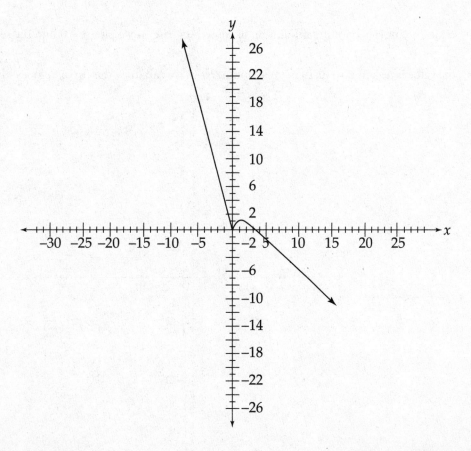

9.   Maximum at (0, 0); Vertical asymptotes at $x = 2$ and $x = -2$; Horizontal asymptote at $y = 3$; No point of inflection.

First, notice that the curve goes through the origin. There are vertical asymptotes at $x = 2$ and $x = -2$. There is a horizontal asymptote at $y = 3$. Next, we take the derivative: $\dfrac{dy}{dx} = \dfrac{(x^2 - 4)(6x) - (3x^2)(2x)}{(x^2 - 4)^2} = \dfrac{-24x}{(x^2 - 4)^2}$. Next, we set the derivative equal to zero to find any critical points. There is a solution at $x = 0$, which means that there is a critical point at (0, 0). Note also that the function is positive for $x < 0$ and negative for $x > 0$, so the curve is increasing for $x < 0$ and decreasing for $x > 0$. Next, we take the second derivative: $\dfrac{d^2 y}{dx^2} = \dfrac{\left(x^2 - 4\right)^2 (-24) - (-24x)\left(4x^3 - 16x\right)}{\left(x^2 - 4\right)^4} = \dfrac{72x^2 + 96}{\left(x^2 - 4\right)^3}$. If we set this equal to zero, we get no solution, which means that there is no point of inflection at the origin. Notice that the second derivative is positive for $x < -2$, negative for $-2 < x < 2$, and positive for $x > 2$. Therefore, the curve is concave up for $x < -2$ and $x > 2$, and concave down for $-2 < x < 2$. Next, we need to determine if each critical point is maximum, minimum, or something else. If we plug $x = 0$ into the second derivative, the value is negative, so (0, 0) is a maximum. Now, we can draw the curve. It looks like the following:

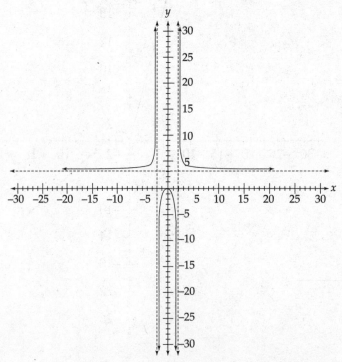

SOLUTIONS TO PRACTICE PROBLEM SET 12

1.   2,000 ft²/s

We are given the rate at which the circumference is increasing, $\dfrac{dC}{dt} = 40$ and are looking for the

rate at which the area is increasing, $\dfrac{dA}{dt}$. Thus, we need to find a way to relate the area of a circle

to its circumference. Recall that the circumference of a circle is $C = 2\pi r$, and the area is $A = \pi r^2$.

We could find C in terms of r and then plug it into the equation for A, or we could work with the

equations separately and then relate them. Let's do both and compare.

Method 1: First, we find C in terms of r: $r = \dfrac{C}{2\pi}$. Now, we plug this in for r in the equa-

tion for A: $A = \pi \left(\dfrac{C}{2\pi}\right)^2 = \dfrac{C^2}{4\pi}$. Next, we take the derivative of the equation with respect

to t: $\dfrac{dA}{dt} = \dfrac{1}{4\pi}(2C)\dfrac{dC}{dt} = \left(\dfrac{C}{2\pi}\right)\dfrac{dC}{dt}$. Next, we plug in $C = 100\pi$ and $\dfrac{dC}{dt} = 40$, and solve:

$\dfrac{dA}{dt} = \left(\dfrac{100\pi}{2\pi}\right)(40) = 2,000$. Therefore, the answer is 2,000 ft²/s.

Method 2: First, we take the derivative of C with respect to t: $\dfrac{dC}{dt} = 2\pi\left(\dfrac{dr}{dt}\right)$. Next, we plug in

$\dfrac{dC}{dt} = 40$ and solve for $\dfrac{dr}{dt}$: $40 = 2\pi\left(\dfrac{dr}{dt}\right)$, so $\dfrac{dr}{dt} = \dfrac{20}{\pi}$. Next, we take the derivative of A with

respect to t: $\dfrac{dA}{dt} = 2\pi r\dfrac{dr}{dt}$. Now, we can plug in for $\dfrac{dr}{dt}$ and r and solve for $\dfrac{dA}{dt}$. Note that when

the circumference is 100π, $r = 50$: $\dfrac{dA}{dt} = 2\pi(50)\left(\dfrac{20}{\pi}\right) = 2,000$. Therefore, the answer is 2,000 ft²/s.

Which method is better? In this case they are about the same. Method 1 is going to be more

efficient if it is easy to solve for one variable in terms of the other, and it is also easy to take the

derivative of the resulting expression. Otherwise, we will prefer to use Method 2 (See Example 3

on page 156).

2. $\dfrac{3}{4}$ in/s

We are given the rate at which the volume is increasing, $\dfrac{dV}{dt} = 27\pi$ and are looking for the rate at

which the radius is increasing, $\dfrac{dr}{dt}$. Thus, we need to find a way to relate the volume of a sphere to its radius. Recall that the volume of a sphere is $V = \frac{4}{3}\pi r^3$. All we have to do is take the derivative

of the equation with respect to t: $\frac{dV}{dt} = \frac{4}{3}\pi\left(3r^2\right)\left(\frac{dr}{dt}\right) = 4\pi r^2 \frac{dr}{dt}$. Now we substitute $\frac{dV}{dt} = 27\pi$

and $r = 3$: $27\pi = 4\pi(3)^2\frac{dr}{dt}$. If we solve for $\frac{dr}{dt}$, we get $\frac{dr}{dt} = \frac{3}{4}$ in/s.

3.   100 km/hr

We are given the rate at which Car A is moving south, $\frac{dA}{dt} = 80$, and the rate at which Car B

is moving west, $\frac{dB}{dt} = 60$, and are looking for the rate at which the distance between them is

increasing, which we'll call $\frac{dC}{dt}$. Note that the directions south and west are at right angles to each

other. Thus, the distance that Car A is from the starting point, which we'll call A, and the distance

that Car B is from the starting point, which we'll call B, are the legs of a right triangle, with C as

the hypotenuse. We can relate the three distances using the Pythagorean theorem. Here, because

A and B are the legs and C is the hypotenuse, $A^2 + B^2 = C^2$. Now, we take the derivative of the

equation with respect to t: $2A\frac{dA}{dt} + 2B\frac{dB}{dt} = 2C\frac{dC}{dt}$. This simplifies to $A\frac{dA}{dt} + B\frac{dB}{dt} = C\frac{dC}{dt}$.

We know that Car A has been driving for 3 hours at 80 km/hr and Car B has been driving for 3

hours at 60 km/hr, so $A = 240$ and $B = 180$. Using the Pythagorean theorem, $240^2 + 180^2 = C^2$, so

$C = 300$. Now we can substitute into our derivative equation: $(240)(80) + (180)(60) = (300)\frac{dC}{dt}$. If

we solve for C, we get $C = 100$ km/hr.

4.   3 m/s

We are given the rate at which the volume of the fluid is increasing, $\frac{dV}{dt} = 108\pi$ and are look-

ing for the rate at which the height of the fluid is increasing, $\frac{dh}{dt}$. Thus, we need to find a way to

relate the volume of a cylinder to its height. We know that the volume of a cylinder is $V = \pi r^2 h$. We

will want to take the derivative of the equation with respect to t. Note that the radius of the tank

doesn't change as the volume changes, therefore r is a constant in this problem, not a variable. We

get $\frac{dV}{dt} = \pi r^2 \frac{dh}{dt}$. Because r is not changing, $\frac{dr}{dt} = 0$, and we would have ended up with the same

derivative.

Now we can plug in $\dfrac{dV}{dt} = 108\pi$ and $r = 6$, and solve for $\dfrac{dh}{dt}$. We get $108\pi = \pi (6)^2 \dfrac{dh}{dt}$. Thus, $\dfrac{dh}{dt} = 3\,\text{m/s}$.

5. $243\sqrt{3}\ \text{in}^2/\text{s}$

We are given the rate at which the sides of the triangle are increasing, $\dfrac{ds}{dt} = 27$ and are looking for the rate at which the area is increasing, $\dfrac{dA}{dt}$. Thus, we need to find a way to relate the area of an equilateral triangle to the length of a side. We know that the area of an equilateral triangle, in terms of its sides, is $A = \dfrac{s^2 \sqrt{3}}{4}$. (If you don't know this formula, memorize it! It will come in very handy in future math problems.) Now we take the derivative of this equation with respect to t: $\dfrac{dA}{dt} = \dfrac{\sqrt{3}}{4}(2s)\left(\dfrac{ds}{dt}\right)$. Next, we plug $\dfrac{ds}{dt} = 27$ and $s = 18$ into the derivative and we get $\dfrac{dA}{dt} = \dfrac{\sqrt{3}}{4}(36)(27) = 243\sqrt{3}\ \text{in}^2/\text{s}$.

6. $-\dfrac{5}{7}\ \text{in/s}$

We are given the rate at which the water is flowing out of the container, $\dfrac{dV}{dt} = -35\pi$ (Why is it negative?), and are looking for the rate at which the depth of the water is dropping, $\dfrac{dh}{dt}$. Thus, we need to find a way to relate the volume of a cone to its height. We know that the volume of a cone is $V = \dfrac{1}{3}\pi r^2 h$. Notice that we have a problem. We have a third variable, r, in the equation. We cannot treat it as a constant the way we did in problem 4 because as the volume of a cone changes, both its height and radius change. But, we also know that in any cone, the ratio of the radius to the height is a constant. Here, when the radius is 21 (because the diameter is 42), the height is 15. Thus, $\dfrac{r}{h} = \dfrac{21}{15} = \dfrac{7}{5}$. We can now isolate r in this equation: $r = \dfrac{7h}{5}$. Now we can plug it into the volume formula to get rid of r: $V = \dfrac{1}{3}\pi \left(\dfrac{7h}{5}\right)^2 h$. This simplifies to $V = \dfrac{49\pi}{75}h^3$. Now we can take the derivative of this equation with respect to t: $\dfrac{dV}{dt} = \dfrac{49\pi}{75}\left(3h^2\right)\dfrac{dh}{dt}$. Next, we plug $\dfrac{dV}{dt} = -35\pi$

and $h = 5$ into the derivative and we get $-35\pi = \dfrac{49\pi}{75}\left(3(5)^2\right)\dfrac{dh}{dt}$. Now we can solve for $\dfrac{dh}{dt}$:

$\dfrac{dh}{dt} = -\dfrac{5}{7}$ in/s .

7. $-\dfrac{25}{6}$ ft/s

We are given the rate at which the length of the rope, R, is changing, $\dfrac{dR}{dt} = -4$ and are looking for

the rate at which the boat, B, is approaching the dock, $\dfrac{dB}{dt}$. The key to this problem is to realize

that the vertical distance from the dock to the bow, the distance from the boat to the dock, and the

length of the rope form a right triangle.

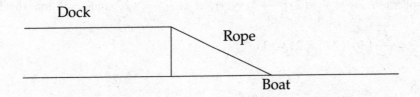

The vertical distance from the dock to the bow is always 7, so, using the Pythagorean theorem, we

get: $7^2 + B^2 = R^2$. Now we can take the derivative of this equation with respect to t: $2B\dfrac{dB}{dt} = 2R\dfrac{dR}{dt}$,

which simplifies to $B\dfrac{dB}{dt} = R\dfrac{dR}{dt}$. We know that $R = 25$ and can use the Pythagorean theorem to

find B: $7^2 + B^2 = 25^2$, so $B = 24$. Now we plug $\dfrac{dR}{dt} = -4$, $R = 25$, and $B = 24$ into the derivative and

we get $24\dfrac{dB}{dt} = 25(-4)$. Now we can solve for $\dfrac{dB}{dt}$: $\dfrac{dB}{dt} = -\dfrac{25}{6}$ ft/s .

8. $\dfrac{4}{3}$ ft/s

We are given the rate at which the woman is walking away from the street lamp and are looking for the rate at which the length of her shadow is changing. It helps to draw a picture of the situation.

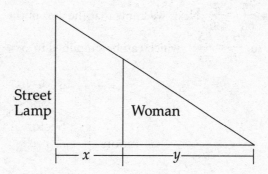

If we label the distance between the woman and the street lamp, x, and the length of the woman's shadow, y, we can use similar triangles to get $\dfrac{y}{6} = \dfrac{x+y}{24}$. We can cross-multiply and simplify.

$24y = 6x + 6y$

$3y = x$

Next, we take the derivative of the equation with respect to t: $3\dfrac{dy}{dt} = \dfrac{dx}{dt}$. Now, we can plug $\dfrac{dx}{dt} = 4$ into the derivative and solve: $\dfrac{dy}{dt} = \dfrac{4}{3}$ ft/s.

9. 380 volts/s

We are given the rate at which the current is decreasing, $\dfrac{dI}{dt} = -4$, and the rate at which the resistance is increasing, $\dfrac{dR}{dt} = 20$, and are looking for the rate at which the voltage is changing. We are also given an equation that relates the three variables: $V = IR$. So, we simply take the derivative of the equation with respect to t (Product Rule!): $\dfrac{dV}{dt} = I\dfrac{dR}{dt} + R\dfrac{dI}{dt}$. Now we can use our equation to find R, which we will need to plug into the derivative equation: $100 = 20R$, so $R = 5$. Next, we plug $\dfrac{dI}{dt} = -4$, $\dfrac{dR}{dt} = 20$, $I = 20$, and $R = 5$ into the derivative equation: $\dfrac{dV}{dt} = (20)(20) + (5)(-4) = 380$ volts/s.

10. $\dfrac{3\pi}{5}$ in^2/min

We know that it takes 60 minutes for the minute hand of a clock to make one complete revolution, so the rate at which the angle, θ, formed by the minute hand and noon increasing, in terms of radians/min, is $\dfrac{d\theta}{dt} = \dfrac{2\pi}{60} = \dfrac{\pi}{30}$. Next, we know that the area of the sector of a circle is proportional to its central angle, so $\dfrac{\theta}{2\pi} = \dfrac{S}{\pi r^2}$, which can be simplified to $S = \dfrac{1}{2} r^2 \theta$.

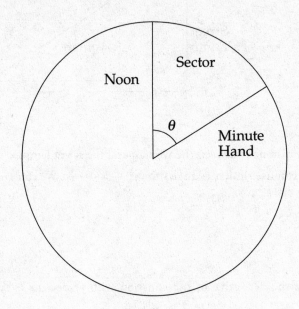

Note that the radius is 6 (the length of the minute hand) and is a constant. Next, we take the derivative of the equation with respect to t: $\dfrac{dS}{dt} = \dfrac{1}{2} r^2 \dfrac{d\theta}{dt}$.

Finally, we substitute $\dfrac{d\theta}{dt} = \dfrac{\pi}{30}$ into the derivative: $\dfrac{dS}{dt} = \dfrac{1}{2}(6)^2 \left(\dfrac{\pi}{30} \right) = \dfrac{3\pi}{5}$ in^2/s .

SOLUTIONS TO PRACTICE PROBLEM SET 13

1. $v(t) = 3t^2 - 18t + 24 \, ; a(t) = 6t - 18$

In order to find the velocity function of the particle, we simply take the derivative of the position function with respect to t: $\dfrac{dx}{dt} = v(t) = 3t^2 - 18t + 24$. In order to find the acceleration function of the particle, we simply take the derivative of the velocity function with respect to t: $\dfrac{dv}{dt} = 6t - 18$.

2. $v(t) = 2\cos(2t) - \sin(t); \ a(t) = -4\sin(2t) - \cos(t)$

In order to find the velocity function of the particle, we simply take the derivative of the position function with respect to t: $\dfrac{dx}{dt} = v(t) = 2\cos(2t) - \sin(t)$. In order to find the acceleration function of the particle, we simply take the derivative of the velocity function with respect to t:

$\dfrac{dv}{dt} = a(t) = -4\sin(2t) - \cos(t)$.

3. $t = 3$

In order to find where the particle is changing direction, we need to find where the velocity of the particle changes sign. The velocity function of the particle is the derivative of the position function:

$\dfrac{dx}{dt} = v(t) = \dfrac{(t^2 + 9)(1) - (t)(2t)}{(t^2 + 9)^2} = \dfrac{9 - t^2}{(t^2 + 9)^2}$. Next, we set the velocity equal to zero. The solutions are: $t = \pm 3$. We can ignore the negative solution because t must be positive. Next, we check the sign of the velocity on either side of $t = 3$. When $0 < t < 3$, the velocity is positive, so the particle is moving to the right. When $t > 3$, the velocity is negative, so the particle is moving to the left. Therefore, the particle is changing direction at $t = 3$.

4. $t = \pi$ and $t = 3\pi$

In order to find where the particle is changing direction, we need to find where the velocity of the particle changes sign. The velocity function of the particle is the derivative of the position function: $\dfrac{dx}{dt} = v(t) = \dfrac{1}{2}\cos\left(\dfrac{t}{2}\right)$. Next, we set the velocity equal to zero. The solutions are: $t = \pi$ and $t = 3\pi$. Actually, there are an infinite number of solutions but remember that we are restricted to $0 < t < 4\pi$. Next, we check the sign of the velocity on the intervals $0 < t < \pi$, $\pi < t < 3\pi$, and

$3\pi < t < 4\pi$. When $0 < t < \pi$, the velocity is positive, so the particle is moving to the right. When $\pi < t < 3\pi$, the velocity is negative, so the particle is moving to the left. When $3\pi < t < 4\pi$, the velocity is positive, so the particle is moving to the right. Therefore, the particle is changing direction at $t = \pi$ and $t = 3\pi$.

5. The distance is 69.

In order to find the distance that the particle travels, we need to look at the position of the particle at $t = 2$ and at $t = 5$. We also need to see if the particle changes direction anywhere on the interval between the two times. If so, we will need to look at the particle's position at those "turning points" as well. The way to find out if the particle is changing direction is to look at the velocity of the particle, which we find by taking the derivative of the position function. We get $\dfrac{dx}{dt} = v(t) = 6t + 2$. If we set the velocity equal to zero, we get $t = -\dfrac{1}{3}$, which is not in the time interval. This means that the velocity doesn't change signs, and thus the particle does not change direction. Now we look at the position of the particle on the interval. At $t = 2$, the particle's position is: $x = 3(2)^2 + 2(2) + 4 = 20$. At $t = 5$, the particle's position is: $x = 3(5)^2 + 2(5) + 4 = 89$. Therefore, the particle travels a distance of 69.

6. The distance is 48.

In order to find the distance that the particle travels, we need to look at the position of the particle at $t = 0$ and at $t = 4$. We also need to see if the particle changes direction anywhere on the interval between the two times. If so, we will need to look at the particle's position at those "turning points" as well. The way to find out if the particle is changing direction is to look at the velocity of the particle, which we find by taking the derivative of the position function. We get $\dfrac{dx}{dt} = v(t) = 2t + 8$. If we set the velocity equal to zero, we get $t = -4$, which is not in the time interval. This means that the velocity doesn't change signs, and thus the particle does not change direction. Now we look at the position of the particle on the interval. At $t = 0$, the particle's position is: $x = (0)^2 + 8(0) = 0$. At $t = 4$, the particle's position is: $x = (4)^2 + 8(4) = 48$. Therefore, the particle travels a distance of 48.

7.   Velocity is 0; acceleration is 0.

This should not be a surprise because $2\sin^2 t + 2\cos^2 t = 2$, so the position is a constant. This means that the particle is not moving and thus has a velocity and acceleration of 0.

8. $t = \dfrac{-8 + \sqrt{70}}{3} \approx 0.122$

In order to find where the particle is changing direction, we need to find where the velocity of the particle changes sign. The velocity function of the particle is the derivative of the position function: $\dfrac{dx}{dt} = v(t) = 3t^2 + 16t - 2$. Next, we set the velocity equal to zero. The solutions are: $t = \dfrac{-8 + \sqrt{70}}{3}$ and $t = \dfrac{-8 - \sqrt{70}}{3}$. We can eliminate the second solution because it is negative and we are restricted to $t > 0$. Next, we check the sign of the velocity on the intervals $0 < t < \dfrac{-8 + \sqrt{70}}{3}$ and $t > \dfrac{-8 + \sqrt{70}}{3}$. When $0 < t < \dfrac{-8 + \sqrt{70}}{3}$, the velocity is negative, so the particle is moving to the left. When $t > \dfrac{-8 + \sqrt{70}}{3}$, the velocity is positive, so the particle is moving to the right. Therefore, the particle is changing direction at $t = \dfrac{-8 + \sqrt{70}}{3}$.

9.   The velocity is never 0, which means that it never changes signs and thus the particle does not change direction.

In order to find where the particle is changing direction, we need to find where the velocity of the particle changes signs. The velocity function of the particle is the derivative of the position function: $\dfrac{dx}{dt} = v(t) = 6t^2 - 12t + 12$. Next, we set the velocity equal to zero. There are no real solutions. If we try a few values, we can see that the velocity is always positive. Therefore, the particle does not change direction.

10.   Distance is $2 + \sin^2 4 \approx 2.573$

In order to find the distance that the particle travels, we need to look at the position of the particle at $t = 0$ and at $t = 2$. We also need to see if the particle changes direction anywhere on the interval between the two times. If so, we will need to look at the particle's position at those "turning points" as well. The way to find out if the particle is changing direction is to look at the velocity of the particle, which we find by taking the derivative of the position function. We get $\dfrac{dx}{dt} = v(t) = 4\sin(2t)\cos(2t)$. If we set the velocity equal to zero, we get $t = \dfrac{\pi}{4}$ and $t = \dfrac{\pi}{2}$.

Actually, there are an infinite number of solutions, but remember that we are restricted to $0 < t < 2$.

Next, we check the sign of the velocity on the intervals $0 < t < \dfrac{\pi}{4}$, $\dfrac{\pi}{4} < t < \dfrac{\pi}{2}$, and $\dfrac{\pi}{2} < t < 2$.

When $0 < t < \dfrac{\pi}{4}$, the velocity is positive, so the particle is moving to the right. When $\dfrac{\pi}{4} < t < \dfrac{\pi}{2}$, the velocity is negative, so the particle is moving to the left. When $\dfrac{\pi}{2} < t < 2$, the velocity is positive, so the particle is moving to the right. Now we look at the position of the particle on the interval. At $t = 0$, the particle's position is: $x = \sin^2(0) = 0$. At $t = \dfrac{\pi}{4}$, the particle's position is $x = \sin^2\left(\dfrac{\pi}{2}\right) = 1$. At $t = \dfrac{\pi}{2}$, the particle's position is: $x = \sin^2(\pi) = 0$. And at $t = 2$, the particle's position is $x = \sin^2(4)$. Therefore, the particle travels a distance of 1 on the interval $0 < t < \dfrac{\pi}{4}$, a distance of 1 in the other direction on the interval $\dfrac{\pi}{4} < t < \dfrac{\pi}{2}$, and a distance of $\sin^2(4)$ on the interval $\dfrac{\pi}{2} < t < 2$. Therefore, the total distance that the particle travels is $2 + \sin^2 4 \approx 2.573$.

SOLUTIONS TO PRACTICE PROBLEM SET 14

1. $f'(x) = \dfrac{4x^3}{x^4 + 8}$

The rule for finding the derivative of $y = \ln u$ is $\dfrac{dy}{dx} = \dfrac{1}{u}\dfrac{du}{dx}$, where u is a function of x. Here $u = x^4 + 8$. Therefore, the derivative is $f'(x) = \dfrac{1}{x^4 + 8}\left(4x^3\right) = \dfrac{4x^3}{x^4 + 8}$.

2. $f'(x) = \dfrac{1}{x} + \dfrac{1}{6 + 2x}$

The rule for finding the derivative of $y = \ln u$ is $\dfrac{dy}{dx} = \dfrac{1}{u}\dfrac{du}{dx}$, where u is a function of x. Before we find the derivative, we can use the laws of logarithms to expand the logarithm. This way, we won't have to use the Product Rule. We get $\ln\left(3x\sqrt{3 + x}\right) = \ln 3 + \ln x + \ln\sqrt{3 + x} = \ln 3 + \ln x + \dfrac{1}{2}\ln(3 + x)$.

Now we can find the derivative: $f'(x) = 0 + \dfrac{1}{x} + \dfrac{1}{2}\dfrac{1}{3 + x} = \dfrac{1}{x} + \dfrac{1}{6 + 2x}$.

3. $f'(x) = \csc x$

The rule for finding the derivative of $y = \ln u$ is $\dfrac{dy}{dx} = \dfrac{1}{u}\dfrac{du}{dx}$, where u is a function of x.

Here $u = \cot x - \csc x$. Therefore, the derivative is $f'(x) = \dfrac{1}{\cot x - \csc x}\left(-\csc^2 x + \csc x \cot x\right)$.

This can be simplified to: $f'(x) = \dfrac{(-\csc x + \cot x)(\csc x)}{\cot x - \csc x} = \csc x$.

4. $f'(x) = \ln(\cos 3x) - 3x \tan 3x - 3x^2$

The rule for finding the derivative of $y = \ln u$ is $\dfrac{dy}{dx} = \dfrac{1}{u}\dfrac{du}{dx}$, where u is a function of x.

Here we will use the Product Rule to find the derivative: $x\left(\dfrac{1}{\cos 3x}\right)(-3\sin 3x) + (1)\ln \cos 3x - 3x^2$.

This simplifies to $f'(x) = \ln(\cos 3x) - 3x \tan 3x - 3x^2$.

5. $f'(x) = \dfrac{2}{x} - \dfrac{x}{5 + x^2}$

The rule for finding the derivative of $y = \ln u$ is $\dfrac{dy}{dx} = \dfrac{1}{u}\dfrac{du}{dx}$, where u is a function of x. Before we find the derivative, we can use the laws of logarithms to expand the logarithm. This way, we won't have to use the Product Rule or the Quotient Rule. We get $\ln\left(\dfrac{5x^2}{\sqrt{5 + x^2}}\right) = \ln 5 + \ln x^2 - \ln\sqrt{5 + x^2} = \ln 5 + 2\ln x - \dfrac{1}{2}\ln(5 + x^2)$. Now we can find the derivative: $f'(x) = 0 + 2\dfrac{1}{x} - \dfrac{1}{2}\dfrac{1}{5 + x^2}(2x) = \dfrac{2}{x} - \dfrac{x}{5 + x^2}$.

6. $f'(x) = e^{x\cos x}\left(\cos x - x \sin x\right)$

The rule for finding the derivative of $y = e^u$ is $\dfrac{dy}{dx} = e^u\dfrac{du}{dx}$, where u is a function of x. Here we will use the Product Rule to find the derivative of the exponent: $f'(x) = e^{x\cos x}\left(\cos x - x \sin x\right)$.

7. $f'(x) = -3e^{-3x}\sin 5x + 5e^{-3x}\cos 5x$

The rule for finding the derivative of $y = e^u$ is $\dfrac{dy}{dx} = e^u\dfrac{du}{dx}$, where u is a function of x. Here we will use the Product Rule to find the derivative: $f'(x) = e^{-3x}(-3)\sin 5x + e^{-3x}(5\cos 5x)$, which we can rearrange to $f'(x) = -3e^{-3x}\sin 5x + 5e^{-3x}\cos 5x$.

8. $f'(x) = e^{\tan 4x} \dfrac{4x \sec^2 4x - 1}{4x^2}$

The rule for finding the derivative of $y = e^u$ is $\dfrac{dy}{dx} = e^u \dfrac{du}{dx}$, where u is a function of x. Here we will

use the Quotient Rule to find the derivative: $f'(x) = \dfrac{(4x)(e^{\tan 4x})(4 \sec^2 4x) - (e^{\tan 4x})(4)}{(4x)^2}$. This

can be simplified to $f'(x) = e^{\tan 4x} \dfrac{4x \sec^2 4x - 1}{4x^2}$.

9. $f'(x) = \pi e^{\pi x} - \pi$

The rule for finding the derivative of $y = e^u$ is $\dfrac{dy}{dx} = e^u \dfrac{du}{dx}$, where u is a function of x and the rule

for finding the derivative of $y = \ln u$ is $\dfrac{dy}{dx} = \dfrac{1}{u} \dfrac{du}{dx}$, where u is a function of x.

We get $f'(x) = e^{\pi x}(\pi) - \dfrac{1}{e^{\pi x}} \pi e^{\pi x}$. This can be simplified to $f'(x) = \pi e^{\pi x} - \pi$. You might have

noticed that $\ln e^{\pi x} = \pi x$, which would have made the derivative a little easier.

10. $f'(x) = \dfrac{3}{x \ln 12}$

The rule for finding the derivative of $y = \log_a u$ is $\dfrac{dy}{dx} = \dfrac{1}{u \ln a} \dfrac{du}{dx}$, where u is a function of x.
Before we find the derivative, we can use the laws of logarithms to expand the logarithm. We get

$f(x) = \log_{12} x^3 = 3 \log_{12} x$. Now we can find the derivative: $f'(x) = \dfrac{3}{x \ln 12}$.

11. $f'(x) = \dfrac{1}{\ln 6}\left(\dfrac{1}{x} + \dfrac{\sec^2 x}{\tan x}\right)$

The rule for finding the derivative of $y = \log_a u$ is $\dfrac{dy}{dx} = \dfrac{1}{u \ln a} \dfrac{du}{dx}$, where u is a func-

tion of x. Before we find the derivative, we can use the laws of logarithms to expand the log-

arithm. This way, we won't have to use the Product Rule or the Quotient Rule. We get

$f(x) = \log_6 (3x \tan x) = \log_6 3 + \log_6 x + \log_6 \tan x$. Now we can find the derivative:

$f'(x) = 0 + \dfrac{1}{x \ln 6} + \dfrac{1}{\tan x \ln 6}(\sec^2 x)$. This can be simplified to $f'(x) = \dfrac{1}{\ln 6}\left(\dfrac{1}{x} + \dfrac{\sec^2 x}{\tan x}\right)$.

12. $$f'(x) = \frac{1}{xe^{4x}\ln 4} - \frac{4\log_4 x}{e^{4x}}$$

The rule for finding the derivative of $y = \log_a u$ is $\dfrac{dy}{dx} = \dfrac{1}{u\ln a}\dfrac{du}{dx}$, and the rule for finding the

derivative of $y = e^u$ is $\dfrac{dy}{dx} = e^u\dfrac{du}{dx}$, where u is a function of x. Here we will use the Quotient

Rule to find the derivative: $f'(x) = \dfrac{\left(e^{4x}\right)\left(\dfrac{1}{x\ln 4}\right) - \left(\log_4 x\right)\left(e^{4x}\right)(4)}{\left(e^{4x}\right)^2}$. This can be simplified to

$$f'(x) = \frac{1}{xe^{4x}\ln 4} - \frac{4\log_4 x}{e^{4x}}.$$

13. $$f'(x) = \frac{3}{2}$$

The rule for finding the derivative of $y = \log_a u$ is $\dfrac{dy}{dx} = \dfrac{1}{u\ln a}\dfrac{du}{dx}$, and the rule for finding the
derivative of $y = a^u$ is $\dfrac{dy}{dx} = a^u(\ln a)\dfrac{du}{dx}$, where u is a function of x. Before we find the derivative,
we can use the laws of logarithms to simplify the logarithm. We get $f(x) = \dfrac{1}{2}\log 10^{3x}$. Now if
you are alert, you will remember that $\log 10^{3x} = 3x$, so this simplifies to $f(x) = \dfrac{1}{2}3x = \dfrac{3}{2}x$. The
derivative is simply $f'(x) = \dfrac{3}{2}$.

14. $$f'(x) = \frac{2\ln x}{x\ln 10}$$

The rule for finding the derivative of $y = \log_a u$ is $\dfrac{dy}{dx} = \dfrac{1}{u\ln a}\dfrac{du}{dx}$, and the rule for finding the
derivative of $y = \ln u$ is $\dfrac{dy}{dx} = \dfrac{1}{u}\dfrac{du}{dx}$, where u is a function of x. Here we will use the Product
Rule to find the derivative: $f'(x) = (\ln x)\left(\dfrac{1}{x\ln 10}\right) + (\log x)\left(\dfrac{1}{x}\right)$. This can be simplified to
$f'(x) = \dfrac{2\ln x}{x\ln 10}$.

15. $$f'(x) = 3e^{3x} - 3^{ex}(e)(\ln 3)$$

The rule for finding the derivative of $y = e^u$ is $\dfrac{dy}{dx} = e^u\dfrac{du}{dx}$, and the rule for finding the derivative
of $y = a^u$ is $\dfrac{dy}{dx} = a^u(\ln a)\dfrac{du}{dx}$, where u is a function of x. We get $f'(x) = 3e^{3x} - 3^{ex}(e)(\ln 3)$.

16. $f'(x) = 10^{\sin x} (\cos x)(\ln 10)$

The rule for finding the derivative of $y = a^u$ is $\dfrac{dy}{dx} = a^u (\ln a)\dfrac{du}{dx}$, where u is a function of x. We get $f'(x) = 10^{\sin x} (\cos x)(\ln 10)$.

17. $f'(x) = 5^{\tan x} (\sec^2 x)\ln 5$

The rule for finding the derivative of $y = a^u$ is $\dfrac{dy}{dx} = a^u (\ln a)\dfrac{du}{dx}$, where u is a function of x. We get $f'(x) = 5^{\tan x} (\sec^2 x)\ln 5$.

18. $f'(x) = \ln 10$

Before we find the derivative, we can use the laws of logarithms to simplify the logarithm. We get $f(x) = \ln(10^x) = x\ln 10$. Now the derivative is simply $f'(x) = \ln 10$.

19. $f'(x) = x^4 5^x (5 + x\ln 5)$

The rule for finding the derivative of $y = a^u$ is $\dfrac{dy}{dx} = a^u (\ln a)\dfrac{du}{dx}$, where u is a function of x. Here we will use the Product Rule to find the derivative: $f'(x) = x^5 (5^x \ln 5) + (5^x)(5x^4)$, which simplifies to $f'(x) = x^4 5^x (5 + x\ln 5)$.

SOLUTIONS TO PRACTICE PROBLEM SET 15

1. $\dfrac{16}{15}$

First, we take the derivative of y: $\dfrac{dy}{dx} = 1 - \dfrac{1}{x^2}$. Next, we find the value of x where $y = \dfrac{17}{4}$:

$\dfrac{17}{4} = x + \dfrac{1}{x}$.

With a little algebra, you should get $x = 4$ or $x = \dfrac{1}{4}$. Because $x > 1$, we can ignore the second

answer. Or, if you are permitted, use your calculator.

Now we can use the formula for the derivative of the inverse of $f(x)$ (see page 178):

$$\frac{d}{dx}f^{-1}(x)\bigg|_{x=c} = \frac{1}{\left[\frac{d}{dy}f(y)\right]_{y=a}}$$, where $f(a)=c$. This formula means that we find the derivative

of the inverse of a function at a value a by taking the reciprocal of the derivative and plugging in

the value of x that makes y equal to a.

$$\frac{1}{\frac{dy}{dx}\bigg|_{x=4}} = \frac{1}{1-\frac{1}{x^2}\bigg|_{x=4}} = \frac{16}{15}$$

2. $-\dfrac{1}{12}$

First, we take the derivative of y: $\dfrac{dy}{dx} = 3 - 15x^2$. Next, we find the value of x where $y = 2$:
$2 = 3x - 5x^3$.

With a little algebra, you should get $x = -1$. Or, if you are permitted, use your calculator.

Now we can use the formula for the derivative of the inverse of $f(x)$ (see page 178):

$$\frac{d}{dx}f^{-1}(x)\bigg|_{x=c} = \frac{1}{\left[\frac{d}{dy}f(y)\right]_{y=a}}$$, where $f(a)=c$. This formula means that we find the derivative

of the inverse of a function at a value a by taking the reciprocal of the derivative and plugging in

the value of x that makes y equal to a.

$$\frac{1}{\frac{dy}{dx}\bigg|_{x=-1}} = \frac{1}{3-15x^2\bigg|_{x=-1}} = -\frac{1}{12}$$

3. $\dfrac{1}{e}$

First, we take the derivative of y: $\dfrac{dy}{dx} = e^x$. Next, we find the value of x where $y = e : e = e^x$. It should be obvious that $x = 1$.

Now we can use the formula for the derivative of the inverse of $f(x)$ (see page 178):

$$\dfrac{d}{dx} f^{-1}(x) \bigg|_{x=c} = \dfrac{1}{\left[\dfrac{d}{dy} f(y)\right]_{y=a}}, \text{ where } f(a) = c.$$ This formula means that we find the derivative

of the inverse of a function at a value a by taking the reciprocal of the derivative and plugging in

the value of x that makes y equal to a.

$$\dfrac{1}{\dfrac{dy}{dx}\bigg|_{x=1}} = \dfrac{1}{e^x\big|_{x=1}} = \dfrac{1}{e}$$

4. $\dfrac{1}{3}$

First, we take the derivative of $f(x)$: $f'(x) = 7x^6 - 10x^4 + 6x^2$. Next, we find the value of x where $f(x) = 1 : x^7 - 2x^5 + 2x^3 = 1$. You should be able to tell by inspection that $x = 1$ is a solution. Or, if you are permitted, use your calculator. Remember that the AP exam won't give you a problem where it is very difficult to solve for the inverse value of y. If the algebra looks difficult, look for an obvious solution, such as $x = 0$ or $x = 1$ or $x = -1$.

Now we can use the formula for the derivative of the inverse of $f(x)$: $\dfrac{d}{dx} f^{-1}(x) \bigg|_{x=c} = \dfrac{1}{\left[\dfrac{d}{dy} f(y)\right]_{y=a}}$,

where $f(a) = c$. This formula means that we find the derivative of the inverse of a function at a

value a by taking the reciprocal of the derivative and plugging in the value of x that makes y equal

to a.

$$\dfrac{1}{\dfrac{dy}{dx}\bigg|_{x=1}} = \dfrac{1}{7x^6 - 10x^4 + 6x^2\big|_{x=1}} = \dfrac{1}{3}$$

5. $\dfrac{1}{4}$

First, we take the derivative of y: $\dfrac{dy}{dx} = 1 + 3x^2$. Next, we find the value of x where $y = -2$: $-2 = x + x^3$. You should be able to tell by inspection that $x = -1$ is a solution. Or, if you are permitted, use your calculator. Remember that the AP exam won't give you a problem where it is very difficult to solve for the inverse value of y. If the algebra looks difficult, look for an obvious solution, such as $x = 0$ or $x = 1$ or $x = -1$.

Now we can use the formula for the derivative of the inverse of $f(x)$: $\left. \dfrac{d}{dx} f^{-1}(x) \right|_{x=c} = \dfrac{1}{\left[\dfrac{d}{dy} f(y) \right]_{y=a}}$,

where $f(a) = c$. This formula means that we find the derivative of the inverse of a function at a

value a by taking the reciprocal of the derivative and plugging in the value of x that makes y equal

to a: $\dfrac{1}{\left. \dfrac{dy}{dx} \right|_{x=-1}} = \dfrac{1}{\left. 1 + 3x^2 \right|_{x=-1}} = \dfrac{1}{4}$.

6. 1

First, we take the derivative of y: $\dfrac{dy}{dx} = 4 - 3x^2$. Next, we find the value of x where $y = 3$: $3 = 4x - x^3$. You should be able to tell by inspection that $x = 1$ is a solution. Or, if you are permitted, use your calculator. Remember that the AP exam won't give you a problem where it is very difficult to solve for the inverse value of y. If the algebra looks difficult, look for an obvious solution, such as $x = 0$ or $x = 1$ or $x = -1$.

Now we can use the formula for the derivative of the inverse of $f(x)$: $\left. \dfrac{d}{dx} f^{-1}(x) \right|_{x=c} = \dfrac{1}{\left[\dfrac{d}{dy} f(y) \right]_{y=a}}$,

where $f(a) = c$. This formula means that we find the derivative of the inverse of a function at a value a by taking the reciprocal of the derivative and plugging in the value of x that makes y equal

to a: $\dfrac{1}{\left.\dfrac{dy}{dx}\right|_{x=1}} = \dfrac{1}{\left.4-3x^2\right|_{x=1}} = 1.$

7.   1

First, we take the derivative of y: $\dfrac{dy}{dx} = \dfrac{1}{x}$. Next, we find the value of x where $y = 0$: $ln\ x = 0$. You

should know that $x = 1$ is the solution. Now we can use the formula for the derivative of the inverse

of $f(x)$: $\left.\dfrac{d}{dx}f^{-1}(x)\right|_{x=c} = \dfrac{1}{\left[\dfrac{d}{dy}f(y)\right]_{y=a}}$, where $f(a) = c$. This formula means that we find the

derivative of the inverse of a function at a value a by taking the reciprocal of the derivative and

plugging in the value of x that makes y equal to a.

$$\dfrac{1}{\left.\dfrac{dy}{dx}\right|_{x=1}} = \dfrac{1}{\left.\dfrac{1}{x}\right|_{x=1}} = 1$$

8. $\dfrac{15}{8}$

First, we take the derivative of y: $\dfrac{dy}{dx} = \dfrac{1}{3}x^{-\frac{2}{3}} + \dfrac{1}{5}x^{-\frac{4}{5}}$. Next, we find the value of x where $y = 2$:
$x^{\frac{1}{3}} + x^{\frac{1}{5}} = 2$. You should be able to tell by inspection that $x = 1$ is a solution. Or, if you are permit-

ted, use your calculator. Remember that the AP exam won't give you a problem where it is very dif-

ficult to solve for the inverse value of y. If the algebra looks difficult, look for an obvious solution,

such as $x = 0$ or $x = 1$ or $x = -1$.

Now we can use the formula for the derivative of the inverse of $f(x)$: $\dfrac{d}{dx}f^{-1}(x)\Big|_{x=c} = \dfrac{1}{\left[\dfrac{d}{dy}f(y)\right]_{y=a}}$,

where $f(a) = c$. This formula means that we find the derivative of the inverse of a function at a

value a by taking the reciprocal of the derivative and plugging in the value of x that makes y equal

to a.

$$\dfrac{1}{\dfrac{dy}{dx}\Big|_{x=1}} = \dfrac{1}{\dfrac{1}{3}x^{-\frac{2}{3}} + \dfrac{1}{5}x^{-\frac{4}{5}}\Big|_{x=1}} = \dfrac{15}{8}$$

SOLUTIONS TO PRACTICE PROBLEM SET 16

1. 5.002

    Recall the differential formula that we use for approximating the value of a function:

    $f(x + \Delta x) \approx f(x) + f'(x)\Delta x$. Here we want to approximate the value of $\sqrt{25.02}$, so we'll use

    $f(x) = \sqrt{x}$ with $x = 25$ and $\Delta x = 0.02$. First, we need to find $f'(x)$: $f'(x) = \dfrac{1}{2\sqrt{x}}$. Now, we

    plug into the formula: $f(x + \Delta x) \approx \sqrt{x} + \dfrac{1}{2\sqrt{x}}\Delta x$. If we plug in $x = 25$ and $\Delta x = 0.02$, we get

    $\sqrt{25 + 0.02} \approx \sqrt{25} + \dfrac{1}{2\sqrt{25}}(0.02)$. If we evaluate this, we get $\sqrt{25.02} \approx 5 + \dfrac{1}{10}(0.02) = 5.002$.

2. 3.999375

    Recall the differential formula that we use for approximating the value of a function:

    $f(x + \Delta x) \approx f(x) + f'(x)\Delta x$. Here we want to approximate the value of $\sqrt{63.97}$,

    so we'll use $f(x) = \sqrt[3]{x}$ with $x = 64$ and $\Delta x = -0.03$. First, we need to find $f'(x)$:

    $f'(x) = \dfrac{1}{3}x^{-\frac{2}{3}} = \dfrac{1}{3\sqrt[3]{x^2}}$. Now, we plug into the formula: $f(x + \Delta x) \approx \sqrt[3]{x} + \dfrac{1}{3\sqrt[3]{x^2}}\Delta x$. If we plug

    in $x = 64$ and $\Delta x = -0.03$, we get $\sqrt[3]{64 - 0.03} \approx \sqrt[3]{64} + \dfrac{1}{3\sqrt[3]{64^2}}(-0.03)$. If we evaluate this, we get

    $\sqrt[3]{63.97} \approx 4 + \dfrac{1}{48}(-0.03) = 3.999375$.

3.   1.802

Recall the differential formula that we use for approximating the value of a function: $f(x + \Delta x) \approx f(x) + f'(x)\Delta x$. Here, we want to approximate the value of tan 61°. Be careful! Whenever we work with trigonometric functions, it is *very* important to work with radians, *not* degrees!

Remember that $60° = \dfrac{\pi}{3}$ radians and $1° = \dfrac{\pi}{180}$ radians, so we'll use $f(x) = \tan x$ with $x = \dfrac{\pi}{3}$ and $\Delta x = \dfrac{\pi}{180}$. First, we need to find $f'(x)$: $f'(x) = \sec^2 x$. Now, we plug into the formula: $f(x + \Delta x) \approx \tan x + \sec^2 x \, \Delta x$. If we plug in $x = \dfrac{\pi}{3}$ and $\Delta x = \dfrac{\pi}{180}$, we get $\tan\left(\dfrac{\pi}{3} + \dfrac{\pi}{180}\right) \approx \tan\left(\dfrac{\pi}{3}\right) + \sec^2\left(\dfrac{\pi}{3}\right)\left(\dfrac{\pi}{180}\right)$. If we evaluate this, we get $\tan\left(\dfrac{61\pi}{180}\right) \approx \sqrt{3} + (2)^2\left(\dfrac{\pi}{180}\right) \approx 1.802$.

4.   997

Recall the differential formula that we use for approximating the value of a function: $f(x + \Delta x) \approx f(x) + f'(x)\Delta x$. Here we want to approximate the value of $(9.99)^3$, so we'll use $f(x) = x^3$ with $x = 10$ and $\Delta x = -0.01$. First, we need to find $f'(x)$: $f'(x) = 3x^2$. Now, we plug into the formula: $f(x + \Delta x) \approx x^3 + 3x^2\Delta x$. If we plug in $x = 10$ and $\Delta x = -0.01$, we get $(10 - 0.01)^3 \approx (10)^3 + 3(10)^2(-0.01)$. If we evaluate this, we get $(9.99)^3 \approx 1{,}000 + 3(10)^2(-0.01) = 997$.

5. ± 2.16 in.^{3}

Recall the formula that we use when we want to approximate the error in a measurement: $dy = f'(x)dx$. Here we want to approximate the error in the volume of a cube when we know that it has a side of length 6 in with an error of ± 0.02 in., where $V(x) = x^3$ (the volume of a cube of side x) with $dx = \pm 0.02$. We find the derivative of the volume: $V'(x) = 3x^2$. Now we can plug into the formula: $dV = 3x^2 dx$. If we plug in $x = 6$ and $dx = \pm 0.02$, we get $dV = 3(6)^2(\pm 0.02) = \pm 2.16$.

6. $\pi \approx 3.142$ mm^3

Recall the formula that we use when we want to approximate the error in a measurement: $dy = f'(x)dx$. Here we want to approximate the increase in the volume of a sphere when we know that it has a radius of length 5 mm with an increase of 0.01 mm, where $V(r) = \dfrac{4}{3}\pi r^3$ (the volume of a sphere of radius r) with $dr = 0.01$. We find the derivative of the volume: $V'(r) = \dfrac{4}{3}\pi\left(3r^2\right) = 4\pi r^2$.

Now we can plug into the formula: $dV = 4\pi r^2 dr$. If we plug in $r = 5$ and $dr = 0.01$, we get

$dV = 4\pi(5)^2(0.01) = \pi \approx 3.142$.

7. -1.732 cm^2

Recall the formula that we use when we want to approximate the error in a measurement: $dy = f'(x)dx$. Here we want to approximate the decrease in the area of an equilateral triangle when we know that it has a side of length 10 cm with a decrease of 0.2 cm, where $A(x) = \dfrac{x^2\sqrt{3}}{4}$ (the area of an equilateral triangle of side x) with $dx = -0.2$. We find the derivative of the area: $A'(x) = \dfrac{2x\sqrt{3}}{4} = \dfrac{x\sqrt{3}}{2}$. Now we can plug into the formula: $dA = \dfrac{x\sqrt{3}}{2}dx$. If we plug in $x = 10$ and $dx = -0.2$, we get $dA = \dfrac{10\sqrt{3}}{2}(-0.2) = -\sqrt{3} \approx -1.732$.

8.   (a) 1.963 m^3;   (b) 15.708 m^3

(a) Recall the formula that we use when we want to approximate the error in a measurement: $dy = f'(x)dx$. Here we want to approximate the error in the volume of a cylinder when we know that it has a diameter of length 5 m (which means that its radius is 2.5 m) and its height is 20 m, with an error in the height of 0.1 m, where $V = \pi r^2 h$ with $dh = 0.1$. Note that the radius is exact, so when we take the derivative we will treat only the height as a variable. We find the derivative of the volume: $V' = \pi r^2$. Now we can plug into the formula: $dV = \pi r^2\, dh$. If we plug in $r = 2.5$, $h = 20$, and $dh = 0.1$, we get $dV = \pi(2.5)^2(0.1) = 0.625\pi \approx 1.963$.

(b) Here we want to approximate the error in the volume of a cylinder when we know that it has a diameter of length 5 m (which means that its radius is 2.5 m) and its height is 20 m, with an error in the diameter of 0.1 m (which means that the error in the radius is 0.05 m), where $V = \pi r^2 h$ with $dr = 0.05$. Note that the height is exact, so when we take the derivative we will treat only the radius as a variable. We find the derivative of the volume: $V' = 2\pi rh$. Now we can plug into the formula: $dV = 2\pi rh\, dh$. If we plug in $r = 2.5$, $h = 20$, and $dr = 0.05$, we get $dV = 2\pi(2.5)(20)(0.05) = 5\pi \approx 15.708$.

SOLUTIONS TO PRACTICE PROBLEM SET 17

1. $-\dfrac{1}{3x^3}+C$

Here we will use the Power Rule, which says that $\int x^n dx = \dfrac{x^{n+1}}{n+1}+C$. The integral is

$\int \dfrac{1}{x^4}dx = \int x^{-4}dx = \dfrac{x^{-3}}{-3}+C = -\dfrac{1}{3x^3}+C$.

2. $10\sqrt{x}+C$

Here we will use the Power Rule, which says that $\int x^n dx = \dfrac{x^{n+1}}{n+1}+C$. The integral is

$\int \dfrac{5}{\sqrt{x}}dx = 5\int x^{-\frac{1}{2}}dx = 5\dfrac{x^{\frac{1}{2}}}{\frac{1}{2}}+C = 10\sqrt{x}+C$.

3. $\dfrac{x^4}{4}-\dfrac{7}{x}+C$

Here we will use the Power Rule, which says that $\int x^n dx = \dfrac{x^{n+1}}{n+1}+C$. First, let's simpli-

fy the integrand: $\int \dfrac{x^5+7}{x^2}\,dx = \int\left(\dfrac{x^5}{x^2}+\dfrac{7}{x^2}\right)dx = \int\left(x^3+7x^{-2}\right)dx$. Now, we can evaluate the

integral: $\int\left(x^3+7x^{-2}\right)dx = \dfrac{x^4}{4}+7\dfrac{x^{-1}}{-1}+C = \dfrac{x^4}{4}-\dfrac{7}{x}+C$.

4. $x^5 - x^3 + x^2 + 6x + C$

Here we will use the Power Rule, which says that $\int x^n dx = \dfrac{x^{n+1}}{n+1}+C$. The integral is

$\int\left(5x^4 - 3x^2 + 2x + 6\right)dx = 5\dfrac{x^5}{5} - 3\dfrac{x^3}{3} + 2\dfrac{x^2}{2} + 6x + C = x^5 - x^3 + x^2 + 6x + C$.

5. $-\dfrac{3}{2x^2}+\dfrac{2}{x}+\dfrac{x^5}{5}+2x^8+C$

Here we will use the Power Rule, which says that $\int x^n dx = \dfrac{x^{n+1}}{n+1}+C$.
The integral is

$\int\left(3x^{-3} - 2x^{-2} + x^4 + 16x^7\right)dx = 3\dfrac{x^{-2}}{-2} - 2\dfrac{x^{-1}}{-1} + \dfrac{x^5}{5} + 16\dfrac{x^8}{8} + C = -\dfrac{3}{2x^2}+\dfrac{2}{x}+\dfrac{x^5}{5}+2x^8+C$.

6.

$$\frac{x^4}{4} - \frac{2x^3}{3} + \frac{x^2}{2} - 2x + C$$

Here we will use the Power Rule, which says that $\int x^n dx = \frac{x^{n+1}}{n+1} + C$. First, let's simplify

the integrand: $\int(1 + x^2)(x - 2)dx = \int(x^3 - 2x^2 + x - 2)dx$. Now, we can evaluate the integral:

$$\int\left(x^3 - 2x^2 + x - 2\right) dx = \frac{x^4}{4} - \frac{2x^3}{3} + \frac{x^2}{2} - 2x + C.$$

7.

$$\frac{3x^{\frac{4}{3}}}{2} + \frac{3x^{\frac{7}{3}}}{7} + C$$

Here we will use the Power Rule, which says that $\int x^n dx = \frac{x^{n+1}}{n+1} + C$. First, let's simplify

the integrand: $\int x^{\frac{1}{3}}(2 + x)\, dx = \int\left(2x^{\frac{1}{3}} + x^{\frac{4}{3}}\right) dx$. Now, we can evaluate the integral:

$$\int\left(2x^{\frac{1}{3}} + x^{\frac{4}{3}}\right) dx = 2\frac{x^{\frac{4}{3}}}{\frac{4}{3}} + \frac{x^{\frac{7}{3}}}{\frac{7}{3}} + C = \frac{3x^{\frac{4}{3}}}{2} + \frac{3x^{\frac{7}{3}}}{7} + C.$$

8.

$$\frac{x^7}{7} + \frac{2x^5}{5} + \frac{x^3}{3} + C$$

Here we will use the Power Rule, which says that $\int x^n dx = \frac{x^{n+1}}{n+1} + C$. First, let's simplify

the integrand: $\int(x^3 + x)^2\, dx = \int(x^6 + 2x^4 + x^2)dx$. Now, we can evaluate the integral:

$$\int\left(x^6 + 2x^4 + x^2\right) dx = \frac{x^7}{7} + \frac{2x^5}{5} + \frac{x^3}{3} + C.$$

9.

$$\frac{x^5}{5} - \frac{2x^3}{3} - \frac{1}{x} + C$$

Here we will use the Power Rule, which says that $\int x^n dx = \frac{x^{n+1}}{n+1} + C$.

First, let's simplify the integrand: $\int \frac{x^6 - 2x^4 + 1}{x^2}\, dx = \int\left(\frac{x^6}{x^2} - 2\frac{x^4}{x^2} + \frac{1}{x^2}\right) dx = \int\left(x^4 - 2x^2 + x^{-2}\right)dx$.

Now, we can evaluate the integral: $\int\left(x^4 - 2x^2 + x^{-2}\right) dx = \frac{x^5}{5} - \frac{2x^3}{3} + \frac{x^{-1}}{-1} + C = \frac{x^5}{5} - \frac{2x^3}{3} - \frac{1}{x} + C.$

10. $$\frac{x^5}{5} - \frac{3x^4}{4} + x^3 - \frac{x^2}{2} + C$$

Here we will use the Power Rule, which says that $\int x^n dx = \frac{x^{n+1}}{n+1} + C$.

First, let's simplify the integrand: $\int x(x-1)^3 dx = \int x(x^3 - 3x^2 + 3x - 1)dx = \int (x^4 - 3x^3 + 3x^2 - x)dx$.

Now, we can evaluate the integral:

$$\int \left(x^4 - 3x^3 + 3x^2 - x\right)dx = \frac{x^5}{5} - \frac{3x^4}{4} + \frac{3x^3}{3} - \frac{x^2}{2} + C = \frac{x^5}{5} - \frac{3x^4}{4} + x^3 - \frac{x^2}{2} + C.$$

11. $\sin x + 5 \cos x + C$

Here we will use the Rules for the Integrals of Trig Functions, namely:

$\int \sin x\, dx = -\cos x + C$ and $\int \cos x\, dx = \sin x + C$. We get $\int (\cos x - 5 \sin x)\, dx = \sin x + 5 \cos x + C$.

12. $\tan x + \sec x + C$

Here we will use the Rules for the Integrals of Trig Functions, namely: $\int \sec^2 x\, dx = \tan x + C$ and $\int (\sec x \tan x)\, dx = \sec x + C$. First, let's expand the integrand: $\int \sec x\,(\sec x + \tan x)\, dx = \int (\sec^2 x + \sec x \tan x)\, dx$. We get $\int (\sec^2 x + \sec x \tan x)\, dx = \tan x + \sec x + C$.

13. $\tan x + \frac{x^2}{2} + C$

Here we will use the Rules for the Integrals of Trig Functions, namely: $\int \sec^2 x\, dx = \tan x + C$ and the Power Rule, which says that $\int x^n dx = \frac{x^{n+1}}{n+1} + C$. We get $\int \left(\sec^2 x + x\right)dx = \tan x + \frac{x^2}{2} + C.$

14. $\sec x + C$

Here we will use the Rules for the Integrals of Trig Functions, namely: $\int (\sec x \tan x)dx = \sec x + C$. First, we need to rewrite the integrand, using trig identities: $\int \frac{\sin x}{\cos^2 x} dx = \int \left(\frac{1}{\cos x} \frac{\sin x}{\cos x}\right) dx = \int (\sec x \tan x)dx$. Now, we can evaluate the integral: $\int (\sec x \tan x)dx = \sec x + C.$

15. $\sin x + 4 \tan x + C$

Here we will use the Rules for the Integrals of Trig Functions, namely: $\int \sec^2 x\, dx = \tan x + C$ and $\int \cos x\, dx = \sin x + C$.

First, we need to rewrite the integrand, using trig identities:

$$\int \frac{\cos^3 x + 4}{\cos^2 x} dx = \int \left(\frac{\cos^3 x}{\cos^2 x} + \frac{4}{\cos^2 x} \right) dx = \int \left(\cos x + 4 \sec^2 x \right) dx.$$ Now, we can evaluate the inte-

gral: $\int(\cos x + 4 \sec^2 x)dx = \sin x + 4 \tan x + C$.

16. $-2 \cos x + C$

Here we will use the Rules for the Integrals of Trig Functions, namely: $\int \sin x\ dx = -\cos x + C$. First, we need to rewrite the integrand, using trig iden-

tities: $\int \frac{\sin 2x}{\cos x} dx = \int \frac{2 \sin x \cos x}{\cos x} dx = \int 2 \sin x\ dx$. Now, we can evaluate the integral:

$\int 2 \sin x\ dx = -2 \cos x + C$.

17. $x + \sin x + C$

Here we will use the Rules for the Integrals of Trig Functions, namely:

$\int \cos x\ dx = \sin x + C$. First, we need to rewrite the integrand, using trig identities:

$\int \left(1 + \cos^2 x \sec x \right) dx = \int \left(1 + \frac{\cos^2 x}{\cos x} \right) dx = \int \left(1 + \cos x \right) dx$. Now, we can evaluate the integral:

$\int(1 + \cos x)dx = x + \sin x + C$.

18. $\tan x - x + C$

Here we will use the Rules for the Integrals of Trig Functions, namely: $\int \sec^2 x\ dx = \tan x + C$. First, we need to rewrite the integrand, using trig identities: $\int(\tan^2 x)dx = \int(\sec^2 x - 1)dx$. Now, we can evaluate the integral: $\int(\sec^2 x - 1)dx = \tan x - x + C$.

19. $-\cos x + C$

Here we will use the Rules for the Integrals of Trig Functions, namely: $\int \sin x\ dx = -\cos x + C$. First, we need to rewrite the integrand, using trig identities: $\int \frac{1}{\csc x} dx = \int \sin x\ dx$. Now we can evaluate the integral: $\int \sin x\ dx = -\cos x + C$.

20. $\dfrac{x^2}{2} - 2\tan x + C$

Here we will use the Rules for the Integrals of Trig Functions, namely:

$\int \sec^2 x \, dx = \tan x + C$. First, we need to rewrite the integrand, using trig identi-

ties: $\int \left(x - \dfrac{2}{\cos^2 x}\right) dx = \int \left(x - 2\sec^2 x\right) dx$. Now, we can evaluate the integral:

$\int \left(x - 2\sec^2 x\right) dx = \dfrac{x^2}{2} - 2\tan x + C$.

SOLUTIONS TO PRACTICE PROBLEM SET 18

1. $\dfrac{\sin^2 2x}{4} + C$

If we let $u = \sin 2x$, then $du = 2\cos 2x \, dx$. We need to substitute for

$\cos 2x \, dx$, so we can divide the du term by 2: $\dfrac{du}{2} = \cos 2x \, dx$. Next we can substitute into the

integral: $\int \sin 2x \cos 2x \, dx = \dfrac{1}{2} \int u \, du$. Now we can integrate: $\dfrac{1}{2} \int u \, du = \dfrac{1}{2}\left(\dfrac{u^2}{2}\right) + C = \dfrac{u^2}{4} + C$.

Last, we substitute back and get $\dfrac{\sin^2 2x}{4} + C$.

2. $-\dfrac{9}{4}\left(10 - x^2\right)^{\frac{2}{3}} + C$

First, pull the constant out of the integrand: $\int \dfrac{3x \, dx}{\sqrt[3]{10 - x^2}} \, dx = 3\int \dfrac{x \, dx}{\sqrt[3]{10 - x^2}} \, dx$. If we let $u = 10 - x^2$,

then $du = -2x \, dx$. We need to substitute for $x \, dx$, so we can divide the du term by -2: $-\dfrac{du}{2} = x \, dx$.

Next we can substitute into the integral: $3\int \dfrac{x \, dx}{\sqrt[3]{10 - x^2}} \, dx = -\dfrac{3}{2} \int u^{-\frac{1}{3}} \, du$. Now we can integrate:

$-\dfrac{3}{2} \int u^{-\frac{1}{3}} \, du = -\dfrac{3}{2} \dfrac{u^{\frac{2}{3}}}{\frac{2}{3}} + C = -\dfrac{9}{4} u^{\frac{2}{3}} + C$. Last, we substitute back and get $-\dfrac{9}{4}\left(10 - x^2\right)^{\frac{2}{3}} + C$.

3. $\dfrac{1}{30}\left(5x^4+20\right)^{\frac{3}{2}}+C$

If we let $u = 5x^4 + 20$, then $du = 20x^3 dx$. We need to substitute for $x^3 dx$, so we can divide the du

term by 20: $\dfrac{du}{20}=x^3 dx$. Next we can substitute into the integral: $\int x^3\sqrt{5x^4+20}\,dx=\dfrac{1}{20}\int u^{\frac{1}{2}}\,du.$

Now we can integrate: $\dfrac{1}{20}\int u^{\frac{1}{2}}\,du=\dfrac{1}{20}\dfrac{u^{\frac{3}{2}}}{\frac{3}{2}}+C=\dfrac{1}{30}u^{\frac{3}{2}}+C$. Last, we substitute back and get

$\dfrac{1}{30}\left(5x^4+20\right)^{\frac{3}{2}}+C$.

4. $-\dfrac{1}{x-1}+C$

If we let $u = x - 1$, then $du = dx$. Next we can substitute into the integral: $\int\dfrac{dx}{\left(x-1\right)^2}=\int u^{-2}\,du$.

Now we can integrate: $\int u^{-2}\,du=\dfrac{u^{-1}}{-1}+C=-\dfrac{1}{u}+C$. Last, we substitute back and get $-\dfrac{1}{x-1}+C$.

5. $-\dfrac{1}{12\left(x^3+3x\right)^4}+C$

If we let $u = x^3 + 3x$, then $du = (3x^2 + 3)\, dx$. We need to substitute for $(x^2 + 1)\, dx$, so we can

divide the du term by 3: $\dfrac{du}{3}=\left(x^2+1\right)dx$. Next we can substitute into the integral:

$\int\left(x^2+1\right)\left(x^3+3x\right)^{-5}dx=\dfrac{1}{3}\int u^{-5}\,du$. Now we can integrate: $\dfrac{1}{3}\int u^{-5}\,du=\dfrac{1}{3}\dfrac{u^{-4}}{-4}+C=-\dfrac{1}{12}\dfrac{1}{u^4}+C$.

Last, we substitute back and get $-\dfrac{1}{12\left(x^3+3x\right)^4}+C$.

6. $-2\cos\sqrt{x}+C$

If we let $u=\sqrt{x}$, then $du=\dfrac{1}{2\sqrt{x}}dx$. We need to substitute for $\dfrac{1}{\sqrt{x}}dx$, so we can multiply the du

term by 2: $2\,du=\dfrac{1}{\sqrt{x}}dx$. Next we can substitute into the integral: $\int\dfrac{1}{\sqrt{x}}\sin\sqrt{x}\,dx=2\int\sin u\,du$.

Now we can integrate: $2\int\sin u\,du=-2\cos u+C$. Last, we substitute back and get $-2\cos\sqrt{x}+C$.

7. $\dfrac{1}{3}\tan\left(x^3\right)+C$

If we let $u = x^3$, then $du = 3x^2 dx$. We need to substitute for $x^2 dx$, so we can divide the du term by

3: $\dfrac{du}{3} = x^2 dx$. Next we can substitute into the integral: $\displaystyle\int x^2 \sec^2\left(x^3\right)dx = \dfrac{1}{3}\int \sec^2 u\, du$. Now we

can integrate: $\dfrac{1}{3}\displaystyle\int \sec^2 u\, du = \dfrac{1}{3}\tan u + C$. Last, we substitute back and get $\dfrac{1}{3}\tan\left(x^3\right)+C$.

8. $-\dfrac{1}{3}\sin\left(\dfrac{3}{x}\right)+C$

If we let $u = \dfrac{3}{x}$, then $du = -\dfrac{3}{x^2}dx$. We need to substitute for $\dfrac{1}{x^2}dx$, so we can divide the du

term by -3: $-\dfrac{du}{3} = \dfrac{1}{x^2}dx$. Next we can substitute into the integral: $\displaystyle\int \dfrac{\cos\left(\dfrac{3}{x}\right)}{x^2}dx = -\dfrac{1}{3}\int \cos u\, du$.

Now we can integrate: $-\dfrac{1}{3}\displaystyle\int \cos u\, du = -\dfrac{1}{3}\sin u + C$. Last, we substitute back and get

$-\dfrac{1}{3}\sin\left(\dfrac{3}{x}\right)+C$.

9. $-\dfrac{1}{4}\left(1-\cos 2x\right)^{-2}+C$

If we let $u = 1 - \cos 2x$, then $du = 2\sin 2x\, dx$. We need to substitute for $\sin 2x\, dx$, so we

can divide the du term by 2: $\dfrac{du}{2} = \sin 2x\, dx$. Next we can substitute into the integral:

$\displaystyle\int \dfrac{\sin 2x}{\left(1-\cos 2x\right)^3}dx = \dfrac{1}{2}\int u^{-3}du$. Now we can integrate: $\dfrac{1}{2}\displaystyle\int u^{-3}du = \dfrac{1}{2}\left(\dfrac{u^{-2}}{-2}\right)+C = -\dfrac{1}{4}\left(\dfrac{1}{u^2}\right)+C$.

Last, we substitute back and get $-\dfrac{1}{4}\left(1-\cos 2x\right)^{-2}+C$.

10. $-\cos(\sin x) + C$

If we let $u = \sin x$, then $du = \cos x\, dx$. Next we can substitute into the integral:
$\int\sin(\sin x)\cos x\, dx = \int\sin u\, du$. Now we can integrate: $\int\sin u\, du = -\cos u + C$. Last, we substitute
back and get $-\cos(\sin x) + C$.

SOLUTIONS TO PRACTICE PROBLEM SET 19

1. $\dfrac{25}{32}$

First, let's draw a picture.

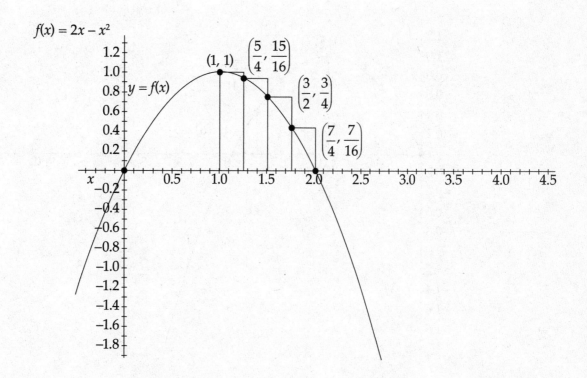

The width of each rectangle is found by taking the difference between the endpoints and dividing by n. Here, the width of each rectangle is $\dfrac{2-1}{4} = \dfrac{1}{4}$.

We find the heights of the rectangles by evaluating $y = 2x - x^2$ at the appropriate endpoints.

$y(1) = 2(1) - (1)^2 = 1$; $\quad y\left(\dfrac{5}{4}\right) = 2\left(\dfrac{5}{4}\right) - \left(\dfrac{5}{4}\right)^2 = \dfrac{15}{16}$; $\quad y\left(\dfrac{3}{2}\right) = 2\left(\dfrac{3}{2}\right) - \left(\dfrac{3}{2}\right)^2 = \dfrac{3}{4}$

and $y\left(\dfrac{7}{4}\right) = 2\left(\dfrac{7}{4}\right) - \left(\dfrac{7}{4}\right)^2 = \dfrac{7}{16}$.

Therefore, the area is $\left(\dfrac{1}{4}\right)\left(1 + \dfrac{15}{16} + \dfrac{3}{4} + \dfrac{7}{16}\right) = \dfrac{25}{32}$.

2. $\dfrac{17}{32}$

First, let's draw a picture.

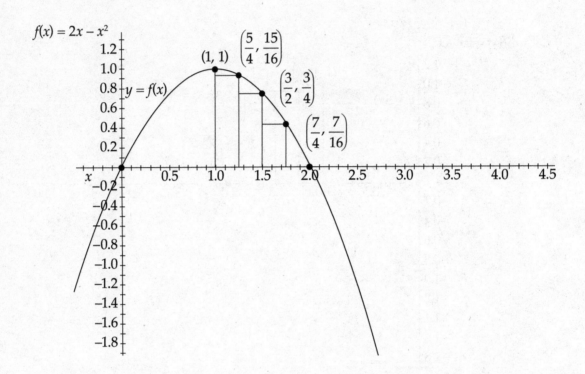

The width of each rectangle is found by taking the difference between the endpoints and dividing by n. Here, the width of each rectangle is $\dfrac{2-1}{4} = \dfrac{1}{4}$.

We find the heights of the rectangles by evaluating $y = 2x - x^2$ at the appropriate endpoints.

$$y\left(\frac{5}{4}\right) = 2\left(\frac{5}{4}\right) - \left(\frac{5}{4}\right)^2 = \frac{15}{16}; \quad y\left(\frac{3}{2}\right) = 2\left(\frac{3}{2}\right) - \left(\frac{3}{2}\right)^2 = \frac{3}{4}; \quad y\left(\frac{7}{4}\right) = 2\left(\frac{7}{4}\right) - \left(\frac{7}{4}\right)^2 = \frac{7}{16};$$

and $y(2) = 2(2) - (2)^2 = 0$.

Therefore, the area is $\left(\dfrac{1}{4}\right)\left(\dfrac{15}{16} + \dfrac{3}{4} + \dfrac{7}{16} + 0\right) = \dfrac{17}{32}$.

3. $\dfrac{21}{32}$

First, let's draw a picture.

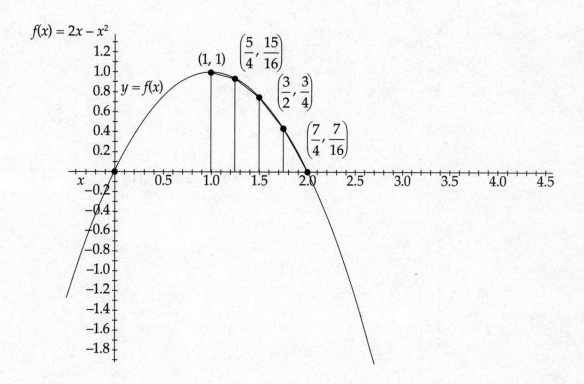

The height of each trapezoid is found by taking the difference between the endpoints and dividing by n. Here, the width of each trapezoid is $\dfrac{2-1}{4} = \dfrac{1}{4}$.

We find the bases of the trapezoids by evaluating $y = 2x - x^2$ at the appropriate endpoints.

$y(1) = 2(1) - (1)^2 = 1$; $y\left(\dfrac{5}{4}\right) = 2\left(\dfrac{5}{4}\right) - \left(\dfrac{5}{4}\right)^2 = \dfrac{15}{16}$; $y\left(\dfrac{3}{2}\right) = 2\left(\dfrac{3}{2}\right) - \left(\dfrac{3}{2}\right)^2 = \dfrac{3}{4}$;

$y\left(\dfrac{7}{4}\right) = 2\left(\dfrac{7}{4}\right) - \left(\dfrac{7}{4}\right)^2 = \dfrac{7}{16}$; and $y(2) = 2(2) - (2)^2 = 0$.

Therefore, the area is $\left(\dfrac{1}{2}\right)\left(\dfrac{1}{4}\right)\left[1 + (2)\left(\dfrac{15}{16}\right) + (2)\left(\dfrac{3}{4}\right) + (2)\left(\dfrac{7}{16}\right) + 0\right] = \dfrac{21}{32}$.

4. $\dfrac{43}{64}$

First, let's draw a picture.

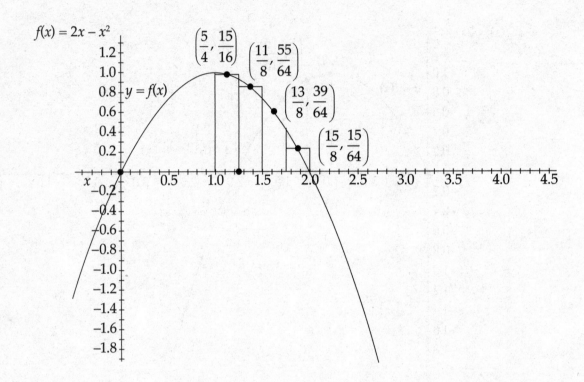

The height of each rectangle is found by taking the difference between the endpoints and dividing by n. Here, the width of each rectangle is $\dfrac{2-1}{4} = \dfrac{1}{4}$.

We find the bases of the rectangles by evaluating $y = 2x - x^2$ at the appropriate endpoints.

$$y\left(\frac{9}{8}\right) = 2\left(\frac{9}{8}\right) - \left(\frac{9}{8}\right)^2 = \frac{63}{64}; \; y\left(\frac{11}{8}\right) = 2\left(\frac{11}{8}\right) - \left(\frac{11}{8}\right)^2 = \frac{55}{64}; \; y\left(\frac{13}{8}\right) = 2\left(\frac{13}{8}\right) - \left(\frac{13}{8}\right)^2 = \frac{39}{64};$$

and $y\left(\dfrac{15}{8}\right) = 2\left(\dfrac{15}{8}\right) - \left(\dfrac{15}{8}\right)^2 = \dfrac{15}{64}$.

Therefore, the area is $\left(\dfrac{1}{4}\right)\left(\dfrac{63}{64} + \dfrac{55}{64} + \dfrac{39}{64} + \dfrac{15}{64}\right) = \dfrac{43}{64}$.

5. $\dfrac{2}{3}$

We will find the exact area by evaluating the integral $\int_1^2 \left(2x - x^2\right) dx$. According to the Fundamental Theorem of Calculus: $\int_1^2 \left(2x - x^2\right) dx = \left(\dfrac{2x^2}{2} - \dfrac{x^3}{3}\right)\Big|_1^2 = \left(x^2 - \dfrac{x^3}{3}\right)\Big|_1^2$.

If we evaluate the integral at the limits, we get $\left[\left(2\right)^2 - \left(\dfrac{2^3}{3}\right)\right] - \left[\left(1\right)^2 - \left(\dfrac{1^3}{3}\right)\right] = \dfrac{2}{3}$.

6.   2

According to the Fundamental Theorem of Calculus: $\int_{-\frac{\pi}{2}}^{\frac{\pi}{2}} \cos x\, dx = \sin x \Big|_{-\frac{\pi}{2}}^{\frac{\pi}{2}} = \sin \dfrac{\pi}{2} - \sin\left(-\dfrac{\pi}{2}\right) = 2$.

7. $\dfrac{968}{5}$

First, let's rewrite the integrand: $\int_1^9 \left(2x\sqrt{x}\right) dx = \int_1^9 \left(2x^{\frac{3}{2}}\right) dx$. Now, according to the Fundamental

Theorem of Calculus: $\int_1^9 \left(2x^{\frac{3}{2}}\right) dx = \left(\dfrac{2x^{\frac{5}{2}}}{\frac{5}{2}}\right)\Big|_1^9 = \left(\dfrac{4x^{\frac{5}{2}}}{5}\right)\Big|_1^9 = \left(\dfrac{4(9)^{\frac{5}{2}}}{5} - \dfrac{4(1)^{\frac{5}{2}}}{5}\right) = \dfrac{968}{5}$.

8. $-\dfrac{161}{20}$

According to the Fundamental Theorem of Calculus:

$$\int_0^1 \left(x^4 - 5x^3 + 3x^2 - 4x - 6\right) dx = \left(\dfrac{x^5}{5} - 5\dfrac{x^4}{4} + x^3 - 2x^2 - 6x\right)\Big|_0^1 =$$

$$\left(\dfrac{(1)^5}{5} - 5\dfrac{(1)^4}{4} + (1)^3 - 2(1)^2 - 6(1)\right) - 0 = -\dfrac{161}{20}$$

9. 16

Recall that the absolute value function must be rewritten as a piecewise function: $|x| = \begin{cases} x; x \geq 0 \\ -x; x < 0 \end{cases}$.

Thus, we need to split the integral into two separate integrals in order to evaluate it:

$\int_{-4}^{4} |x| dx = \int_{-4}^{0} (-x) dx + \int_{0}^{4} x dx$. Now, according to the Fundamental Theorem of Calculus:

$\int_{-4}^{0} (-x) dx = \left(-\frac{x^2}{2} \right) \Big|_{-4}^{0} = (0) - \left(-\frac{(-4)^2}{2} \right) = 8$ and $\int_{0}^{4} (x) dx = \left(\frac{x^2}{2} \right) \Big|_{0}^{4} = \left(\frac{(4)^2}{2} \right) - (0) = 8$. There-

fore, the answer is 8 + 8 = 16.

10. 0

According to the Fundamental Theorem of Calculus:

$$\int_{-\frac{\pi}{2}}^{\frac{\pi}{2}} \sin x\, dx = -\cos x \Big|_{-\frac{\pi}{2}}^{\frac{\pi}{2}} = -\cos\frac{\pi}{2} + \cos\left(-\frac{\pi}{2} \right) = 0$$

11. Recall that the formula for finding the area under the curve using the left end-

points is: $\left(\frac{b-a}{n} \right) [y_0 + y_1 + y_2 + ... + y_{n-1}]$. This formula assumes that the x-val-

ues are evenly spaced, but they aren't here, so we will replace the values of $\left(\frac{b-a}{n} \right)$

with the appropriate widths of each rectangle. The width of the first rectangle is

1 − 0 = 1; the second width is 3 − 1 = 2; the third is 5 − 3 = 2; the fourth is 9 − 5 = 4;

and the fifth is 14 − 9 = 5. We find the height of each rectangle by evaluating $g(x)$

at the appropriate value of x, the left endpoint of each interval on the x-axis. Here

$y_0 = 10$, $y_1 = 8$, $y_2 = 11$, $y_3 = 17$, and $y_4 = 20$. Therefore, we can approximate the integral with:

$\int_{0}^{14} g(x) dx = (1)(10) + (2)(8) + (2)(11) + (4)(17) + (5)(20) = 216$.

SOLUTIONS TO PRACTICE PROBLEM SET 20

1. $2\sqrt{\dfrac{2}{\pi}}$

We find the average value of the function, $f(x)$, on the interval $[a, b]$ using the formula

$f(c) = \dfrac{1}{b-a}\displaystyle\int_a^b f(x)\,dx$. Here we are looking for the average value of $f(x) = 4x\cos x^2$ on the interval

$\left[0, \sqrt{\dfrac{\pi}{2}}\,\right]$. Using the formula, we need to find: $\dfrac{1}{\sqrt{\dfrac{\pi}{2}}-0}\displaystyle\int_0^{\sqrt{\frac{\pi}{2}}}\left(4x\cos x^2\right)dx = \sqrt{\dfrac{2}{\pi}}\displaystyle\int_0^{\sqrt{\frac{\pi}{2}}}\left(4x\cos x^2\right)dx$.

We will need to use u-substitution to evaluate the integral. Let $u = x^2$ and $du = 2x\,dx$. We need

to substitute for $4x\,dx$, so we multiply by 2 to get $2\,du = 4x\,dx$. Now we can substitute into the

integral: $\int(4x\cos x^2)dx = 2\int\cos u\,du = 2\sin u$. If we substitute back, we get $2\sin x^2$. Now we can

evaluate the integral: $\sqrt{\dfrac{2}{\pi}}\displaystyle\int_0^{\sqrt{\frac{\pi}{2}}}\left(4x\cos x^2\right)dx = \sqrt{\dfrac{2}{\pi}}\left(2\sin x^2\right)\Big|_0^{\sqrt{\frac{\pi}{2}}} = \sqrt{\dfrac{2}{\pi}}\left(2\sin\dfrac{\pi}{2}-2\sin 0\right)=2\sqrt{\dfrac{2}{\pi}}$.

2. $\dfrac{8}{3}$

We find the average value of the function, $f(x)$, on the interval $[a, b]$ using the formula

$f(c) = \dfrac{1}{b-a}\displaystyle\int_a^b f(x)\,dx$. Here we are looking for the average value of $f(x) = \sqrt{x}$

on the interval $[0, 16]$. Using the formula, we need to find $\dfrac{1}{16-0}\displaystyle\int_0^{16}\left(\sqrt{x}\right)dx$. We get

$\dfrac{1}{16}\displaystyle\int_0^{16}\left(\sqrt{x}\right)dx = \dfrac{1}{16}\left(\dfrac{x^{\frac{3}{2}}}{\dfrac{3}{2}}\right)\Bigg|_0^{16} = \dfrac{1}{24}\left(x^{\frac{3}{2}}\right)\Bigg|_0^{16} = \dfrac{1}{24}\left((16)^{\frac{3}{2}}-0\right)=\dfrac{8}{3}$.

3. $\dfrac{2\sqrt{2}}{3}$

We find the average value of the function, $f(x)$, on the interval $[a,\ b]$ using the

formula $\quad f(c)=\dfrac{1}{b-a}\displaystyle\int_{a}^{b}f(x)dx.$ Here we are looking for the average

value of $f(x)=\sqrt{1-x}$ on the interval $[-1,\ 1]$. Using the formula, we need to find:

$\dfrac{1}{1-(-1)}\displaystyle\int_{-1}^{1}\left(\sqrt{1-x}\right)dx$. We get

$$\frac{1}{2}\int_{-1}^{1}(1-x)^{\frac{1}{2}}dx=\frac{1}{2}\left(-\frac{(1-x)^{\frac{3}{2}}}{\frac{3}{2}}\right)\Bigg|_{-1}^{1}=-\frac{1}{3}(1-x)^{\frac{3}{2}}\Bigg|_{-1}^{1}=-\frac{1}{3}\left(0-2^{\frac{3}{2}}\right)=\frac{2\sqrt{2}}{3}.$$

4. 1

We find the average value of the function, $f(x)$, on the interval $[a,\ b]$ using the formula

$f(c)=\dfrac{1}{b-a}\displaystyle\int_{a}^{b}f(x)dx$. Here we are looking for the average value of $f(x)=2|x|$ on the interval

$[-1,\ 1]$. Using the formula, we need to find: $\dfrac{1}{1-(-1)}\displaystyle\int_{-1}^{1}\left(2|x|\right)dx$. Recall that the absolute value

function must be rewritten as a piecewise function: $|x|=\begin{cases}x;x\geq 0\\-x;x<0\end{cases}$. Thus, we need to split the

integral into two separate integrals in order to evaluate it: $\dfrac{1}{1-(-1)}\displaystyle\int_{-1}^{1}\left(2|x|\right)dx=\int_{-1}^{0}(-x)dx+\int_{0}^{1}x\,dx.$

We get $\displaystyle\int_{-1}^{0}(-x)dx+\int_{0}^{1}x\,dx=\left(-\frac{x^{2}}{2}\right)\Bigg|_{-1}^{0}+\left(\frac{x^{2}}{2}\right)\Bigg|_{-1}^{0}=\left(0-\left(-\frac{1}{2}\right)\right)+\left(\frac{1}{2}-0\right)=1.$

5. $\sin^{2}x$

We find the derivative of an integral using the Second Fundamental Theorem of Calculus:

$\dfrac{d}{dx}\displaystyle\int_{a}^{x}f(t)dt=f(x)$. We get $\dfrac{d}{dx}\displaystyle\int_{1}^{x}\sin^{2}t\,dt=\sin^{2}x$.

6. $27x^2 - 9x$

We find the derivative of an integral using the Second Fundamental Theorem of Calculus: $\dfrac{d}{dx}\displaystyle\int_a^x f(t)\,dt = f(x)$.   We   get $\dfrac{d}{dx}\displaystyle\int_1^{3x}\left(t^2 - t\right)dt = 3\left[(3x)^2 - (3x)\right] = 27x^2 - 9x$.   Don't

forget that because the upper limit is a function, we need to multiply the answer by the

derivative of that function.

7. $2x^3$

We find the derivative of an integral using the Second Fundamental Theorem of Calculus: $\dfrac{d}{dx}\displaystyle\int_a^x f(t)\,dt = f(x)$. Normally, we would need to rewrite the absolute value function as a

piecewise function, but notice that we are evaluating the absolute value over an interval where

all values will be positive. Thus, we can ignore the absolute value and rewrite the integral as

$\dfrac{d}{dx}\displaystyle\int_0^{x^2}|t|\,dt = \dfrac{d}{dx}\displaystyle\int_0^{x^2} t\,dt$. We get $\dfrac{d}{dx}\displaystyle\int_0^{x^2} t\,dt = \left(x^2\right)(2x) = 2x^3$. Don't forget that because the up-

per limit is a function, we need to multiply the answer by the derivative of that function.

8. $-2\cos x$

We find the derivative of an integral using the Second Fundamental Theorem of Calculus: $\dfrac{d}{dx}\displaystyle\int_a^x f(t)\,dt = f(x)$. We get $\dfrac{d}{dx}\displaystyle\int_1^x -2\cos t\,dt = -2\cos x$.

SOLUTIONS TO PRACTICE PROBLEM SET 21

1. $\ln |\tan x| + C$

   Whenever we have an integral in the form of a quotient, we check to see if the solution

   is a logarithm. A clue is whether the numerator is the derivative of the denominator, as

   it is here. Let's use u-substitution. If we let $u = \tan x$, then $du = \sec^2 x\ dx$. If we substitute

   into the integrand, we get $\int \dfrac{\sec^2 x}{\tan x} dx = \int \dfrac{du}{u}$. Recall that $\int \dfrac{du}{u} = \ln |u| + C$. Substituting back, we

   get $\int \dfrac{\sec^2 x}{\tan x} dx = \ln |\tan x| + C$.

2. $-\ln |1 - \sin x| + C$

   Whenever we have an integral in the form of a quotient, we check to see if the

   solution is a logarithm. A clue is whether the numerator is the derivative of the

   denominator, as it is here. Let's use u-substitution. If we let $u = 1 - \sin x$, then

   $du = -\cos x\ dx$. If we substitute into the integrand, we get $\int \dfrac{\cos x}{1 - \sin x} dx = -\int \dfrac{du}{u}$.
   Recall that $\int \dfrac{du}{u} = \ln |u| + C$. Substituting back, we get $\int \dfrac{\cos x}{1 - \sin x} dx = -\ln |1 - \sin x| + C$.

3. $\ln |\ln x| + C$

   Whenever we have an integral in the form of a quotient, we check to see if the

   solution is a logarithm. A clue is whether the numerator is the derivative of the

   denominator, as it is here. Let's use u-substitution. If we let $u = \ln x$, then $du = \dfrac{1}{x} dx$.

   If we substitute into the integrand, we get $\int \dfrac{1}{x \ln x} dx = \int \dfrac{du}{u}$. Recall that

   $\int \dfrac{du}{u} = \ln |u| + C$. Substituting back, we get $\int \dfrac{1}{x \ln x} dx = \ln |\ln x| + C$.

4. $\sin(\ln x) + C$

Recall that $\dfrac{d}{dx}\ln x = \dfrac{1}{x}$. Here, we can use u-substitution to get rid of the log in the

integrand. If we let $u = \ln x$, then $du = \dfrac{1}{x}dx$. If we substitute into the integrand, we get

$\displaystyle\int\dfrac{1}{x}\cos(\ln x)\,dx = \int\cos u\ du = \sin u + C.$ Substituting back we get $\displaystyle\int\dfrac{1}{x}\cos(\ln x)\,dx = \sin(\ln x) + C.$

5. $-\ln|\cos x| - x + C$

First, let's rewrite the integrand.

$\displaystyle\int\dfrac{\sin x - \cos x}{\cos x}\,dx = \int\left(\dfrac{\sin x}{\cos x} - \dfrac{\cos x}{\cos x}\right)dx = \int\left(\dfrac{\sin x}{\cos x} - 1\right)dx = \int\left(\dfrac{\sin x}{\cos x}\right)dx - \int dx.$ Whenever

we have an integral in the form of a quotient, we check to see if the solution is a logarithm. A

clue is whether the numerator is the derivative of the denominator, as it is in the first integral.

Let's use u-substitution. If we let $u = \cos x$, then $du = -\sin x\ dx$. If we substitute into the in-

tegrand, we get $\displaystyle\int\left(\dfrac{\sin x}{\cos x}\right)dx = -\int\dfrac{du}{u}.$ Recall that $\displaystyle\int\dfrac{du}{u} = \ln|u| + C.$ Substituting back, we get

$\displaystyle\int\left(\dfrac{\sin x}{\cos x}\right)dx = -\ln|\cos x| + C.$ The second integral is simply $\int dx = x + C.$ Therefore, the integral is

$\displaystyle\int\left(\dfrac{\sin x}{\cos x}\right)dx - \int dx = -\ln|\cos x| - x + C.$

6. $\ln\left(1+2\sqrt{x}\right)+C$

Whenever we have an integral in the form of a quotient, we check to see if the solution is a logarithm. A clue is whether the numerator is the derivative of the denominator, as it is here.

Let's use u-substitution. If we let $u=1+2\sqrt{x}$, then $du=\dfrac{1}{\sqrt{x}}dx$. If we substitute into the integrand, we get $\displaystyle\int\dfrac{1}{\sqrt{x}\left(1+2\sqrt{x}\right)}dx=\int\dfrac{du}{u}$. Recall that $\displaystyle\int\dfrac{du}{u}=\ln|u|+C$. Substituting back, we get $\displaystyle\int\dfrac{1}{\sqrt{x}\left(1+2\sqrt{x}\right)}dx=\ln\left(1+2\sqrt{x}\right)+C$. Notice that we don't need the absolute value bars because $1+2\sqrt{x}$ is never negative.

7. $\ln(1 + e^x) + C$

Whenever we have an integral in the form of a quotient, we check to see if the solution is a logarithm. A clue is whether the numerator is the derivative of the denominator, as it is here.

Let's use u-substitution. If we let $u = 1 + e^x$, then $du = e^x dx$. If we substitute into the integrand, we get $\displaystyle\int\dfrac{e^x}{1+e^x}dx=\int\dfrac{du}{u}$. Recall that $\displaystyle\int\dfrac{du}{u}=\ln|u|+C$. Substituting back, we get $\displaystyle\int\dfrac{e^x}{1+e^x}dx=\ln\left(1+e^x\right)+C$. Notice that we don't need the absolute value bars because $1 + e^x$ is never negative.

8. $\dfrac{1}{10}e^{5x^2-1}+C$

Recall that $\int e^u du = e^u + C$. Let's use u-substitution. If we let $u = 5x^2 - 1$, then $du = 10x\,dx$. But we need to substitute for $x\,dx$, so if we divide du by 10, we get $\dfrac{1}{10}du = x\,dx$. Now, if we substitute into the integrand, we get $\displaystyle\int xe^{5x^2-1}dx=\dfrac{1}{10}\int e^u du=\dfrac{1}{10}e^u+C$. Substituting back, we get $\displaystyle\int xe^{5x^2-1}dx=\dfrac{1}{10}e^{5x^2-1}+C$.

9. $\sin(2 + e^x) + C$

Recall that $\int e^u \, du = e^u + C$. Let's use u-substitution. If we let $u = 2 + e^x$, then $du = e^x \, dx$. If we substitute into the integrand, we get $\int e^x \cos(2 + e^x) dx = \int \cos u \, du = \sin u + C$. Substituting back, we get $\int e^x \cos(2 + e^x) \, dx = \sin(2 + e^x) + C$.

10. $\ln|e^x - e^{-x}| + C$

Recall that $\int e^u \, du = e^u + C$. Let's use u-substitution. If we let $u = e^x - e^{-x}$, then $du = e^x + e^{-x} dx$. If we substitute into the integrand, we get $\int \dfrac{e^x + e^{-x}}{e^x - e^{-x}} \, dx = \int \dfrac{du}{u}$. Recall that $\int \dfrac{du}{u} = \ln|u| + C$. Substituting back, we get $\int \dfrac{e^x + e^{-x}}{e^x - e^{-x}} \, dx = \ln|e^x - e^{-x}| + C$.

11. $-\dfrac{4^{-x^2}}{\ln 16} + C$

Recall that $\int a^u \, du = \dfrac{1}{\ln a} a^u + C$. Let's use u-substitution. If we let $u = -x^2$, then $du = -2x \, dx$. But we need to substitute for $x \, dx$, so if we divide du by -2, we get $-\dfrac{1}{2} du = x \, dx$. Now, if we substitute into the integrand, we get $\int x 4^{-x^2} \, dx = -\dfrac{1}{2} \int 4^u \, du = -\dfrac{1}{2}\left(\dfrac{1}{\ln 4}\right) 4^u + C$. Substituting back, we get $\int x 4^{-x^2} \, dx = -\dfrac{1}{2}\left(\dfrac{1}{\ln 4}\right) 4^{-x^2} + C = -\dfrac{1}{\ln 16} 4^{-x^2} + C$.

12. $\dfrac{7^{\sin x}}{\ln 7} + C$

Recall that $\int a^u \, du = \dfrac{1}{\ln a} a^u + C$. Let's use u-substitution. If we let $u = \sin x$, then $du = \cos x \, dx$. If we substitute into the integrand, we get $\int 7^{\sin x} \cos x \, dx = \int 7^u \, du = \dfrac{1}{\ln 7} 7^u + C$. Substituting back, we get $\int 7^{\sin x} \cos x \, dx = \dfrac{7^{\sin x}}{\ln 7} + C$.

SOLUTIONS TO PRACTICE PROBLEM SET 22

1. $\dfrac{32}{3}$

We find the area of a region bounded by $f(x)$ above and $g(x)$ below at all points of the interval $[a, b]$ using the formula $\int_a^b \left[f(x) - g(x) \right] dx$. Here $f(x) = 2$ and $g(x) = x^2 - 2$. First, let's make a sketch of the region.

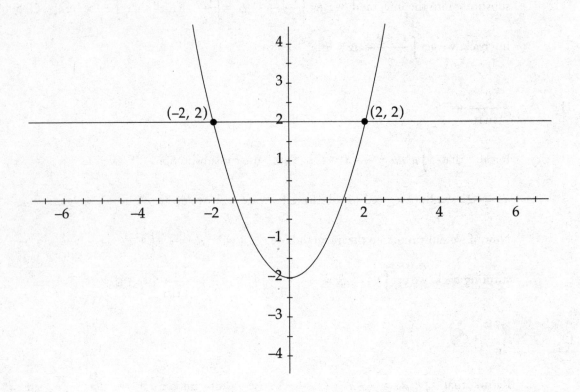

Next, we need to find where the two curves intersect, which will be the endpoints of the region.

We do this by setting the two curves equal to each other. We get $x^2 - 2 = 2$. The solutions are

$(-2, 0)$ and $(2, 0)$. Therefore, in order to find the area of the region, we need to evaluate the integral

$\int_{-2}^2 \left(2 - \left(x^2 - 2 \right) \right) dx = \int_{-2}^2 \left(4 - x^2 \right) dx$. We get

$$\int_{-2}^2 \left(4 - x^2 \right) dx = \left(4x - \frac{x^3}{3} \right) \Bigg|_{-2}^2 = \left(4(2) - \frac{(2)^3}{3} \right) - \left(4(-2) - \frac{(-2)^3}{3} \right) = \frac{32}{3}.$$

2. $\dfrac{8}{3}$

We find the area of a region bounded by $f(x)$ above and $g(x)$ below at all points of the interval $[a, b]$ using the formula $\int_a^b \left[f(x) - g(x) \right] dx$. Here $f(x) = 4x - x^2$ and $g(x) = x^2$.

First, let's make a sketch of the region.

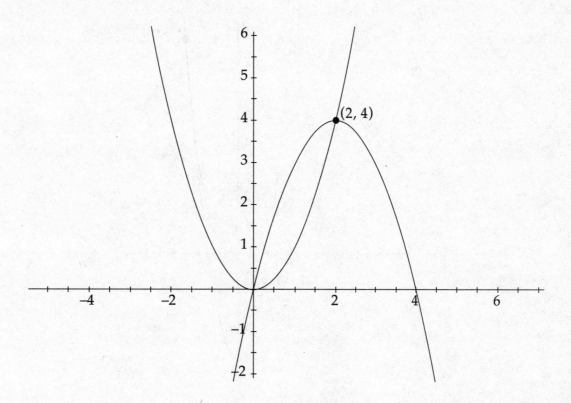

Next, we need to find where the two curves intersect, which will be the endpoints of the region. We do this by setting the two curves equal to each other. We get

$4x - x^2 = x^2$. The solutions are $(0, 0)$ and $(2, 4)$. Therefore, in order to find the area of the region, we need to evaluate the integral $\int_0^2 \left(\left(4x - x^2 \right) - x^2 \right) dx = \int_0^2 \left(4x - 2x^2 \right) dx$. We get

$$\int_0^2 \left(4x - 2x^2 \right) dx = \left(2x^2 - \frac{2x^3}{3} \right) \Bigg|_0^2 = \left(2(2)^2 - \frac{2(2)^3}{3} \right) - 0 = \frac{8}{3}.$$

3. $\dfrac{27}{4}$

We find the area of a region bounded by $f(x)$ above and $g(x)$ below at all points of the interval $[a, b]$ using the formula $\int_a^b \left[f(x) - g(x) \right] dx$. Here $f(x) = x^3$ and $g(x) = 3x^2 - 4$.

First, let's make a sketch of the region.

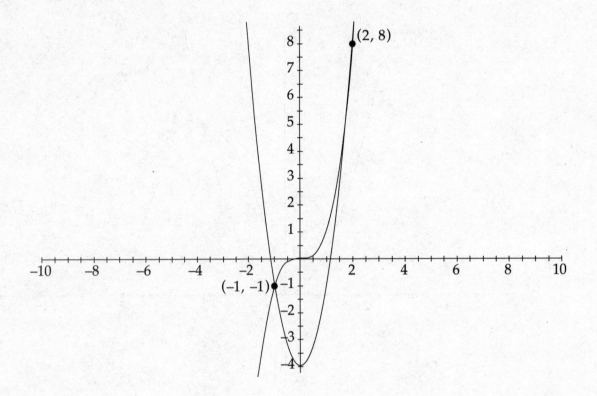

Next, we need to find where the two curves intersect, which will be the endpoints of the region. We do this by setting the two curves equal to each other. We get $3x^2 - 4 = x^3$.

The solutions are $(-1, -1)$ and $(2, 8)$. Therefore, in order to find the area of the region, we need to evaluate the integral $\int_{-1}^{2} \left(x^3 - \left(3x^2 - 4 \right) \right) dx = \int_{-1}^{2} \left(x^3 - 3x^2 + 4 \right) dx$. We get

$$\int_{-1}^{2} \left(x^3 - 3x^2 + 4 \right) dx = \left. \left(\frac{x^4}{4} - x^3 + 4x \right) \right|_{-1}^{2} = \left(\frac{(2)^4}{4} - (2)^3 + 4(2) \right) - \left(\frac{(-1)^4}{4} - (-1)^3 + 4(-1) \right) = \frac{27}{4}.$$

4.   36

We find the area of a region bounded by $f(x)$ above and $g(x)$ below at all points of the interval $[a, b]$ using the formula $\int_a^b \left[f(x) - g(x) \right] dx$. Here $f(x) = 2x - 5$ and $g(x) = x^2 - 4x - 5$. First, let's make a sketch of the region.

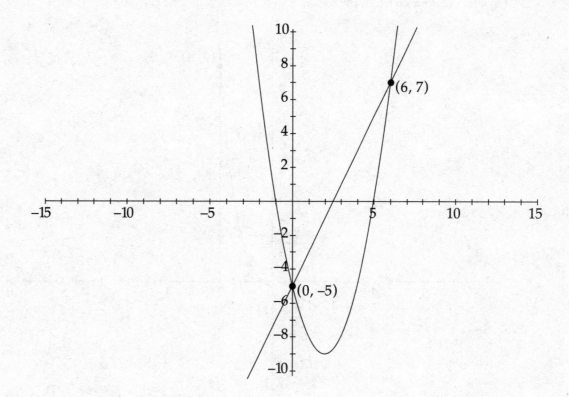

Next, we need to find where the two curves intersect, which will be the endpoints of the region. We do this by setting the two curves equal to each other. We get $x^2 - 4x - 5 = 2x - 5$. The solutions are $(0, -5)$ and $(6, 7)$. Therefore, in order to find the area of the region, we need to evaluate the integral $\int_0^6 \left[(2x - 5) - (x^2 - 4x - 5) \right] dx = \int_0^6 (-x^2 + 6x) dx$. We get

$$\int_0^6 (-x^2 + 6x) dx = \left(-\frac{x^3}{3} + 3x^2 \right) \Bigg|_0^6 = \left(-\frac{(6)^3}{3} + 3(6)^2 \right) - 0 = 36.$$

5. $\dfrac{17}{4}$

We find the area of a region bounded by $f(x)$ above and $g(x)$ below at all points of the interval $[a, b]$

using the formula $\displaystyle\int_a^b \left[f(x) - g(x) \right] dx$.

First, let's make a sketch of the region.

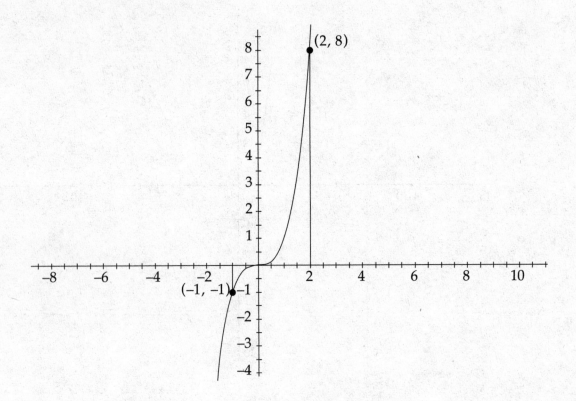

Notice that in the region from $x = -1$ to $x = 0$ the top curve is $f(x) = 0$ (the x-axis), and the

bottom curve is $g(x) = x^3$, but from $x = 0$ to $x = 2$ the situation is reversed, so the top curve is

$f(x) = x^3$, and the bottom curve is $g(x) = 0$. Thus, we split the region into two pieces and find the

area by evaluating two integrals and adding the answers: $\displaystyle\int_{-1}^0 \left(0 - x^3\right) dx$ and $\displaystyle\int_0^2 \left(x^3 - 0\right) dx$. We get

$\displaystyle\int_{-1}^0 \left(0 - x^3\right) dx = \left(-\dfrac{x^4}{4} \right)\Bigg|_{-1}^0 = 0 - \left(-\dfrac{(-1)^4}{4} \right) = \dfrac{1}{4}$ and $\displaystyle\int_0^2 \left(x^3 - 0\right) dx = \left(\dfrac{x^4}{4} \right)\Bigg|_0^2 = \dfrac{(2)^4}{4} - 0 = 4$. There-

fore, the area of the region is $\dfrac{17}{4}$.

6. $\dfrac{9}{2}$

We find the area of a region bounded by $f(y)$ on the right and $g(y)$ on the left at all points of the interval $[c, d]$ using the formula $\displaystyle\int_c^d \left[f(y) - g(y) \right] dy$. Here, $f(y) = y + 2$ and $g(y) = y^2$.

First, let's make a sketch of the region.

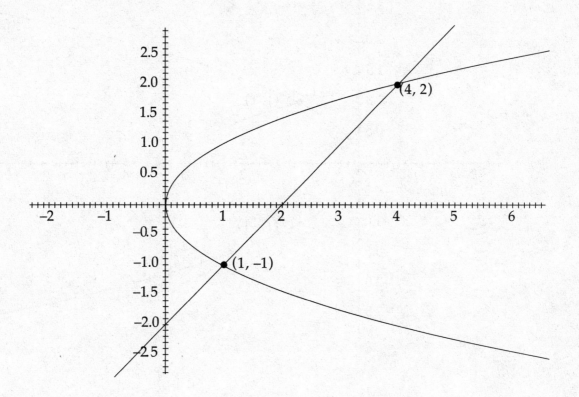

Next, we need to find where the two curves intersect, which will be the endpoints of the region.

We do this by setting the two curves equal to each other. We get $y^2 = y + 2$. The solutions are $(4, 2)$

and $(1, -1)$. Therefore, in order to find the area of the region, we need to evaluate the integral

$\displaystyle\int_{-1}^{2} \left(y + 2 - y^2 \right) dy$. We get

$$\int_{-1}^{2} \left(y + 2 - y^2 \right) dy = \left(\frac{y^2}{2} + 2y - \frac{y^3}{3} \right) \Bigg|_{-1}^{2} = \left(\frac{(2)^2}{2} + 2(2) - \frac{(2)^3}{3} \right) - \left(\frac{(-1)^2}{2} + 2(-1) - \frac{(-1)^3}{3} \right) = \frac{9}{2}.$$

7.   4

We find the area of a region bounded by $f(y)$ on the right and $g(y)$ on the left at all points of the interval $[c, d]$ using the formula $\int_c^d \left[f(y) - g(y) \right] dy$. Here, $f(y) = 3 - 2y^2$ and $g(y) = y^2$.

First, let's make a sketch of the region.

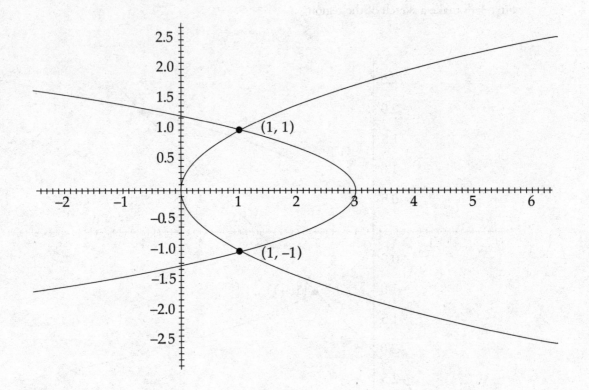

Next, we need to find where the two curves intersect, which will be the endpoints of the region. We do this by setting the two curves equal to each other. We get $y^2 = 3 - 2y^2$. The solutions are $(1,1)$ and $(1, -1)$. Therefore, in order to find the area of the region, we need to evaluate the integral $\int_{-1}^1 \left(3 - 2y^2 - y^2 \right) dy = \int_{-1}^1 \left(3 - 3y^2 \right) dy$. We get

$$\int_{-1}^1 \left(3 - 3y^2 \right) dy = \left(3y - y^3 \right) \Big|_{-1}^1 = \left(3(1) - (1)^3 \right) - \left(3(-1) - (-1)^3 \right) = 4.$$

8. $\dfrac{37}{12}$

We find the area of a region bounded by $f(y)$ on the right and $g(y)$ on the left at all points of the interval $[c, d]$ using the formula $\int_{c}^{d}\left[f(y) - g(y)\right]dy$. Here $f(y) = y^3 - y^2$ and $g(y) = 2y$.

First, let's make a sketch of the region.

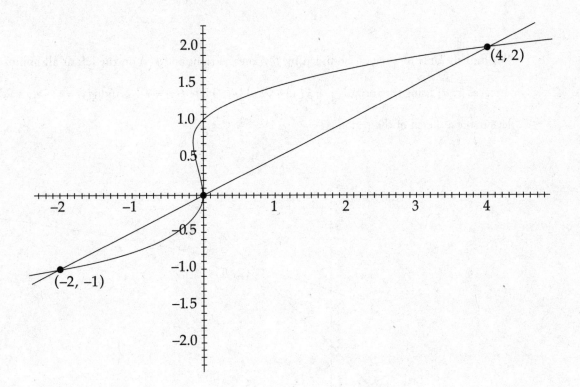

Next, we need to find where the two curves intersect, which will be the endpoints of the region. We do this by setting the two curves equal to each other. We get $y^3 - y^2 = 2y$. The solutions are $(4, 2)$, $(0, 0)$, and $(-2, -1)$. Notice that in the region from $y = -1$ to $y = 0$, the right curve is $f(y) = y^3 - y^2$ and the left curve is $g(y) = 2y$, but from $y = 0$ to $y = 2$ the situation is reversed, so the right curve is $f(y) = 2y$ and the left curve is $g(y) = y^3 - y^2$. Thus, we split the region into two pieces and find the area by evaluating two integrals and adding the answers: $\int_{-1}^{0}\left(y^3 - y^2 - 2y\right)dy$ and $\int_{0}^{2}\left(2y - \left(y^3 - y^2\right)\right)dy = \int_{0}^{2}\left(2y - y^3 + y^2\right)dy$.

We get $\int_{-1}^{0}(y^3 - y^2 - 2y)dy = \left(\dfrac{y^4}{4} - \dfrac{y^3}{3} - y^2\right)\Bigg|_{-1}^{0} = 0 - \left(\dfrac{(-1)^4}{4} - \dfrac{(-1)^3}{3} - (-1)^2\right) = \dfrac{5}{12}$ and

$\int_{0}^{2}(2y - y^3 + y^2)dy = \left(y^2 - \dfrac{y^4}{4} + \dfrac{y^3}{3}\right)\Bigg|_{0}^{2} = \dfrac{8}{3}$. Therefore, the area of the region is $\dfrac{5}{12} + \dfrac{8}{3} = \dfrac{37}{12}$.

9. $\dfrac{9}{2}$

We find the area of a region bounded by $f(y)$ on the right and $g(y)$ on the left at all points of the interval $[c, d]$ using the formula $\int_{c}^{d}\left[f(y) - g(y)\right]dy$. Here $f(y) = y - 2$ and $g(y) = y^2 - 4y + 2$. First, let's make a sketch of the region.

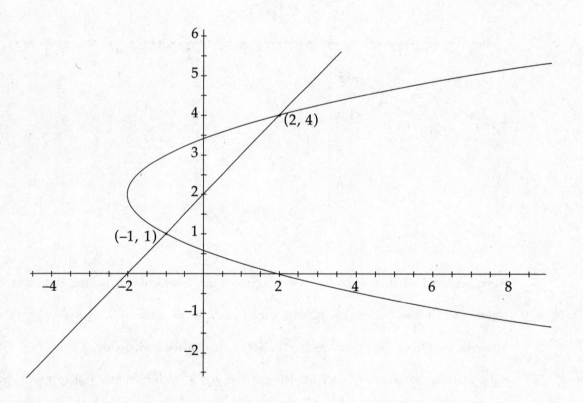

Next, we need to find where the two curves intersect, which will be the endpoints of the region. We do this by setting the two curves equal to each other. We get $y^2 - 4y + 2 = y - 2$. The solutions are $(2, 4)$ and $(-1, 1)$. Therefore, in order to find the area of the region, we need to evaluate the integral $\int_1^4 \left[(y-2) - \left(y^2 - 4y + 2 \right) \right] dy = \int_1^4 \left[\left(-y^2 + 5y - 4 \right) \right] dy$. We get $\int_1^4 \left[\left(-y^2 + 5y - 4 \right) \right] dy =$

$$\left(-\frac{y^3}{3} + \frac{5y^2}{2} - 4y \right)\Bigg|_1^4 = \left(-\frac{(4)^3}{3} + \frac{5(4)^2}{2} - 4(4) \right) - \left(-\frac{(1)^3}{3} + \frac{5(1)^2}{2} - 4(1) \right) = \frac{9}{2}.$$

10. $\dfrac{12}{5}$

We find the area of a region bounded by $f(y)$ on the right and $g(y)$ on the left at all points of the interval $[c, d]$ using the formula $\int_c^d \left[f(y) - g(y) \right] dy$. Here $f(y) = 2 - y^4$ and $g(y) = y^{\frac{2}{3}}$. First, let's make a sketch of the region.

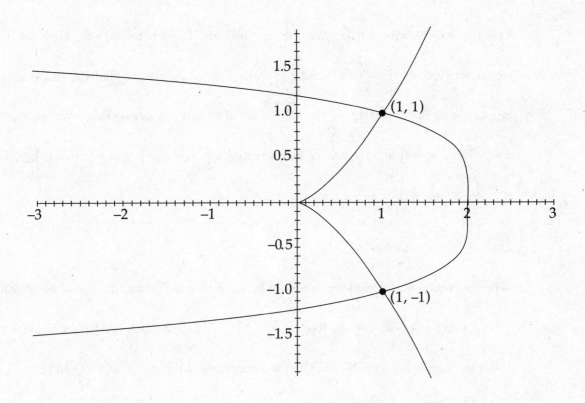

Next, we need to find where the two curves intersect, which will be the end-points of the region. We do this by setting the two curves equal to each other. We get $2 - y^4 = y^{\frac{2}{3}}$. The solutions are $(1, 1)$ and $(1, -1)$. Therefore, in order to find the area of the region, we need to evaluate the integral $\int_{-1}^{1}\left(2-y^4-y^{\frac{2}{3}}\right)dy$. We get

$$\int_{-1}^{1}\left(2-y^4-y^{\frac{2}{3}}\right)dy=\left(2y-\frac{y^5}{5}-\frac{3}{5}y^{\frac{5}{3}}\right)\Bigg|_{-1}^{1}=\left(2(1)-\frac{(1)^5}{5}-\frac{3}{5}(1)^{\frac{5}{3}}\right)-\left(2(-1)-\frac{(-1)^5}{5}-\frac{3}{5}(-1)^{\frac{5}{3}}\right)=\frac{12}{5}.$$

SOLUTIONS TO PRACTICE PROBLEM SET 23

1. $36\,\pi$

When the region we are revolving is defined between a curve $f(x)$ and the x-axis, we can find the volume using disks. We use the formula $V=\pi\int_{a}^{b}\left[f(x)\right]^2dx$. Here we have a region between $f(x)=\sqrt{9-x^2}$ and the x-axis. First, we need to find the endpoints of the region. We do this by setting $f(x)=\sqrt{9-x^2}$ equal to zero and solving for x. We get $x=-3$ and $x=3$. Thus, we will find the volume by evaluating $\pi\int_{-3}^{3}\left(\sqrt{9-x^2}\right)^2dx=\pi\int_{-3}^{3}\left(9-x^2\right)dx$. We get

$$\pi\int_{-3}^{3}\left(9-x^2\right)dx=\pi\left(9x-\frac{x^3}{3}\right)\Bigg|_{-3}^{3}=36\pi.$$

2. $2\,\pi$

When the region we are revolving is defined between a curve $f(x)$ and the x-axis, we can find the volume using disks. We use the formula $V=\pi\int_{a}^{b}\left[f(x)\right]^2dx$. Here we have a region between $f(x)=\sec x$ and the x-axis. We are given the endpoints of the region: $x=-\frac{\pi}{4}$ and $x=\frac{\pi}{4}$. Thus, we will find the volume by evaluating $\pi\int_{-\frac{\pi}{4}}^{\frac{\pi}{4}}\sec^2 x\,dx$. We get $\pi\int_{-\frac{\pi}{4}}^{\frac{\pi}{4}}\sec^2 x\,dx=\pi\left(\tan x\right)\Big|_{-\frac{\pi}{4}}^{\frac{\pi}{4}}=2\pi$.

3. $\dfrac{16\pi}{15}$

When the region we are revolving is defined between a curve $f(y)$ and the y-axis, we can find the volume using disks. We use the formula $V = \pi \int_a^b \left[f(y) \right]^2 dy$ (see page 261 and note that when $g(y) = 0$ we get disks instead of washers). Here we have a region between $f(y) = 1 - y^2$ and the y-axis. First, we need to find the endpoints of the region. We do this by setting $f(y) = 1 - y^2$ equal to zero and solving for y. We get $y = -1$ and $y = 1$. Thus, we will find the volume by evaluating $\pi \int_{-1}^1 \left(1 - y^2\right)^2 dy = \pi \int_{-1}^1 \left(1 - 2y^2 + y^4\right) dy$. We get $\pi \int_{-1}^1 \left(1 - 2y^2 + y^4\right) dy = \pi \left(y - \dfrac{2y^3}{3} + \dfrac{y^5}{5} \right)\Big|_{-1}^1 = \dfrac{16\pi}{15}$.

4. 2π

When the region we are revolving is defined between a curve $f(y)$ and the y-axis, we can find the volume using disks. We use the formula $V = \pi \int_a^b \left[f(y) \right]^2 dy$ (see page 261 and note that when $g(y) = 0$ we get disks instead of washers). Here we have a region between $f(y) = \sqrt{5} y^2$ and the y-axis. We are given the endpoints of the region: $y = -1$ and $y = 1$. Thus, we will find the volume by evaluating $\pi \int_{-1}^1 \left(\sqrt{5} y^2\right)^2 dy = \pi \int_{-1}^1 \left(5 y^4\right) dy$. We get $\pi \int_{-1}^1 \left(5 y^4\right) dy = \pi \left(y^5 \right)\Big|_{-1}^1 = 2\pi$.

5. $\dfrac{16\pi}{5}$

When the region we are revolving is defined between a curve $f(x)$ and $g(x)$, we can find the volume using cylindrical shells. We use the formula $V = 2\pi \int_a^b x \left[f(x) - g(x) \right] dx$. Here we have a region between $f(x) = x^3$ and the line $x = 2$ that we are revolving around the line $x = 2$. If we use vertical slices, then we will need to use cylindrical shells to find the volume. If we use horizontal slices, then we will need to use washers to find the volume. Here we will use cylindrical shells. (Try doing it yourself using washers. You should get the same answer but it is much harder!) First, we need to find the endpoints of the region. We get the left endpoint by setting $f(x) = x^3$ equal to zero and solving for x. We get $x = 0$. The right endpoint is simply $x = 2$. Next, note that we are not revolving around the x-axis but around the line $x = 2$. Thus, the radius of each shell is not x but rather $2 - x$. The height of each shell is simply $f(x) - g(x) = x^3 - 0 = x^3$. Therefore, we will find the volume by evaluating

$2\pi \int_0^2 \left[(2-x)(x^3) \right] dx = 2\pi \int_0^2 (2x^3 - x^4) dx$. We get $2\pi \int_0^2 (2x^3 - x^4) dx = 2\pi \left(\dfrac{x^4}{2} - \dfrac{x^5}{5} \right) \Big|_0^2 = \dfrac{16\pi}{5}$.

6. 8π

When the region we are revolving is defined between a curve $f(x)$ and $g(x)$, we can find the volume using cylindrical shells. We use the formula $V = 2\pi \int_a^b x\left[f(x) - g(x) \right] dx$. Here we have $f(x) = x$ and $g(x) = -\dfrac{x}{2}$. Thus, the height of each shell is $f(x) - g(x) = x - \left(-\dfrac{x}{2} \right) = \dfrac{3x}{2}$, and the radius is simply x. The endpoints of our region are $x = 0$ and $x = 2$. Therefore, we will find the volume by evaluating: $2\pi \int_0^2 \left(x \left(\dfrac{3x}{2} \right) \right) dx = 2\pi \int_0^2 \left(\dfrac{3x^2}{2} \right) dx = 3\pi \int_0^2 x^2 dx$. We get $3\pi \int_0^2 x^2 dx = 3\pi \left(\dfrac{x^3}{3} \right) \Big|_0^2 = 8\pi$.

7. $\dfrac{7\pi}{15}$

When the region we are revolving is defined between a curve $f(x)$ and $g(x)$, we can find the volume using cylindrical shells. We use the formula $V = 2\pi \int_a^b x\left[f(x) - g(x) \right] dx$. Here we have $f(x) = \sqrt{x}$ and $g(x) = 2x - 1$. Thus, the height of each shell is $f(x) - g(x) = \sqrt{x} - (2x - 1) = \sqrt{x} - 2x + 1$, and the radius is simply x. The left endpoint of our region is $x = 0$, and we find the right endpoint by finding where $f(x) = \sqrt{x}$ intersects $g(x) = 2x - 1$. We get $x = 1$. Therefore, we will find the

volume by evaluating: $2\pi \int_0^1 \left(x \left(\sqrt{x} - 2x + 1 \right) \right) dx = 2\pi \int_0^1 \left(x^{\frac{3}{2}} - 2x^2 + x \right) dx$. We get

$$2\pi \int_0^1 \left(x^{\frac{3}{2}} - 2x^2 + x \right) dx = 2\pi \left(\frac{2x^{\frac{5}{2}}}{5} - \frac{2x^3}{3} + \frac{x^2}{2} \right) \Bigg|_0^1 = \frac{7\pi}{15}.$$

8. $\dfrac{128\pi}{5}$

When the region we are revolving is defined between a curve $f(y)$ and $g(y)$, we can find the

volume using cylindrical shells. We use the formula $V = 2\pi \int_a^b y \left[f(y) - g(y) \right] dy$. Here we have

$f(y) = \sqrt{y}$ (which we get by solving $y = x^2$ for x) and $g(y) = 0$ (the y-axis). We can easily see

that the top endpoint is $y = 4$ and the bottom one is $y = 0$. Therefore, we will find the volume by

evaluating: $2\pi \int_0^4 \left(y\sqrt{y} \right) dy = 2\pi \int_0^4 \left(y^{\frac{3}{2}} \right) dy$. We get $2\pi \int_0^4 \left(y^{\frac{3}{2}} \right) dy = 2\pi \left(\frac{2y^{\frac{5}{2}}}{5} \right) \Bigg|_0^4 = \frac{128\pi}{5}$.

9. $\dfrac{256\pi}{5}$

When the region we are revolving is defined between a curve $f(x)$ and $g(x)$, we can find the

volume using cylindrical shells. We use the formula $V = 2\pi \int_a^b x \left[f(x) - g(x) \right] dx$. Here we

have $f(x) = 2\sqrt{x}$ and $g(x) = 0$. Thus, the height of each shell is $f(x) - g(x) = 2\sqrt{x}$, and the

radius is simply x. We can easily see that the left endpoint is $x = 0$ and that the right endpoint is

$x = 4$. Therefore, we will find the volume by evaluating: $2\pi \int_0^4 \left(x \left(2\sqrt{x} \right) \right) dx = 2\pi \int_0^4 \left(2x^{\frac{3}{2}} \right) dx$. We

get $2\pi \int_0^4 \left(2x^{\frac{3}{2}} \right) dx = 4\pi \left(\frac{2x^{\frac{5}{2}}}{5} \right) \Bigg|_0^4 = \frac{256\pi}{5}$.

10. $\dfrac{896\pi}{15}$

When the region we are revolving is defined between a curve $f(x)$ and $g(x)$, we can

find the volume using cylindrical shells. We use the formula $V = 2\pi \int_{a}^{b} x\left[f(x) - g(x)\right]dx$. Here

we have $f(x) = 2\sqrt{2x}$ (which we get by solving $y^2 = 8x$ for y and taking the top half above

the x-axis) and $g(x) = -2\sqrt{2x}$ (the bottom half). Thus, the height of each shell is

$f(x) - g(x) = 4\sqrt{2x}$. We can easily see that the left endpoint is $x = 0$ and that the right

endpoint is $x = 2$. Next, note that we are not revolving around the x-axis but around the line

$x = 4$. Thus, the radius of each shell is not x but rather $4 - x$. Therefore, we will find

the volume by evaluating: $2\pi \int_{0}^{2}\left[(4-x)4\sqrt{2x}\right]dx = 8\pi\sqrt{2}\int_{0}^{2}\left(4x^{\frac{1}{2}} - x^{\frac{3}{2}}\right)dx$. We get

$8\pi\sqrt{2}\int_{0}^{2}\left(4x^{\frac{1}{2}} - x^{\frac{3}{2}}\right)dx = 8\pi\sqrt{2}\left(\dfrac{8x^{\frac{3}{2}}}{3} - \dfrac{2x^{\frac{5}{2}}}{5}\right)\Bigg|_{0}^{2} = \dfrac{896\pi}{15}.$

11. $\dfrac{256}{3}$

To find the volume of a solid with a cross-section of an isosceles right triangle, we integrate

the area of the square ($side^2$) over the endpoints of the interval. Here the sides of the squares

are found by $f(x) - g(x) = \sqrt{16 - x^2} - 0$, and the intervals are found by setting

$y = \sqrt{16 - x^2}$ equal to zero. We get $x = -4$ and $x = 4$. Thus, we find the volume by evaluating the

integral $\int_{-4}^{4} \left(\sqrt{16 - x^2} \right)^2 dx = \int_{-4}^{4} \left(16 - x^2 \right) dx$. We get $\int_{-4}^{4} \left(16 - x^2 \right) dx = \left(16x - \frac{x^3}{3} \right) \Big|_{-4}^{4} = \frac{256}{3}$.

12. $\dfrac{128}{15}$

To find the volume of a solid with a cross-section of a square, we integrate the area

of the isosceles right triangle $\dfrac{hypotenuse^2}{4}$ over the endpoints of the interval. Here,

the hypotenuses of the triangles are found by $f(x) - g(x) = 4 - x^2$, and the intervals

are found by setting $y = x^2$ equal to $y = 4$. We get $x = -2$ and $x = 2$. Thus, we find the

volume by evaluating the integral $\int_{-2}^{2} \frac{\left(4 - x^2 \right)^2}{4} dx = \int_{-2}^{2} \frac{\left(16 - 8x^2 + x^4 \right)}{4} dx = \int_{-2}^{2} \left(4 - 2x^2 + \frac{x^4}{4} \right) dx$.

We get $\int_{-2}^{2} \left(4 - 2x^2 + \frac{x^4}{4} \right) dx = \left(4x - \frac{2x^3}{3} + \frac{x^5}{20} \right) \Big|_{-2}^{2} = \frac{128}{15}$.

SOLUTIONS TO PRACTICE PROBLEM SET 24

1. $y = \sqrt[4]{\dfrac{28x^3}{3} - 236}$

We solve this differential equation by separation of variables. We want to get all of the y variables on one side of the equals sign and all of the x variables on the other side. We can do this easily by cross multiplying. We get $y^3 dy = 7x^2 dx$. Next, we integrate both sides.

$$\int y^3 dy = \int 7x^2 dx$$

$$\frac{y^4}{4} = 7\frac{x^3}{3} + C$$

$$y^4 = \frac{28x^3}{3} + C$$

Now we solve for C. We plug in $x = 3$ and $y = 2$.

$16 = 252 + C$

$C = -236$

Therefore, $y^4 = \dfrac{28x^3}{3} - 236$.

Now we isolate y. We get the equation $y = \sqrt[4]{\dfrac{28x^3}{3} - 236}$.

2. $y = 6e^{\frac{5x^3}{3}}$

We solve this differential equation by separation of variables. We want to get all of the y variables on one side of the equals sign and all of the x variables on the other side. We can do this easily by dividing both sides by y and multiplying both sides by dx. We get $\dfrac{dy}{y} = 5x^2 dx$. Next, we integrate both sides.

$$\int \frac{dy}{y} = \int 5x^2 dx$$

$$\ln y = \frac{5x^3}{3} + C_0$$

Now we isolate y: $y = e^{\frac{5x^3}{3} + C}$. We can rewrite this as $y = e^{\frac{5x^3}{3}}\left(e^{C_0}\right) = Ce^{\frac{5x^3}{3}}$.

Note that we are using the letter C in the last equation. This is to distinguish it from the C_0 in the first equation. Now we solve for C. We plug in $x = 0$ and $y = 6$: $6 = Ce^0 = C$. Therefore, the equation is $y = 6e^{\frac{5x^3}{3}}$.

3. $y = \sqrt{2\tan^{-1}x + 4}$

We solve this differential equation by separation of variables. We want to get all of the y variables on one side of the equals sign and all of the x variables on the other side. First, we factor the y out of the denominator of the right hand expression: $\dfrac{dy}{dx} = \dfrac{1}{y + x^2 y} = \dfrac{1}{y(1 + x^2)}$. Next, we multiply both sides by y and by dx. We get $y\,dy = \dfrac{dx}{1 + x^2}$. Next, we integrate both sides.

$$\int y\,dy = \int \frac{dx}{1 + x^2}$$

$$\frac{y^2}{2} = \tan^{-1}x + C$$

Now we isolate y: $y^2 = 2\tan^{-1}x + C$.

$$y = \sqrt{2\tan^{-1}x + C}$$

Now we solve for C. We plug in $x = 0$ and $y = 2$.

$$2 = \sqrt{2\tan^{-1}0 + C}$$

$$C = 4$$

Therefore, the equation is $y = \sqrt{2\tan^{-1}x + 4}$.

4. $y = \sqrt[3]{3e^x - 2}$

We solve this differential equation by separation of variables. We want to get all of the y variables on one side of the equals sign and all of the x variables on the other side. We can do this easily by cross multiplying. We get $y^2\,dy = e^x dx$. Next, we integrate both sides.

$$\int y^2\,dy = \int e^x dx$$

$$\frac{y^3}{3} = e^x + C$$

Now we isolate y.

$$y^3 = 3e^x + C$$

$$y = \sqrt[3]{3e^x + C}$$

Now we solve for C. We plug in $x = 0$ and $y = 1$.

$$1 = \sqrt[3]{3e^0 + C}$$

$$C = -2$$

Therefore, the equation is $y = \sqrt[3]{3e^x - 2}$.

5. $y = 2x^2$

We solve this differential equation by separation of variables. We want to get all of the y variables on one side of the equals sign and all of the x variables on the other side. We can do this easily by dividing both sides by y^2 and multiplying both sides by dx. We get

$$\frac{dy}{y^2} = \frac{dx}{x^3}$$. Next, we integrate both sides.

$$\int \frac{dy}{y^2} = \int \frac{dx}{x^3}$$

$$-\frac{1}{y} = -\frac{1}{2x^2} + C$$

Now we isolate y.

$$\frac{1}{y} = \frac{1}{2x^2} + C$$

Now we solve for C. We plug in $x = 1$ and $y = 2$.

$$\frac{1}{2} = \frac{1}{2(1)^2} + C$$

$$C = 0$$

Therefore, the equation is $\dfrac{1}{y} = \dfrac{1}{2x^2} + C$, which can be rewritten as $y = 2x^2$.

6. $y = \sin^{-1}(-\cos x)$

We solve this differential equation by separation of variables. We want to get all of the y variables on one side of the equals sign and all of the x variables on the other side. We can do this easily by cross multiplying. We get $\cos y\, dy = \sin x\, dx$. Next, we integrate both sides.

$$\int \cos y\, dy = \int \sin x\, dx$$

$$\sin y = -\cos x + C$$

Now we solve for C. We plug in $x = 0$ and $y = \dfrac{3\pi}{2}$.

$$\sin\frac{3\pi}{2} = -\cos 0 + C$$

$$-1 = -1 + C$$

$$C = 0$$

Now we isolate y to get the equation $y = \sin^{-1}(-\cos x)$.

7.   20,000 (approximately)

The phrase "exponential growth" means that we can represent the situation with the differential equation $\dfrac{dy}{dt} = ky$, where k is a constant and y is the population at a time t. Here we are also told that $y = 4{,}000$ at $t = 0$ and $y = 6{,}500$ at $t = 3$. We solve this differential equation by separation of variables. We want to get all of the y variables on one side of the equals sign and all of the t variables on the other side. We can do this easily by dividing both sides by y and multiplying both sides by dt. We get $\dfrac{dy}{y} = k\,dt$. Next, we integrate both sides.

$$\int \frac{dy}{y} = \int k\,dt$$

$$\ln y = kt + C_0$$
$$y = Ce^{kt}$$

Next, we plug in $y = 4{,}000$ and $t = 0$ to solve for C.

$$4{,}000 = Ce^0$$

$$C = 4{,}000$$

Now, we have $y = 4{,}000e^{kt}$. Next, we plug in $y = 6{,}500$ and $t = 3$ to solve for k.

$$6{,}500 = 4{,}000e^{3k}$$

$$k = \frac{1}{3}\ln\frac{13}{8} \approx 0.162$$

Therefore, our equation for the population of bacteria, y, at time, t, is $y \approx 4{,}000e^{0.162t}$, or if we want an exact solution, it is $y = 4{,}000\left(\dfrac{13}{8}\right)^{\frac{t}{3}}$. Finally, we can solve for the population at time $t = 10$: $y \approx 4{,}000e^{.162(10)} \approx 20{,}212$ or $y = 4{,}000\left(\dfrac{13}{8}\right)^{\frac{10}{3}} \approx 20{,}179$. (Notice how even with an "exact" solution, we still have to round the answer. And, if we are concerned with significant figures, the population can be written as 20,000.)

8.   45 m

Because acceleration is the derivative of velocity, we can write $\dfrac{dv}{dt} = -9$. We are also told that $v = 18$ and $h = 45$ when $t = 0$. We solve this differential equation by separation of variables. We want to get all of the v variables on one side of the equals sign and all of the t variables on the other side. We can do this easily by multiplying both sides by dt: $dv = -9\,dt$. Next, we integrate both sides:

$$\int dv = \int -9\,dt$$

$$v = -9 + C_0$$

Now we can solve for C_0 by plugging in $v = 18$ and $t = 0$: $18 = -9\,(0) + C_0$ so $C_0 = 18$. Thus, our equation for the velocity of the rock is $v = -9t + 18$. Next, because the height of the rock is the derivative of the velocity, we can write $\dfrac{dh}{dt} = -9t + 18$. We again separate the variables by multiplying both sides by dt: $dh = (-9t + 18)\,dt$. We integrate both sides:

$$\int dh = \int (-9t + 18)\,dt$$

$$h = -\frac{9t^2}{2} + 18t + C_1$$

Next, we plug in $h = 45$ and $t = 0$ to solve for C_1.

$$45 = -\frac{9(0)^2}{2} + 18(0) + C_1, \text{ so } C_1 = 45$$

Therefore, the equation for the height of the rock, h, at time, t, is $h = -\dfrac{9t^2}{2} + 18t + 45$.

Finally, we can solve for the height of the rock at time $t = 4$: $h = -\dfrac{9t^2}{2} + 18t + 45$, so $h = -\dfrac{9(4)^2}{2} + 18(4) + 45 = 45$ m.

9. $\dfrac{1{,}125\pi}{2} \approx 1{,}800 \text{ ft}^3$

We can express this situation with the differential equation $\dfrac{dV}{dt} = kV$ where k is a constant and V is the volume of the sphere at time t. We are also told that $V = 36\pi$ when $t = 0$ and $V = 90\pi$ when $t = 1$. We solve this differential equation by separation of variables. We want to get all of the V variables on one side of the equals sign and all of the t variables on the other side. We can do this easily by dividing both sides by V and multiplying both sides by dt. We get $\dfrac{dV}{V} = k\, dt$. Next, we integrate both sides.

$\displaystyle \int \dfrac{dV}{V} = \int k\, dt$

$\ln V = kt + C_0$

$V = Ce^{kt}$

Next, we plug in $V = 36\pi$ and $t = 0$ to solve for C: $36\pi = Ce^0$ so $C = 36\pi$. This gives us the equation $V = 36\pi e^{kt}$. Next, we plug in $V = 90\pi$ and $t = 1$ to solve for k: $90\pi = 36\pi e^k$.

$k = \ln \dfrac{5}{2} \approx 0.916$

Therefore, the equation for the volume of the sphere, V, at time, t, is $V \approx 36\pi e^{0.916t}$, or if we want an exact solution, it is $V = 36\pi \dfrac{5^t}{2}$. Finally, we can solve for the volume at time $t = 3$: $V \approx 36\pi e^{.916(3)} \approx 1{,}766 \text{ ft}^3$ or $V = \dfrac{1{,}125\pi}{2} \text{ ft}^3$. (And, if we are concerned with significant figures, the volume can be written as 1,800.)

10.   8,900 grams (approximately)

We can express this situation with the differential equation $\frac{dm}{dt} = -km$, where m is the mass at time t. We are also given that $m = 10,000$ when $t = 0$ and $m = 5,000$ when $t = 5,750$. We solve this differential equation by separation of variables. We want to get all of the m variables on one side of the equals sign and all of the t variables on the other side. We can do this easily by dividing both sides by m and multiplying both sides by dt. We get $\frac{dm}{m} = -k\,dt$. Next, we integrate both sides.

$$\int \frac{dm}{m} = \int -k\,dt$$

$$\ln m = -kt + C_0$$

$$m = Ce^{-kt}$$

Now we can solve for C by plugging in $m = 10,000$ and $t = 0$: $10,000 = Ce^0$, so $C = 10,000$. This gives us the equation $m = 10,000e^{-kt}$. Next, we can solve for k by plugging in $m = 5,000$ and $t = 5,750$.

$$5,000 = 10,000e^{-5,750k}$$

$$\frac{1}{2} = e^{-5,750k}$$

$$-\frac{1}{5,750}\ln\frac{1}{2} = k \approx 0.000121$$

Therefore, the equation for the mass of the element, m, at time, t, is $m \approx 10,000e^{-0.000121t}$, or if we want an exact solution, it is $m = 10,000\left(\frac{1}{2}\right)^{\frac{t}{5,750}}$.

Finally, we can solve for the mass of the element at time $t = 1,000$:

$m \approx 10,000e^{-.000121(1,000)} \approx 8,860\,\text{gms}$ or $m = 10,000\left(\frac{1}{2}\right)^{\frac{1,000}{5,750}} \approx 8,864\,\text{gms}$. (And, if we are concerned with significant figures, the mass can be written as 8,900.)

11.

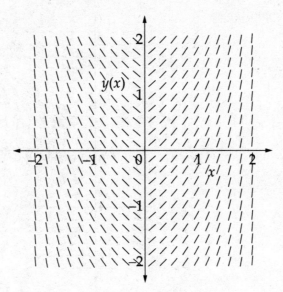

12.

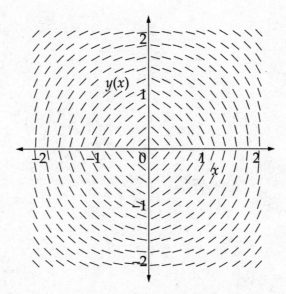

13.

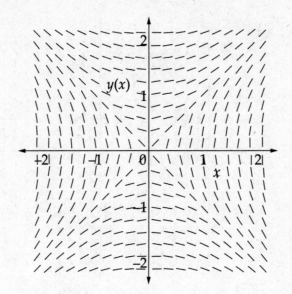

Part V
Practice Tests

Chapter 20
Practice Test 1

AP® Calculus AB Exam

SECTION I: Multiple-Choice Questions

DO NOT OPEN THIS BOOKLET UNTIL YOU ARE TOLD TO DO SO.

At a Glance

Total Time
1 hour and 45 minutes
Number of Questions
45
Percent of Total Grade
50%
Writing Instrument
Pencil required

Instructions

Section I of this examination contains 45 multiple-choice questions. Fill in only the ovals for numbers 1 through 45 on your answer sheet.

CALCULATORS MAY NOT BE USED IN THIS PART OF THE EXAMINATION.

Indicate all of your answers to the multiple-choice questions on the answer sheet. No credit will be given for anything written in this exam booklet, but you may use the booklet for notes or scratch work. After you have decided which of the suggested answers is best, completely fill in the corresponding oval on the answer sheet. Give only one answer to each question. If you change an answer, be sure that the previous mark is erased completely. Here is a sample question and answer.

Sample Question   Sample Answer

Chicago is a
(A) state
(B) city
(C) country
(D) continent
(E) village

Use your time effectively, working as quickly as you can without losing accuracy. Do not spend too much time on any one question. Go on to other questions and come back to the ones you have not answered if you have time. It is not expected that everyone will know the answers to all the multiple-choice questions.

About Guessing

Many candidates wonder whether or not to guess the answers to questions about which they are not certain. Multiple choice scores are based on the number of questions answered correctly. Points are not deducted for incorrect answers, and no points are awarded for unanswered questions. Because points are not deducted for incorrect answers, you are encouraged to answer all multiple-choice questions. On any questions you do not know the answer to, you should eliminate as many choices as you can, and then select the best answer among the remaining choices.

CALCULUS AB

SECTION I, Part A

Time—55 Minutes

Number of questions—28

A CALCULATOR MAY NOT BE USED ON THIS PART OF THE EXAMINATION

Directions: Solve each of the following problems, using the available space for scratchwork. After examining the form of the choices, decide which is the best of the choices given and fill in the corresponding oval on the answer sheet. No credit will be given for anything written in the test book. Do not spend too much time on any one problem.

In this test: Unless otherwise specified, the domain of a function f is assumed to be the set of all real numbers x for which $f(x)$ is a real number.

1. If $f(x) = 5x^{\frac{4}{3}}$, then $f'(8) =$

   (A) 10

   (B) $\dfrac{40}{3}$

   (C) 40

   (D) 80

   (E) $\dfrac{160}{3}$

GO ON TO THE NEXT PAGE.

2. $\displaystyle\lim_{x\to\infty}\frac{5x^2-3x+1}{4x^2+2x+5}$ is

   (A) 0

   (B) $\dfrac{4}{5}$

   (C) $\dfrac{3}{11}$

   (D) $\dfrac{5}{4}$

   (E) ∞

3. If $f(x)=\dfrac{3x^2+x}{3x^2-x}$, then $f'(x)$ is

   (A) 1

   (B) $\dfrac{6x^2+1}{6x^2-1}$

   (C) $\dfrac{-6}{(3x-1)^2}$

   (D) $\dfrac{-2x^2}{(x^2-x)^2}$

   (E) $\dfrac{36x^3-2x}{(x^2-x)^2}$

GO ON TO THE NEXT PAGE.

4. If the function f is continuous for all real numbers and if $f(x) = \dfrac{x^2 - 7x + 12}{x - 4}$ when $x \neq 4$, then $f(4) =$

   (A) 1

   (B) $\dfrac{8}{7}$

   (C) -1

   (D) 0

   (E) undefined

5. If $x^2 - 2xy + 3y^2 = 8$, then $\dfrac{dy}{dx} =$

   (A) $\dfrac{8 + 2y - 2x}{6y - 2x}$

   (B) $\dfrac{3y - x}{y - x}$

   (C) $\dfrac{2x - 2y}{6y - 2x}$

   (D) $\dfrac{1}{3}$

   (E) $\dfrac{y - x}{3y - x}$

GO ON TO THE NEXT PAGE.

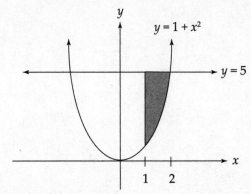

6. Which of the following integrals correctly corresponds to the area of the shaded region in the figure above ?

(A) $\int_{1}^{2} (x^2 - 4)\, dx$

(B) $\int_{1}^{2} (4 - x^2)\, dx$

(C) $\int_{1}^{5} (x^2 - 4)\, dx$

(D) $\int_{1}^{2} (x^2 + 4)\, dx$

(E) $\int_{1}^{5} (4 - x^2)\, dx$

GO ON TO THE NEXT PAGE.

7. If $f(x) = \sec x + \csc x$, then $f'(x) =$

    (A) 0
    (B) $\sec^2 x + \csc^2 x$
    (C) $\csc x - \sec x$
    (D) $\sec x \tan x + \csc x \cot x$
    (E) $\sec x \tan x - \csc x \cot x$

---

8. An equation of the line normal to the graph of $y = \sqrt{(3x^2 + 2x)}$ at $(2, 4)$ is

    (A) $-4x + y = 20$
    (B) $4x + 7y = 20$
    (C) $-7x + 4y = 2$
    (D) $7x + 4y = 30$
    (E) $4x + 7y = 36$

GO ON TO THE NEXT PAGE.

9. $\displaystyle\int_{-1}^{1} \frac{4}{1+x^2}\,dx =$

(A) 0
(B) π
(C) 1
(D) 2π
(E) 2

10. If $f(x) = \cos^2 x$, then $f''(\pi) =$

(A) -2
(B) 0
(C) 1
(D) 2
(E) 2π

GO ON TO THE NEXT PAGE.

11. If $f(x) = \dfrac{5}{x^2+1}$ and $g(x) = 3x$, then $g(f(2)) =$

   (A) -3

   (B) $\dfrac{5}{37}$

   (C) 3

   (D) 5

   (E) $\dfrac{37}{5}$

12. $\displaystyle\int x\sqrt{5x^2-4}\,dx =$

   (A) $\dfrac{1}{10}(5x^2-4)^{\frac{3}{2}} + C$

   (B) $\dfrac{1}{15}(5x^2-4)^{\frac{3}{2}} + C$

   (C) $-\dfrac{1}{5}(5x^2-4)^{-\frac{1}{2}} + C$

   (D) $\dfrac{20}{3}(5x^2-4)^{\frac{3}{2}} + C$

   (E) $\dfrac{3}{20}(5x^2-4)^{\frac{3}{2}} + C$

GO ON TO THE NEXT PAGE.

13. The slope of the line tangent to the graph of $3x^2 + 5 \ln y = 12$ at $(2, 1)$ is

   (A) $-\dfrac{12}{5}$

   (B) $\dfrac{12}{5}$

   (C) $\dfrac{5}{12}$

   (D) 12

   (E) -7

---

14. The equation $y = 2 - 3 \sin \dfrac{\pi}{4}(x - 1)$ has a fundamental period of

   (A) $\dfrac{1}{8}$

   (B) $\dfrac{\pi}{4}$

   (C) $\dfrac{4}{\pi}$

   (D) 8

   (E) 2π

GO ON TO THE NEXT PAGE.

15. If $f(x) = \begin{cases} x^2 + 5 \text{ if } x < 2 \\ 7x - 5 \text{ if } x \ge 2 \end{cases}$, for all real numbers x, which of the following must be true?

    I. $f(x)$ is continuous everywhere.
   II. $f(x)$ is differentiable everywhere.
  III. $f(x)$ has a local minimum at $x = 2$.

(A) I only
(B) I and II only
(C) II and III only
(D) I and III only
(E) I, II, and III

16. For what value of x does the function $f(x) = x^3 - 9x^2 - 120x + 6$ have a local minimum?

(A) 10
(B) 4
(C) 3
(D) −4
(E) −10

GO ON TO THE NEXT PAGE.

17. The acceleration of a particle moving along the x-axis at time t is given by $a(t) = 4t - 12$. If the velocity is 10 when $t = 0$ and the position is 4 when $t = 0$, then the particle is changing direction at

   (A) $t = 1$

   (B) $t = 3$

   (C) $t = 5$

   (D) $t = 1$ and $t = 5$

   (E) $t = 1$, $t = 3$, and $t = 5$

18. The average value of the function $f(x) = (x - 1)^2$ on the interval from $x = 1$ to $x = 5$ is

   (A) $-\dfrac{16}{3}$

   (B) $\dfrac{16}{3}$

   (C) $\dfrac{64}{3}$

   (D) $\dfrac{66}{3}$

   (E) $\dfrac{256}{3}$

GO ON TO THE NEXT PAGE.

19. $\int (e^{3\ln x} + e^{3x})\, dx =$

   (A) $3 + \dfrac{e^{3x}}{3} + C$

   (B) $\dfrac{x^4}{4} + 3e^{3x} + C$

   (C) $\dfrac{e^{x^4}}{4} + 3e^{3x} + C$

   (D) $\dfrac{e^{x^4}}{4} + \dfrac{e^{3x}}{3} + C$

   (E) $\dfrac{x^4}{4} + \dfrac{e^{3x}}{3} + C$

---

20. If $f(x) = (x^2 + x + 11)\sqrt{(x^3 + 5x + 121)}$, then $f'(0) =$

   (A) $\dfrac{5}{2}$

   (B) $\dfrac{27}{2}$

   (C) 22

   (D) $22 + \dfrac{2}{\sqrt{5}}$

   (E) $\dfrac{247}{2}$

GO ON TO THE NEXT PAGE.

21. If $f(x) = 5^{3x}$, then $f'(x) =$

   (A) $5^{3x}(\ln 125)$

   (B) $\dfrac{5^{3x}}{3\ln 5}$

   (C) $3(5^{2x})$

   (D) $3(5^{3x})$

   (E) $3x(5^{3x-1})$

22. A solid is generated when the region in the first quadrant enclosed by the graph of $y = (x^2 + 1)^3$, the line $x = 1$, the x-axis, and the y-axis is revolved about the x-axis. Its volume is found by evaluating which of the following integrals?

   (A) $\pi \displaystyle\int_1^8 (x^2 + 1)^3\, dx$

   (B) $\pi \displaystyle\int_1^8 (x^2 + 1)^6\, dx$

   (C) $\pi \displaystyle\int_0^1 (x^2 + 1)^3\, dx$

   (D) $\pi \displaystyle\int_0^1 (x^2 + 1)^6\, dx$

   (E) $2\pi \displaystyle\int_0^1 (x^2 + 1)^6\, dx$

GO ON TO THE NEXT PAGE.

23. $\displaystyle\lim_{x \to 0} 4\, \frac{\sin x \cos x - \sin x}{x^2} =$

   (A) 2

   (B) $\dfrac{40}{3}$

   (C) ∞

   (D) 0

   (E) undefined

24. If $\dfrac{dy}{dx} = \dfrac{(3x^2 + 2)}{y}$ and $y = 4$ when $x = 2$, then when $x = 3$, $y =$

   (A) 18

   (B) $\pm\sqrt{66}$

   (C) 58

   (D) $\pm\sqrt{74}$

   (E) $\pm\sqrt{58}$

25. $\int \dfrac{dx}{9+x^2} =$

    (A) $3\tan^{-1}\left(\dfrac{x}{3}\right)+C$

    (B) $\dfrac{1}{3}\tan^{-1}\left(\dfrac{x}{3}\right)+C$

    (C) $\dfrac{1}{9}\tan^{-1}\left(\dfrac{x}{3}\right)+C$

    (D) $\dfrac{1}{3}\tan^{-1}(x)+C$

    (E) $\dfrac{1}{9}\tan^{-1}(x)+C$

26. If $f(x) = \cos^3 (x + 1)$, then $f'(\pi) =$

    (A) $-3 \cos^2 (\pi + 1) \sin (\pi + 1)$
    (B) $3 \cos^2 (\pi + 1)$
    (C) $3 \cos^2 (\pi + 1) \sin (\pi + 1)$
    (D) $3\pi \cos^2 (\pi + 1)$
    (E) 0

GO ON TO THE NEXT PAGE.

27. $\int x\sqrt{x+3}\,dx =$

(A) $\frac{2}{3}(x)^{\frac{3}{2}}+6(x)^{\frac{1}{2}}+C$

(B) $\frac{2(x+3)^{\frac{3}{2}}}{3}+C$

(C) $\frac{2}{5}(x+3)^{\frac{5}{2}}-2(x+3)^{\frac{3}{2}}+C$

(D) $\frac{3(x+3)^{\frac{3}{2}}}{2}+C$

(E) $\frac{4x^2(x+3)^{\frac{3}{2}}}{3}+C$

---

28. If $f(x) = \ln(\ln(1-x))$, then $f'(x) =$

(A) $-\dfrac{1}{\ln(1-x)}$

(B) $\dfrac{1}{(1-x)\ln(1-x)}$

(C) $\dfrac{1}{(1-x)^2}$

(D) $-\dfrac{1}{(1-x)\ln(1-x)}$

(E) $-\dfrac{1}{\ln(1-x)^2}$

END OF PART A, SECTION I

IF YOU FINISH BEFORE TIME IS CALLED, YOU MAY CHECK YOUR WORK ON PART A ONLY.

DO NOT GO ON TO PART B UNTIL YOU ARE TOLD TO DO SO.

CALCULUS AB

SECTION I, Part B

Time—50 Minutes

Number of questions—17

A GRAPHING CALCULATOR IS REQUIRED FOR SOME QUESTIONS ON THIS PART OF THE EXAMINATION

Directions: Solve each of the following problems, using the available space for scratchwork. After examining the form of the choices, decide which is the best of the choices given and fill in the corresponding oval on the answer sheet. No credit will be given for anything written in the test book. Do not spend too much time on any one problem.

In this test:

1. The **exact** numerical value of the correct answer does not always appear among the choices given. When this happens, select from among the choices the number that best approximates the exact numerical value.

2. Unless otherwise specified, the domain of a function f is assumed to be the set of all real numbers x for which $f(x)$ is a real number.

29. $\int_0^{\frac{\pi}{4}} \sin x \, dx + \int_{-\frac{\pi}{4}}^{0} \cos x \, dx =$

   (A) $-\sqrt{2}$

   (B) -1

   (C) 0

   (D) 1

   (E) $\sqrt{2}$

GO ON TO THE NEXT PAGE.

30. Boats A and B leave the same place at the same time. Boat A heads due north at 12 km/hr. Boat B heads due east at 18 km/hr. After 2.5 hours, how fast is the distance between the boats increasing (in km/hr)?

    (A) 21.63
    (B) 31.20
    (C) 75.00
    (D) 9.84
    (E) 54.08

---

31. $\displaystyle\lim_{h \to 0} \frac{\tan\left(\dfrac{\pi}{6} + h\right) - \tan\left(\dfrac{\pi}{6}\right)}{h} =$

    (A) $\dfrac{\sqrt{3}}{3}$

    (B) $\dfrac{4}{3}$

    (C) $\sqrt{3}$

    (D) 0

    (E) $\dfrac{3}{4}$

GO ON TO THE NEXT PAGE.

32. If $\int_{30}^{100} f(x)\,dx = A$ and $\int_{50}^{100} f(x)\,dx = B$, then $\int_{30}^{50} f(x)\,dx =$

(A) $A + B$
(B) $A - B$
(C) 0
(D) $B - A$
(E) 20

33. If $f(x) = 3x^2 - x$, and $g(x) = f^{-1}(x)$, then $g'(10)$ could be

(A) 59

(B) $\dfrac{1}{59}$

(C) $\dfrac{1}{10}$

(D) 11

(E) $\dfrac{1}{11}$

GO ON TO THE NEXT PAGE.

34. The graph of $y = x^3 - 5x^2 + 4x + 2$ has a local minimum at

    (A) (0.46, 2.87)
    (B) (0.46, 0)
    (C) (2.87, –4.06)
    (D) (4.06, 2.87)
    (E) (1.66, –0.59)

---

35. The volume generated by revolving about the y-axis the region enclosed by the graphs $y = 9 - x^2$ and $y = 9 - 3x$, for $0 \leq x \leq 2$, is

    (A) -8π
    (B) 4π
    (C) 8π
    (D) 24π
    (E) 48π

GO ON TO THE NEXT PAGE.

36. The average value of the function $f(x) = \ln^2 x$ on the interval $[2, 4]$ is

    (A) −1.204
    (B) 1.204
    (C) 2.159
    (D) 2.408
    (E) 8.636

---

37. $\dfrac{d}{dx}\displaystyle\int_0^{3x} \cos(t)\, dt =$

    (A) $\sin 3x$
    (B) $-3 \sin 3x$
    (C) $\cos 3x$
    (D) $3 \sin 3x$
    (E) $3 \cos 3x$

GO ON TO THE NEXT PAGE.

38. If the definite integral $\int_1^3 (x^2+1)\, dx$ is approximated by using the Trapezoid Rule with $n = 4$, the error is

   (A) 0

   (B) $\dfrac{7}{3}$

   (C) $\dfrac{1}{12}$

   (D) $\dfrac{65}{6}$

   (E) $\dfrac{97}{3}$

---

39. The radius of a sphere is increasing at a rate proportional to itself. If the radius is 4 initially, and the radius is 10 after two seconds, what will the radius be after three seconds?

   (A) 62.50
   (B) 13.00
   (C) 15.81
   (D) 16.00
   (E) 25.00

GO ON TO THE NEXT PAGE.

40. Use differentials to approximate the change in the volume of a sphere when the radius is increased from 10 to 10.02 cm.

    (A) 4,213.973
    (B) 1,261.669
    (C) 1,256.637
    (D) 25.233
    (E) 25.133

41. $\int \ln 2x \, dx =$

    (A) $\dfrac{\ln 2x}{x} + C$

    (B) $\dfrac{\ln 2x}{2x} + C$

    (C) $x \ln x - x + C$

    (D) $x \ln 2x - x + C$

    (E) $2x \ln 2x - 2x + C$

GO ON TO THE NEXT PAGE.

42. If the function $f(x)$ is differentiable and $f(x) = \begin{cases} ax^3 - 6x; & \text{if } x \le 1 \\ bx^2 + 4; & x > 1 \end{cases}$, then $a =$

   (A) 0
   (B) 1
   (C) −14
   (D) −24
   (E) 26

43. Two particles leave the origin at the same time and move along the y-axis with their respective positions determined by the functions $y_1 = \cos 2t$ and $y_2 = 4\sin t$ for $0 < t < 6$. For how many values of t do the particles have the same acceleration?

   (A) 0
   (B) 1
   (C) 2
   (D) 3
   (E) 4

GO ON TO THE NEXT PAGE.

44. Find the distance traveled (to three decimal places) in the first four seconds, for a particle whose velocity is given by $v(t) = 7e^{-t^2}$, where t stands for time.

    (A) 0.976
    (B) 6.204
    (C) 6.359
    (D) 12.720
    (E) 7.000

45. $\int \tan^6 x \sec^2 x \, dx =$

    (A) $\dfrac{\tan^7 x}{7} + C$

    (B) $\dfrac{\tan^7 x}{7} + \dfrac{\sec^3 x}{3} + C$

    (C) $\dfrac{\tan^7 x \sec^3 x}{21} + C$

    (D) $7 \tan^7 x + C$

    (E) $\dfrac{2}{7} \tan^7 x \sec x + C$

STOP
END OF PART B, SECTION I
IF YOU FINISH BEFORE TIME IS CALLED, YOU MAY CHECK YOUR WORK ON PART B ONLY.
DO NOT GO ON TO SECTION II UNTIL YOU ARE TOLD TO DO SO.

SECTION II
GENERAL INSTRUCTIONS

You may wish to look over the problems before starting to work on them, since it is not expected that everyone will be able to complete all parts of all problems. All problems are given equal weight, but the parts of a particular problem are not necessarily given equal weight.

A GRAPHING CALCULATOR IS REQUIRED FOR SOME PROBLEMS OR PARTS OF PROBLEMS ON THIS SECTION OF THE EXAMINATION.

- You should write all work for each part of each problem in the space provided for that part in the booklet. Be sure to write clearly and legibly. If you make an error, you may save time by crossing it out rather than trying to erase it. Erased or crossed-out work will not be graded.

- Show all your work. You will be graded on the correctness and completeness of your methods as well as your answers. Correct answers without supporting work may not receive credit.

- Justifications require that you give mathematical (noncalculator) reasons and that you clearly identify functions, graphs, tables, or other objects you use.

- You are permitted to use your calculator to solve an equation, find the derivative of a function at a point, or calculate the value of a definite integral. However, you must clearly indicate the setup of your problem, namely the equation, function, or integral you are using. If you use other built-in features or programs, you must show the mathematical steps necessary to produce your results.

- Your work must be expressed in standard mathematical notation rather than calculator syntax. For example, $\int_1^5 x^2\, dx$ may not be written as fnInt (X², X, 1, 5).

- Unless otherwise specified, answers (numeric or algebraic) need not be simplified. If your answer is given as a decimal approximation, it should be correct to three places after the decimal point.

- Unless otherwise specified, the domain of a function f is assumed to be the set of all real numbers x for which $f(x)$ is a real number.

SECTION II, PART A
Time—30 minutes
Number of problems—2

A graphing calculator is required for some problems or parts of problems.

During the timed portion for Part A, you may work only on the problems in Part A.

On Part A, you are permitted to use your calculator to solve an equation, find the derivative of a function at a point, or calculate the value of a definite integral. However, you must clearly indicate the setup of your problem, namely the equation, function, or integral you are using. If you use other built-in features or programs, you must show the mathematical steps necessary to produce your results.

GO ON TO THE NEXT PAGE.

1. A particle moves along the *x*-axis so that its acceleration at any time $t > 0$ is given by $a(t) = 12t - 18$. At time $t = 1$, the velocity of the particle is $v(1) = 0$ and the position is $x(1) = 9$.

    (a)   Write an expression for the velocity of the particle $v(t)$.

    (b)   At what values of t does the particle change direction?

    (c)   Write an expression for the position $x(t)$ of the particle.

    (d)   Find the total distance traveled by the particle from $t = \dfrac{3}{2}$ to $t = 6$.

2. Let R be the region enclosed by the graphs of $y = 2 \ln x$ and $y = \dfrac{x}{2}$, and the lines $x = 2$ and $x = 8$.

    (a)   Find the area of R.

    (b)   Set up, <u>but do not integrate</u>, an integral expression, in terms of a single variable, for the volume of the solid generated when R is revolved about the *x*-axis.

    (c)   Set up, <u>but do not integrate</u>, an integral expression, in terms of a single variable, for the volume of the solid generated when R is revolved about the line $x = -1$.

GO ON TO THE NEXT PAGE.

SECTION II, PART B
Time—1 hour
Number of problems—4

No calculator is allowed for these problems.

During the timed portion for Part B, you may continue to work on the problems in Part A without the use of any calculator.

---

3. Consider the equation $x^2 - 2xy + 4y^2 = 64$.

   (a) Write an expression for the slope of the curve at any point (x, y).

   (b) Find the equation of the tangent lines to the curve at the point $x = 2$.

   (c) Find $\dfrac{d^2 y}{dx^2}$ at $(0, 4)$.

---

4. Water is draining at the rate of 48π ft³/second from the vertex at the bottom of a conical tank whose diameter at its base is 40 feet and whose height is 60 feet.

   (a) Find an expression for the volume of water in the tank, in terms of its radius, at the surface of the water.

   (b) At what rate is the radius of the water in the tank shrinking when the radius is 16 feet?

   (c) How fast is the height of the water in the tank dropping at the instant that the radius is 16 feet?

GO ON TO THE NEXT PAGE.

5. Let f be the function given by $f(x) = 2x^4 - 4x^2 + 1$.

    (a)   Find an equation of the line tangent to the graph at $(-2, 17)$.

    (b)   Find the x- and y-coordinates of the relative maxima and relative minima. Verify your answer.

    (c)   Find the x- and y-coordinates of the points of inflection. Verify your answer.

---

6. Let $F(x) = \int_0^x \left[\cos\left(\dfrac{t}{2} \right) + \left(\dfrac{3}{2} \right) \right] dt$ on the closed interval $[0, 4\pi]$.

    (a)   Approximate $F(2\pi)$ using four inscribed rectangles.

    (b)   Find $F'(2\pi)$.

    (c)   Find the average value of $F'(x)$ on the interval $[0, 4\pi]$.

STOP

END OF EXAM

Chapter 21
Practice Test 1
Answers and
Explanations

ANSWER KEY TO SECTION I

1. B	11. C	21. A	31. B	41. D
2. D	12. B	22. D	32. B	42. C
3. C	13. A	23. D	33. E	43. D
4. A	14. D	24. E	34. C	44. B
5. E	15. A	25. B	35. C	45. A
6. B	16. A	26. A	36. B	
7. E	17. D	27. C	37. E	
8. E	18. B	28. D	38. C	
9. D	19. E	29. D	39. C	
10. A	20. B	30. A	40. E	

ANSWERS AND EXPLANATIONS TO SECTION I

1. **B** If $f(x) = 5x^{\frac{4}{3}}$, then $f''(8) =$

   We need to use basic differentiation to solve this problem.

   Step 1: $f'(x) = \dfrac{4}{3}\left(5x^{\frac{1}{3}}\right)$

   Step 2: Now all we have to do is plug in 8 for x and simplify.

   $$\frac{4}{3}\left(5\left(8^{\frac{1}{3}}\right)\right) = \frac{4}{3}(5(2)) = \frac{40}{3}$$

2. **D** $\displaystyle\lim_{x\to\infty} \dfrac{5x^2 - 3x + 1}{4x^2 + 2x + 5}$ is

   Step 1: To solve this problem, you need to remember how to evaluate limits. Always do limit problems on the first pass. Whenever we have a limit of a polynomial fraction where $x \to \infty$, we divide the numerator and the denominator, separately, by the highest power of x in the fraction.

   $$\lim_{x\to\infty} \frac{5x^2 - 3x + 1}{4x^2 + 2x + 5} = \lim_{x\to\infty} \frac{\dfrac{5x^2}{x^2} - \dfrac{3x}{x^2} + \dfrac{1}{x^2}}{\dfrac{4x^2}{x^2} + \dfrac{2x}{x^2} + \dfrac{5}{x^2}}$$

   Step 2: Simplify $\displaystyle\lim_{x\to\infty} \dfrac{5 - \dfrac{3}{x} + \dfrac{1}{x^2}}{4 + \dfrac{2}{x} + \dfrac{5}{x^2}}$.

   Step 3: Now take the limit. Remember that the $\displaystyle\lim_{x\to\infty} \dfrac{k}{x^n} = 0$, if $n > 0$, where k is a constant. Thus, we get

   $$\lim_{x\to\infty} \frac{5 - \dfrac{3}{x} + \dfrac{1}{x^2}}{4 + \dfrac{2}{x} + \dfrac{5}{x^2}} = \lim_{x\to\infty} \frac{5 - 0 + 0}{4 + 0 + 0} = \frac{5}{4}$$

3. **C**   If $f(x) = \dfrac{3x^2 + x}{3x^2 - x}$, then $f'(x)$ is

Step 1: We need to use the Quotient Rule to evaluate this derivative. Remember, the derivative of

$\dfrac{u}{v} = \dfrac{v\dfrac{du}{dx} - u\dfrac{dv}{dx}}{v^2}$. But, before we take the derivative, we should factor an x out of the top and bottom and cancel, simplifying the quotient.

$$f(x) = \frac{3x^2 + x}{3x^2 - x} = \frac{x(3x+1)}{x(3x-1)} = \frac{3x+1}{3x-1}$$

Step 2: Now take the derivative.

$$f'(x) = \frac{(3x-1)(3) - (3x+1)(3)}{(3x-1)^2}$$

Step 3: Simplify.

$$\frac{9x - 3 - 9x - 3}{(3x-1)^2} = \frac{-6}{(3x-1)^2}$$

4. **A**   If the function f is continuous for all real numbers and if $f(x) = \dfrac{x^2 - 7x + 12}{x - 4}$ when $x \neq 4$, then $f(4) =$

This problem is testing your knowledge of continuity.

Step 1: Notice that if we plug 4 into the numerator and denominator we get $\dfrac{0}{0}$, which is undefined. So, the first thing that we should do is factor the numerator. What we are looking for is a common factor in the numerator and denominator. If we find a common factor, we can cancel the factors and simplify the problem.

We get $f(x) = \dfrac{x^2 - 7x + 12}{x - 4} = \dfrac{(x-3)(x-4)}{x-4} = (x-3)$.

Step 2: Now we plug in 4 for x and we get 1.

5. **E**   If $x^2 - 2xy + 3y^2 = 8$, then $\dfrac{dy}{dx} =$

Whenever we have a polynomial where the x's and y's are not easily separated, we need to use implicit differentiation to find the derivative.

Step 1: Take the derivative of everything with respect to x.

$$2x\frac{dx}{dx} - 2\left(x\frac{dy}{dx} + y\frac{dx}{dx}\right) + 6y\frac{dy}{dx} = 0$$

Remember that $\frac{dx}{dx} = 1$!

Step 2: Simplify, and then put all of the terms containing $\frac{dy}{dx}$ on one side, and all of the other terms on the other side.

$$2x - 2x\frac{dy}{dx} - 2y + 6y\frac{dy}{dx} = 0$$

$$-2x\frac{dy}{dx} + 6y\frac{dy}{dx} = 2y - 2x$$

Factor out the $\frac{dy}{dx}$, and then isolate it.

$$\frac{dy}{dx}(6y - 2x) = 2y - 2x$$

$$\frac{dy}{dx} = \frac{2y - 2x}{(6y - 2x)} = \frac{y - x}{3y - x}$$

6. **B**

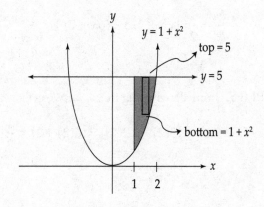

$$A = \int_{1}^{2}(5 - (1 + x^2))\,dx$$

Which of the following integrals correctly corresponds to the area of the region in the figure above between the curve $y = 1 + x^2$ and the line $y = 5$ from $x = 1$ to $x = 2$?

We use integrals to find the area between two curves. If the top curve of a region is $f(x)$ and the bottom curve of a region is $g(x)$, from $x = a$ to $x = b$, then the area is found by the following integral:

$$\int_a^b \left[(f(x) - g(x)) \right] dx$$

Step 1: The top curve here is the line $y = 5$, the bottom curve is $y = 1 + x^2$, and the region extends from the line $x = 1$ to the line $x = 2$. Thus, the integral for the area is

$$\int_1^2 \left[(5) - (1 + x^2) \right] dx = \int_1^2 (4 - x^2) \, dx$$

7. **E** If $f(x) = \sec x + \csc x$, then $f'(x) =$

This question is testing whether you know your derivatives of trigonometric functions. If you do, this is an easy problem.

Step 1: The derivative of $\sec x$ is $\sec x \tan x$, and the derivative of $\csc x$ is $-\csc x \cot x$. That makes the derivative here $\sec x \tan x - \csc x \cot x$.

8. **E** An equation of the line normal to the graph of $y = \sqrt{(3x^2 + 2x)}$ at $(2, 4)$ is

Here we do everything that we normally do for finding the equations of tangent lines, except that we use the negative reciprocal of the slope to find the normal line. This is because the normal line is perpendicular to the tangent line.

Step 1: First, find the slope of the tangent line.

$$\frac{dy}{dx} = \frac{1}{2}(3x^2 + 2x)^{-\frac{1}{2}}(6x + 2)$$

Step 2: DON'T SIMPLIFY. Immediately plug in $x = 2$. We get

$$\frac{dy}{dx} = \frac{1}{2}(3x^2 + 2x)^{-\frac{1}{2}}(6x + 2) = \frac{1}{2}(3(2)^2 + 2(2))^{-\frac{1}{2}}(6(2) + 2) = \frac{1}{2}(16)^{-\frac{1}{2}}(14) = \frac{7}{4}$$

This means that the slope of the tangent line at $x = 2$ is $\dfrac{7}{4}$, so the slope of the normal line is $-\dfrac{4}{7}$.

Step 3: Then the equation of the normal line is $(y - 4) = -\dfrac{4}{7}(x - 2)$.

Step 4: Multiply through by 7 and simplify.

$$7y - 28 = -4x + 8$$

$$4x + 7y = 36$$

9. **D** $\displaystyle\int_{-1}^{1} \frac{4}{1 + x^2}\, dx =$

You should recognize this integral as one of the inverse trigonometric integrals.

Step 1: As you should recall, $\displaystyle\int \frac{dx}{1 + x^2} = \tan^{-1}(x) + C$. The 4 is no big deal. Just multiply the integral by 4 to get $4\tan^{-1}(x)$. Then we just have to evaluate the limits of integration.

Step 2: $4\tan^{-1}(x)\Big|_{-1}^{1} = 4\tan^{-1}(1) - 4\tan^{-1}(-1) = 4\left(\dfrac{\pi}{4}\right) - 4\left(-\dfrac{\pi}{4}\right) = 2\pi$

10. **A** If $f(x) = \cos^2 x$, then $f''(\pi) =$

This problem is just asking us to find a higher order derivative of a trigonometric function.

Step 1: The first derivative requires the Chain Rule.

$f(x) = \cos^2 x$
$f'(x) = 2(\cos x)(-\sin x) = -2\cos x \sin x$

Step 2: The second derivative requires the Product Rule.

$f'(x) = -2\cos x \sin x$
$f''(x) = -2(\cos x \cos x - \sin x \sin x) = -2(\cos^2 x - \sin^2 x)$

Step 3: Now plug in π for x and simplify.

$-2(\cos^2(\pi) - \sin^2(\pi)) = -2(1 - 0) = -2$

11. **C** If $f(x) = \dfrac{5}{x^2 + 1}$ and $g(x) = 3x$, then $g(f(2)) =$

Step 1: To find $g(f(x))$, all you need to do is to replace all of the x's in $g(x)$ with $f(x)$'s.

$$g(f(x)) = 3f(x) = 3\left(\frac{5}{x^2 + 1}\right) = \frac{15}{x^2 + 1}$$

Step 2: Now all we have to do is plug in 2 for x.

$$g(f(2)) = \frac{15}{2^2 + 1} = 3$$

12. **B** $\int x\sqrt{5x^2 - 4}\, dx =$

Any time we have an integral with an x factor whose power is one less than another x factor, we can try to do the integral with u-substitution. This is our favorite technique for doing integration and the most important one to master.

Step 1: Let $u = 5x^2 - 4$ and $du = 10x\, dx$, so $\dfrac{1}{10}\, du = x\, dx$.

Then we can rewrite the integral as

$$\int x\sqrt{5x^2 - 4}\, dx = \frac{1}{10}\int u^{\frac{1}{2}}\, du$$

Step 2: Now this becomes a basic integral.

$$\frac{1}{10}\int u^{\frac{1}{2}}\, du = \frac{1}{10}\left(\frac{u^{\frac{3}{2}}}{\frac{3}{2}}\right) + C = \frac{1}{15}u^{\frac{3}{2}} + C$$

Step 3: Reverse the substitution and we get $\dfrac{1}{15}(5x^2 - 4)^{\frac{3}{2}} + C$

13. **A** The slope of the line tangent to the graph of $3x^2 + 5\ln y = 12$ at $(2, 1)$ is

This is another equation of a tangent line problem, combined with implicit differentiation. Often, the AP exam has more than one tangent line problem, so make sure that you can do these well!

By the way, do you remember the derivative of $\ln(f(x))$? It is $\dfrac{f'(x)}{f(x)}$.

Step 1: First, we take the derivative of the equation.

$$6x\frac{dx}{dx} + \frac{5}{y}\frac{dy}{dx} = 0$$

Step 2: Next, we simplify and solve for $\dfrac{dy}{dx}$.

$$6x + \frac{5}{y}\frac{dy}{dx} = 0$$

$$\frac{dy}{dx} = \frac{-6xy}{5}$$

Step 3: Now, we plug in 2 for x and 1 for y to get the slope of the tangent line.

$$\frac{dy}{dx} = \frac{-6(2)(1)}{5} = \frac{-12}{5}$$

14.　**D**　The equation $y = 2 - 3 \sin\frac{\pi}{4}(x-1)$ has a fundamental period of

The AP people expect you to remember a lot of your trigonometry, so if you're rusty, review the unit in the Appendix.

Step 1: In an equation of the form $f(x) = A \sin B (x \pm C) \pm D$, you should know four components. The amplitude of the equation is $|A|$, the horizontal or phase shift is $\pm C$, the vertical shift is $\pm D$, and the fundamental period is $\frac{2\pi}{B}$.

The same is true for $f(x) = A \cos B (x \pm C) \pm D$.

Step 2: All we have to do is plug into the formula for the period.

$$\frac{2\pi}{B} = \frac{2\pi}{\frac{\pi}{4}} = 8$$

15.　**A**　If $f(x) = \begin{cases} x^2 + 5 & \text{if } x < 2 \\ 7x - 5 & \text{if } x \geq 2 \end{cases}$, for all real numbers x, which of the following must be true?

I.　$f(x)$ is continuous everywhere.

II.　$f(x)$ is differentiable everywhere.

III.　$f(x)$ has a local minimum at $x = 2$.

This problem is testing your knowledge of the rules of continuity and differentiability. While the more formal treatment is located in the unit on continuity, here we'll go directly to a shortcut to the right answer. This type of function is called a piecewise function because it is broken into two or more pieces, depending on the value of x that one is looking at.

Step 1: If a piecewise function is continuous at a point a, then when you plug a into each of the pieces of the function, you should get the same answer. The function consists of a pair of polynomials (remember that all polynomials are continuous!), where the only point that might be a problem is $x = 2$. So here we'll plug 2 into both pieces of the function to see if we get the same value. If we do, then the function is continuous. If we don't, then it's discontinuous. At $x = 2$, the upper piece is equal to 9 and the lower piece is also equal to 9. So the function is continuous everywhere, and **I** is true. You should then eliminate answer choice (C).

Step 2: If a piecewise function is differentiable at a point a, then when you plug a into each of the derivatives of the pieces of the function, you should get the same answer. It is the same idea as in Step 1. So here we will plug 2 into the derivatives of both pieces of the function to see if we get the same value. If we do, then the function is differentiable. If we don't, then it is non-differentiable at $x = 2$.

The derivative of the upper piece is $2x$, and at $x = 2$, the derivative is 4.

The derivative of the lower piece is 7 everywhere.

Because the two derivatives are not equal, the function is not differentiable everywhere, and **II** is false. You should then eliminate answer choices (B) and (E).

Step 3: The slope of the function to the left of $x = 2$ is 4. The slope of the function to the right of $x = 2$ is 7. If the slope of a continuous function has the same sign on either side of a point, then the function cannot have a local minimum or maximum at that point. So **III** is false because of what we found in Step 2. You should then eliminate answer choice (D).

16. **A**   For what value of x does the function $f(x) = x^3 - 9x^2 - 120x + 6$ have a local minimum?

This problem requires you to know how to find maxima/minima. This is a part of curve sketching and is one of the most important parts of differential calculus. A function has *critical points* where the derivative is zero or undefined (which is never a problem when the function is an ordinary polynomial). After finding the critical points, we test them to determine whether they are maxima or minima or something else.

Step 1: First, as usual, take the derivative and set it equal to zero.

$$f'(x) = 3x^2 - 18x - 120$$
$$3x^2 - 18x - 120 = 0$$

Step 2: Find the values of x that make the derivative equal to zero. These are the critical points.

$$3x^2 - 18x - 120 = 0$$
$$x^2 - 6x - 40 = 0$$
$$(x - 10)(x + 4) = 0$$
$$x = \{10, -4\}$$

Step 3: In order to determine whether a critical point is a maximum or a minimum, we need to take the second derivative.

$$f''(x) = 6x - 18$$

Step 4: Now, we plug the critical points from Step 2 into the second derivative. If it yields a negative value, then the point is a maximum. If it yields a positive value, then the point is a minimum. If it yields zero, it is neither, and is most likely a point of inflection.

$$6(10) - 18 = 42$$
$$6(-4) - 18 = -42$$

Therefore, 10 is a minimum.

17. **D** The acceleration of a particle moving along the x-axis at time t is given by $a(t) = 4t - 12$. If the velocity is 10 when $t = 0$ and the position is 4 when $t = 0$, then the particle is changing direction at

Step 1: Because acceleration is the derivative of velocity, if we know the acceleration of a particle, we can find the velocity by integrating the acceleration with respect to t.

$$v = \int a \, dt = \int (4t - 12) \, dt = 2t^2 - 12t + C$$

Next, because the velocity is 10 at $t = 0$, we can plug in 0 for t and solve for the constant.

$$2(0)^2 - 12(0) + C = 10$$

Therefore, $C = 10$ and the velocity, $v(t)$, is $2t^2 - 12t + 10$.

Step 2: In order to find when the particle is changing direction we need to know when the velocity is equal to zero, so we set $v(t) = 0$ and solve for t.

$$2t^2 - 12t + 10 = 0$$
$$t^2 - 6t + 5 = 0$$
$$(t - 5)(t - 1) = 0$$
$$t = \{1, 5\}$$

Now, provided that the acceleration is not also zero at $t = \{1, 5\}$, the particle will be changing direction at those times. The acceleration is found by differentiating the equation for velocity with respect to time: $a(t) = 4t - 12$. This is not zero at either $t = 1$ or $t = 5$. Therefore, the particle is changing direction when $t = 1$ and $t = 5$.

18. **B** The average value of the function $f(x) = (x - 1)^2$ on the interval from $x = 1$ to $x = 5$ is

Step 1: If you want to find the average value of $f(x)$ on an interval $[a, b]$, you need to evaluate the integral $\dfrac{1}{b - a}\displaystyle\int_a^b f(x)\, dx$.

So here we would evaluate the integral $\dfrac{1}{5 - 1}\displaystyle\int_1^5 (x - 1)^2\, dx$.

Step 2: $\dfrac{1}{5 - 1}\displaystyle\int_1^5 (x - 1)^2\, dx = \dfrac{1}{4}\displaystyle\int_1^5 (x^2 - 2x + 1)\, dx$

$$= \dfrac{1}{4}\left(\dfrac{x^3}{3} - x^2 + x\right)\Bigg|_1^5 = \dfrac{1}{4}\left[\left(\dfrac{5^3}{3} - 5^2 + 5\right) - \left(\dfrac{1}{3} - 1 + 1\right)\right]$$

$$= \dfrac{1}{4}\left(\dfrac{125}{3} - 20 - \dfrac{1}{3}\right) = \dfrac{64}{12} = \dfrac{16}{3}$$

19. **E** $\displaystyle\int (e^{3\ln x} + e^{3x})\, dx =$

This problem requires that you know your rules of exponential functions.

Step 1: First of all, $e^{3\ln x} = e^{\ln x^3} = x^3$. So we can rewrite the integral as

$$\int (e^{3\ln x} + e^{3x})\, dx = \int (x^3 + e^{3x})\, dx$$

Step 2: The rule for the integral of an exponential function is $\displaystyle\int e^k\, dx = \dfrac{1}{k}e^{kx} + C$.

Now we can do the integral $\displaystyle\int (x^3 + e^{3x})\, dx = \dfrac{x^4}{4} + \dfrac{1}{3}e^{3x} + C$.

20. **B** If $f(x) = \sqrt{(x^3 + 5x + 121)}(x^2 + x + 11)$, then $f'(0) =$

This problem is just a complicated derivative, requiring you to be familiar with the Chain Rule and the Product Rule.

Step 1: $f'(x) = \dfrac{1}{2}(x^3 + 5x + 121)^{-\frac{1}{2}}(3x^2 + 5)(x^2 + x + 11) + (x^3 + 5x + 121)^{\frac{1}{2}}(2x + 1)$

Step 2: Whenever a problem asks you to find the value of a complicated derivative at a particular point, NEVER simplify the derivative. Immediately plug in the value for x and do arithmetic instead of algebra.

$$f'(0) = \frac{1}{2}(0^3 + 5(0) + 121)^{-\frac{1}{2}}(3(0)^2 + 5)((0)^2 + (0) + 11) + ((0)^3 + 5(0) + 121)^{\frac{1}{2}}(2(0) + 1)$$

$$= \frac{1}{2}(121)^{-\frac{1}{2}}(5)(11) + (121)^{\frac{1}{2}}(1) = \frac{5}{2} + 11 = \frac{27}{2}$$

21.　**A**　If $f(x) = 5^{3x}$, then $f'(x) =$

This problem requires you to know how to find the derivative of an exponential function. The rule is: If a function is of the form $a^{f(x)}$, its derivative is $a^{f(x)}(\ln a) f'(x)$. Now all we have to do is follow the rule!

Step 1: $f(x) = 5^{3x}$

$\qquad f'(x) = 5^{3x}(\ln 5)(3)$

Step 2: If you remember your rules of logarithms, $3\ln 5 = \ln(5^3) = \ln 125$.

So we can rewrite the answer to $f'(x) = 5^{3x}(\ln 5)(3) = 5^{3x}\ln 125$.

22.　**D**　A solid is generated when the region in the first quadrant enclosed by the graph of $y = (x^2 + 1)^3$, the line $x = 1$, the y-axis and the x-axis, is revolved about the x-axis. Its volume is found by evaluating which of the following integrals?

This problem requires you to know how to find the volume of a solid of revolution.

If you have a region between two curves, from $x = a$ to $x = b$, then the volume generated when the region is revolved around the x-axis is: $\pi\int_a^b \left[f(x)^2 - g(x)^2 \right] dx$, if $f(x)$ is above $g(x)$ throughout the region.

Step 1: First, we have to determine what the region looks like. The curve looks like the following:

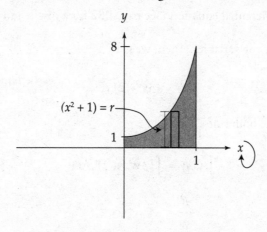

The shaded region is the part that we are interested in. Notice that the curve is always above the x-axis (which is $g(x)$). Now we just follow the formula.

$$\pi\int_0^1\left[\left[(x^2+1)^3\right]^2-[0]^2\right]dx=\pi\int_0^1(x^2+1)^6\,dx$$

23.   D $\displaystyle\lim_{x\to0}4\frac{\sin x\cos x-\sin x}{x^2}=$

This problem requires us to evaluate the limit of a trigonometric function.

There are two important trigonometric limits to memorize.

$$\lim_{x\to0}\frac{\sin x}{x}=1\text{ and }\lim_{x\to0}\frac{1-\cos x}{x}=0$$

Step 1: The first step that we always take when evaluating the limit of a trigonometric function is to rearrange the function so that it looks like some combination of the limits above. We can do this by factoring $\sin x$ out of the numerator.

Now we can break this into limits that we can easily evaluate.

$$\lim_{x\to0}4\frac{\sin x\cos x-\sin x}{x^2}=4\lim_{x\to0}\left(\frac{\sin x}{x}\right)\left(\frac{\cos x-1}{x}\right)$$

$$\left(\text{Note that }\lim_{x\to0}\frac{1-\cos x}{x}=-\lim_{x\to0}\frac{\cos x-1}{x}=0\right)$$

Step 2: Now if we take the limit as $x\to0$ we get $4(1)(0)=0$.

24.   E   If $\dfrac{dy}{dx}=\dfrac{(3x^2+2)}{y}$ and $y=4$ when $x=2$, then when $x=3$, $y=$

This is a very basic differential equation. See page 282 for a discussion of *separation of variables*.

Step 1: First, separate the variables. Then, we get

$$y\,dy=(3x^2+2)\,dx$$

Step 2: Now integrate both sides.

$$\int y\,dy=\int(3x^2+2)\,dx$$

$$\frac{y^2}{2}=x^3+2x+C$$

Notice how we use only one constant. All we have to do now is solve for C. We do this by plugging in 2 for x and 4 for y.

$$\frac{16}{2} = 2^3 + 4 + C$$

$$C = -4$$

So we can rewrite the equation as $\frac{y^2}{2} = x^3 + 2x - 4$.

Step 3: Now, if we plug in 3 for x, we can find y.

$$\frac{y^2}{2} = 27 + 6 - 4$$

$$y^2 = 58$$

$$y = \pm\sqrt{58}$$

25. **B** $\displaystyle\int \frac{dx}{9 + x^2} =$

This is another inverse trigonometric integral.

Step 1: We know that $\displaystyle\int \frac{dx}{1 + x^2} = \tan^{-1}(x) + C$.

(See problem 9 if you're not sure of this.) The trick here is to get the denominator of the fraction in the integrand to be of the correct form. If we factor 9 out of the denominator, we get

$$\int \frac{dx}{9 + x^2} = \int \frac{dx}{9\left(1 + \dfrac{x^2}{9}\right)} = \frac{1}{9}\int \frac{dx}{1 + \dfrac{x^2}{9}} = \frac{1}{9}\int \frac{dx}{1 + \left(\dfrac{x}{3}\right)^2}$$

Step 2: Now if we use u-substitution we will be able to evaluate this integral.

Let $u = \dfrac{x}{3}$ and $du = \dfrac{1}{3}\, dx$ or $3\, du = dx$. Then we have

$$\frac{1}{9}\int \frac{dx}{1 + \left(\dfrac{x}{3}\right)^2} = \frac{1}{9}\int \frac{3\, du}{1 + u^2} = \frac{1}{3}\int \frac{du}{1 + u^2} = \frac{1}{3}\tan^{-1}(u) + C$$

Step 3: Now all we have to do is reverse the u-substitution and we're done.

$$\frac{1}{3}\tan^{-1}(u) + C = \frac{1}{3}\tan^{-1}\left(\frac{x}{3}\right) + C$$

26. **A**   If $f(x) = \cos^3(x + 1)$, then $f'(\pi) =$

Think of $\cos^3(x + 1)$ as $[\cos(x + 1)]^3$.

Step 1: First, we take the derivative of the outside function and ignore the inside functions. The derivative of u^3 is $3u^2$.

We get $\dfrac{d}{dx}[u]^3 = 3[u]^2$.

Step 2: Next, we take the derivative of the cosine term and multiply. The derivative of $\cos u$ is $-\sin u$.

$$\frac{d}{dx}[\cos(u)]^3 = -3[\cos(u)]^2\sin(u)$$

Step 3: Finally, we take the derivative of $x + 1$ and multiply. The derivative of $x + 1$ is 1.

$$\frac{d}{dx}[\cos(x + 1)]^3 = -3[\cos(x + 1)]^2\sin(x + 1)$$

27. **C** $\displaystyle\int x\sqrt{x + 3}\ dx =$

We can do this integral with u-substitution.

Step 1: Let $u = x + 3$. Then $du = dx$ and $u - 3 = x$.

Step 2: Substituting, we get

$$\int x\sqrt{x + 3}\ dx = \int (u - 3)u^{\frac{1}{2}}\ du$$

Why is this better than the original integral, you might ask? Because now we can distribute and the integral becomes easy.

Step 3: When we distribute, we get

$$\int (u - 3)u^{\frac{1}{2}}\ du = \int \left(u^{\frac{3}{2}} - 3u^{\frac{1}{2}}\right)\ du$$

Step 4: Now we can integrate.

$$\int\left(u^{\frac{3}{2}} - 3u^{\frac{1}{2}}\right)du = \frac{2}{5}u^{\frac{5}{2}} - 3\left(\frac{2}{3}u^{\frac{3}{2}}\right) + C$$

Step 5: Substituting back, we get

$$\frac{2}{5}u^{\frac{5}{2}} - 3\left(\frac{2}{3}u^{\frac{3}{2}}\right) + C = \frac{2}{5}(x+3)^{\frac{5}{2}} - 2(x+3)^{\frac{3}{2}} + C$$

28. **D** If $f(x) = \ln(\ln(1-x))$, then $f'(x) =$

Here, we use the Chain Rule.

Step 1: First, take the derivative of the outside function.

The derivative of $\ln u$ is $\dfrac{du}{u}$.

We get

$$\frac{d}{dx}\ln(\ln(u)) = \frac{1}{\ln(u)}$$

Step 2: Now we take the derivative of the function in the denominator. Once again, the function is $\ln u$.

We get

$$\frac{d}{dx}\ln(\ln(1-x)) = \frac{1}{\ln(1-x)}\left(\frac{-1}{1-x}\right) = \frac{-1}{(1-x)\ln(1-x)}$$

29. **D** $\displaystyle\int_0^{\frac{\pi}{4}} \sin x\ dx + \int_{-\frac{\pi}{4}}^0 \cos x\ dx =$

These are a pair of basic trigonometric integrals. You should have memorized several trigonometric integrals, particularly $\displaystyle\int \sin x\ dx = -\cos x + C$ and $\displaystyle\int \cos x\ dx = \sin x + C$.

Step 1: $\displaystyle\int_0^{\frac{\pi}{4}} \sin x\ dx + \int_{-\frac{\pi}{4}}^0 \cos x\ dx = -\cos x\Big|_0^{\frac{\pi}{4}} + \sin x\Big|_{-\frac{\pi}{4}}^0$

Step 2: Now we evaluate the limits of integration, and we're done.

$$-\cos x\Big|_0^{\frac{\pi}{4}} + \sin x\Big|_{-\frac{\pi}{4}}^0 = \left[\left(-\cos\frac{\pi}{4}\right) - (-\cos(0))\right] + \left[(\sin(0)) - \left(\sin\left(-\frac{\pi}{4}\right)\right)\right] = -\frac{1}{\sqrt{2}} + 1 + 0 + \frac{1}{\sqrt{2}} = 1$$

30. **A**   Boats A and B leave the same place at the same time. Boat A heads due north at 12 km/hr. Boat B heads due east at 18 km/hr. After 2.5 hours, how fast is the distance between the boats increasing (in km/hr)?

Step 1: The boats are moving at right angles to each other and are thus forming a right triangle with the distance between them forming the hypotenuse.

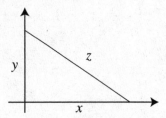

Whenever we see right triangles in related rates problems, we look to use the Pythagorean theorem. Call the distance that Boat A travels y and the distance that Boat B travels x.

Then the rate at which Boat A goes north is $\dfrac{dy}{dt}$, and the rate at which Boat B travels is $\dfrac{dx}{dt}$. The distance between the two boats is z, and we are looking for how fast z is growing, which is $\dfrac{dz}{dt}$. Now we can use the Pythagorean theorem to set up the relationship: $x^2 + y^2 = z^2$.

Step 2: Differentiating both sides we obtain

$$2x\frac{dx}{dt} + 2y\frac{dy}{dt} = 2z\frac{dz}{dt} \text{ or } x\frac{dx}{dt} + y\frac{dy}{dt} = z\frac{dz}{dt}$$

Step 3: After 2.5 hours, Boat A has traveled 30 km and Boat B has traveled 45 km. Because of the Pythagorean theorem, we also know that when $y = 30$ and $x = 45$, $z = 54.08$.

Step 4: Now we plug everything into the equation from Step 2 and solve for $\dfrac{dz}{dt}$.

$$(45)(18) + (30)(12) = (54.08)\frac{dz}{dt}$$

$$1170 = (54.08)\frac{dz}{dt}$$

$$21.63 = \frac{dz}{dt}$$

31. **B** $\lim\limits_{h \to 0} \dfrac{\tan\left(\dfrac{\pi}{6} + h\right) - \tan\left(\dfrac{\pi}{6}\right)}{h} =$

This may *appear* to be a limit problem, but it is *actually* testing to see whether you know the definition of the derivative.

Step 1: You should recall that the definition of the derivative says

$$\lim_{h \to 0} \frac{f(x + h) - f(x)}{h} = f'(x)$$

Thus, if we replace $f(x)$ with tan (x), we can rewrite the problem as

$$\lim_{h \to 0} \frac{\tan(x + h) - \tan(x)}{h} = \left[\tan(x)\right]'$$

Step 2: The derivative of tan x is sec^2 x. Thus,

$$\lim_{h \to 0} \frac{\tan\left(\dfrac{\pi}{6} + h\right) - \tan\left(\dfrac{\pi}{6}\right)}{h} = \sec^2\left(\frac{\pi}{6}\right)$$

Step 3: Because $\sec\left(\dfrac{\pi}{6}\right) = \dfrac{2}{\sqrt{3}}$, $\sec^2\left(\dfrac{\pi}{6}\right) = \dfrac{4}{3}$.

Note: If you had trouble with this problem, you should review the units on the definition of the derivative and derivatives of trigonometric functions.

32. **B** If $\displaystyle\int_{30}^{100} f(x)\, dx = A$ and $\displaystyle\int_{50}^{100} f(x)\, dx = B$, then $\displaystyle\int_{30}^{50} f(x)\, dx =$

This question is testing your knowledge of the rules of definite integrals.

Step 1: Generally speaking, $\displaystyle\int_a^b f(x)\, dx + \int_b^c f(x)\, dx = \int_a^c f(x)\, dx$.

So here, $\displaystyle\int_{30}^{50} f(x)\, dx + \int_{50}^{100} f(x)\, dx = \int_{30}^{100} f(x)\, dx$.

If we substitute $\displaystyle\int_{30}^{100} f(x)\, dx = A$ and $\displaystyle\int_{50}^{100} f(x)\, dx = B$, we get $\displaystyle\int_{30}^{50} f(x)\, dx + B = A$.

33. **E** If $f(x) = 3x^2 - x$, and $g(x) = f^{-1}(x)$, then $g'(10)$ could be

This problem requires you to know how to find the derivative of an inverse function.

Step 1: The rule for finding the derivative of an inverse function is

$$\text{If } y = f(x) \text{ and if } g(x) = f^{-1}(x) \text{ then } g'(x) = \frac{1}{f'(y)}$$

Step 2: In order to use the formula, we need to find the derivative of f and the value of x that corresponds to $y = 10$.

First, $f'(x) = 6x - 1$. Second, when $y = 10$ we get $10 = 3x^2 - x$.

If we solve this for x, we get $x = 2$ and $x = -\dfrac{5}{3}$, but we'll use 2—it's easier.

Step 3: Plugging into the formula, we get $\dfrac{1}{f'(y)} = \dfrac{1}{(6)(2) - 1} = \dfrac{1}{11}$.

Note: There was another possible answer using $x = -\dfrac{5}{3}$, but that doesn't give us one of the answer choices. Generally, the AP exam sticks to the easier answer. They are testing whether you know what to do and are usually NOT trying to trick you.

34. **C** The graph of $y = x^3 - 5x^2 + 4x + 2$ has a local minimum at

This is another maxima/minima question.

Step 1: Take the derivative of the function and set it equal to zero.

$$f'(x) = 3x^2 - 10x + 4 = 0$$

Step 2: Use the quadratic formula to solve for x. You should get $x = \{2.87, 0.46\}$.

Step 3: Now take the second derivative of the function.

$$f''(x) = 6x - 10$$

Step 4: Plug each of the critical values from Step 2 into the second derivative. If you get a positive value, the point is a minimum. If you get a negative value, the point is a maximum. If you get zero, the point is probably a point of inflection (don't worry about that here).

$$f''(2.87) = 7.21$$
$$f''(.46) = -7.21$$

So 2.87 is the x-coordinate of the minimum. To find the y-coordinate, just plug 2.87 into $f(x)$ and you get -4.06.

35. **C** The volume generated by revolving about the y-axis the region enclosed by the graphs $y = 9 - x^2$ and $y = 9 - 3x$, for $0 \le x \le 2$, is

This is another volume of a solid of revolution problem. As you should have noticed by now, these are very popular on the AP exam and show up in both the multiple-choice section and in the free-response section. If you are not good at these, go back and review the unit carefully. You cannot afford to get these wrong! The good thing about *this* volume problem is that it is in the calculator part of the multiple-choice section, so you can use a graphing calculator to assist you.

Step 1: First, graph the two curves on the same set of axes. The graph should look like the following:

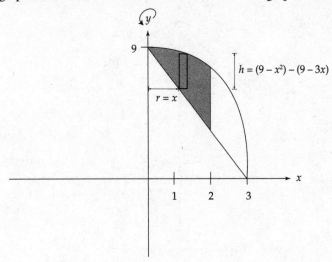

$$v = 2\pi \int_0^2 x\left[(9 - x^2) - (9 - 3x)\right]$$

Graphing Calculator (TI-83 and TI-84)

Press the Y= button, and enter the following values to the list:

$Y_1 = 9 - x^2$

$Y_2 = 9 - 3x$

Press 2nd and MODE to view the home screen.

Press MATH and select 9: fnInt from the list, then enter the following:

fnInt(2π X($Y_1 - Y_2$), X, 0, 2)

- Press VARS and go to Y-VARS to select the variables Y_1 and Y_2.

The result is 25.13274123. Divide that by π to get 8, answer (C).

Step 2: We are being asked to rotate this region around the y-axis, and both of the functions are in terms of x, so we should use the method of shells. We use this method whenever we take a vertical slice of a region and rotate it around an axis parallel to the slice (review the unit if you are not sure what it means). This will give us a region that looks like the following:

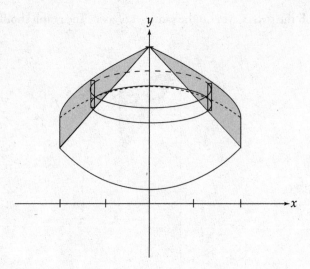

Step 3: The formula for the method of shells says that if you have a region between two curves, $f(x)$ and $g(x)$ from $x = a$ to $x = b$, then the volume generated when the region is revolved around the y-axis is: $2\pi \int_a^b x\left[f(x) - g(x) \right] dx$ if $f(x)$ is above $g(x)$ throughout the region. Thus, our integral is

$$2\pi \int_0^2 x\left[(9 - x^2) - (9 - 3x) \right] dx$$

We can simplify this integral to $2\pi \int_0^2 x(3x - x^2)\, dx = 2\pi \int_0^2 (3x^2 - x^3)\, dx$.

Step 4: Evaluate the integral.

$$2\pi \int_0^2 (3x^2 - x^3)\, dx = 2\pi \left(x^3 - \frac{x^4}{4} \right)\Bigg|_0^2 = 8\pi$$

36. **B**   The average value of the function $f(x) = \ln^2 x$ on the interval $[2, 4]$ is

This problem requires you to be familiar with the Mean Value Theorem for integrals, which we use to find the average value of a function.

Step 1: If you want to find the average value of $f(x)$ on an interval $[a, b]$, you need to evaluate the integral $\dfrac{1}{b - a}\int_a^b f(x)\, dx$. So here we evaluate the integral $\dfrac{1}{2}\int_2^4 \ln^2 x\, dx$.

You have to do this integral on your calculator because you do not know how to evaluate this integral analytically unless you are very good with integration by parts!

Use **fnint.** Divide this by 2 and you will get 1.204.

37. **E** $\dfrac{d}{dx} \displaystyle\int_0^{3x} \cos(t)\, dt =$

This problem is testing your knowledge of the Second Fundamental Theorem of Calculus. The theorem states that $\dfrac{d}{dx} \displaystyle\int_a^u f(t)\, dt = f(u)\dfrac{du}{dx}$, where a is a constant and u is a function of x. So all we have to do is follow the theorem: $\dfrac{d}{dx} \displaystyle\int_0^{3x} \cos(t)\, dt = 3\cos 3x$.

38. **C** If the definite integral $\displaystyle\int_1^3 (x^2 + 1)\, dx$ is approximated by using the Trapezoid Rule with $n = 4$, the error is

This problem will require you to be familiar with the Trapezoid Rule. This is very easy to do on the calculator, and some of you may even have written programs to evaluate this. Even if you haven't, the formula is easy. The area under a curve from $x = a$ to $x = b$, divided into n intervals, is approximated by the Trapezoid Rule, and it is

$$\left(\frac{1}{2}\right)\left(\frac{b-a}{n}\right)\left[y_0 + 2y_1 + 2y_2 + 2y_3 \ldots + 2y_{n-2} + 2y_{n-1} + y_n\right]$$

This formula may look scary, but it actually is quite simple, and the AP exam never uses a very large value for n anyway.

Step 1: $\dfrac{b-a}{n} = \dfrac{3-1}{4} = \dfrac{1}{2}$. Plugging into the formula, we get

$$\frac{1}{4}\left[(1^2 + 1) + 2(1.5^2 + 1) + 2(2^2 + 1) + 2(2.5^2 + 1) + (3^2 + 1)\right]$$

This is easy to plug into your calculator and you will get 10.75 or $\dfrac{43}{4}$.

Step 2: In order to find the error, we now need to know the actual value of the integral.

$$\int_1^3 (x^2 + 1)\, dx = \frac{x^3}{3} + x\Big|_1^3 = \frac{32}{3}$$

Step 3: The error is $\dfrac{43}{4} - \dfrac{32}{3} = \dfrac{1}{12}$.

39. **C** The radius of a sphere is increasing at a rate proportional to itself. If the radius is 4 initially, and the radius is 10 after two seconds, then what will the radius be after three seconds?

This is not a related-rate problem; this is a differential equation! It just happens to involve a rate.

Step 1: If we translate the first sentence into an equation we get $\dfrac{dR}{dt} = kR$.

Put all of the terms that contain an R on the left of the equals sign, and all of the terms that contain a t on the right-hand side.

$$\frac{dR}{R} = k\, dt$$

Step 2: Integrate both sides.

$$\int \frac{dR}{R} = k \int dt$$

Step 3: If we solve this for R, we get $R = Ce^{kt}$ (see the unit on differential equations).

Now we need to solve for C and k. First, we solve for C by plugging in the information that the radius is 4 initially. This means that $R = 4$ when $t = 0$.

$$\text{If } 4 = Ce^0, \text{ then } C = 4$$

Next, we solve for k by plugging in the information that $R = 10$ when $t = 2$.

$$10 = 4e^{2k}$$

$$\frac{5}{2} = e^{2k}$$

$$\ln \frac{5}{2} = 2k$$

$$\frac{1}{2} \ln \frac{5}{2} = k$$

Step 4: Now we have our final equation: $R = 4e^{\left(\frac{1}{2} \ln \frac{5}{2}\right)t}$.

If we plug in $t = 3$ we get $R = 4e^{\left(\frac{1}{2} \ln \frac{5}{2}\right)(3)} \approx 15.811$.

40. **E** Use differentials to approximate the change in the volume of a sphere when the radius is increased from 10 to 10.02 cm.

The volume of a sphere is $V = \frac{4}{3}\pi R^3$. Using differentials, the change will be: $dV = 4\pi R^2\, dR$.

Substitute in $R = 10$ and $dR = 0.02$, and we get

$$dV = 4\pi(10^2)(0.02)$$

$$dV = 8\pi \approx 25.133 \text{ cm}^3$$

41. **D** $\int \ln 2x\, dx =$

This is a simple integral that we do using integration by parts. You should memorize that $\int \ln(ax)\, dx = x \ln(ax) - x + C$, which makes this integral easy.

Step 1: The formula for integration by parts is $\int u\, dv = uv - \int v\, du$.

The trick is that we have to let $dv = dx$.

$$\text{Let } u = \ln 2x \text{ and } dv = dx$$

$$du = \frac{2}{2x}\, dx = \frac{1}{x}\, dx \text{ and } v = x$$

Plugging in to the formula we get

$$\int \ln 2x\, dx = x \ln 2x - \int dx = x \ln 2x - x + C$$

42. **C** For the function $f(x) = \begin{cases} ax^3 - 6x; \text{ if } x \le 1 \\ bx^2 + 4; \ x > 1 \end{cases}$ to be continuous and differentiable, $a =$

This question is testing your knowledge of the rules of continuity, where we also discuss differentiability.

Step 1: If the function is continuous, then plugging 1 into the top and bottom pieces of the function should yield the same answer.
$$a(1^3) - 6(1) = b(1^2) + 4$$
$$a - 6 = b + 4$$

Step 2: If the function is differentiable, then plugging 1 into the derivatives of the top and bottom pieces of the function should yield the same answer.

$$3a(1^2) - 6 = 2b(1)$$
$$3a - 6 = 2b$$

Step 3: Now we have a pair of simultaneous equations. If we solve them, we get $a = -14$.

43. **D** Two particles leave the origin at the same time and move along the y-axis with their respective positions determined by the functions $y_1 = \cos 2t$ and $y_2 = 4\sin t$ for $0 < t < 6$. For how many values of t do the particles have the same acceleration?

If you want to find acceleration, all you have to do is take the second derivative of the position functions.

Step 1:
$$\frac{dy_1}{dx} = -2\sin 2t \text{ and } \frac{dy_2}{dx} = 4\cos t$$

$$\frac{d^2 y_1}{dx^2} = -4\cos 2t \text{ and } \frac{d^2 y_2}{dx^2} = -4\sin t$$

Step 2: Now all we have to do is to graph both of these equations on the same set of axes on a calculator. You should make the window from $x = 0$ to $x = 7$ (leave yourself a little room so that you can see the whole range that you need). You should get a picture that looks like the following:

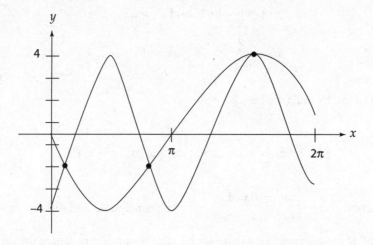

Where the graphs intersect, the acceleration is the same. There are three points of intersection.

44. **B** Find the distance traveled (to three decimal places) in the first four seconds, for a particle whose velocity is given by $v(t) = 7e^{-t^2}$, where t stands for time.

Step 1: If we want to find the distance traveled, we take the integral of velocity from the starting time to the finishing time. Therefore, we need to evaluate $\int_0^4 7e^{-t^2}\, dt$.

Step 2: But we have a problem! We can't take the integral of e^{-t^2}. This means that the AP exam wants you to find the answer using your calculator.

Rounded to three decimal places, the answer is 6.204.

45. **A** $\int \tan^6 x \sec^2 x \, dx =$

We can do this integral with u-substitution.

Step 1: Let $u = \tan x$. Then $du = \sec^2 x \, dx$.

Step 2: Substituting, we get $\int \tan^6 x \sec^2 x \, dx = \int u^6 \, du$.

Step 3: This is an easy integral: $\int u^6 \, du = \dfrac{u^7}{7} + C$.

Step 4: Substituting back, we get $\dfrac{\tan^7 x}{7} + C$.

ANSWERS AND EXPLANATIONS TO SECTION II

1.   A particle moves along the x-axis so that its acceleration at any time $t > 0$ is given by $a(t) = 12t - 18$. At time $t = 1$, the velocity of the particle is $v(1) = 0$ and the position is $x(1) = 9$.

(a)   Write an expression for the velocity of the particle $v(t)$.

Step 1: We know that the derivative of velocity with respect to time is acceleration, so the integral of acceleration with respect to time is velocity.

$$\int a(t) \, dt = v(t)$$
$$\int 12t - 18 \, dt = 6t^2 - 18t + C = v(t)$$

If we plug in the information that at time $t = 1$, $v(1) = 0$, we can solve for C.

$$6(1)^2 - 18(1) + C = 0$$
$$-12 + C = 0$$
$$C = 12$$

This means that the velocity of the particle is $6t^2 - 18t + 12$.

(b)   At what values of t does the particle change direction?

When a particle is in motion, it changes direction at the time when its velocity is zero. (As long as acceleration is not also zero.) So all we have to do is set velocity equal to zero and solve for t.

$$6t^2 - 18t + 12 = 0$$
$$t^2 - 3t + 2 = 0$$
$$(t - 2)(t - 1) = 0$$
$$t = 1, 2$$

(c) Write an expression for the position $x(t)$ of the particle.

We know that the derivative of position with respect to time is velocity, so the integral of velocity with respect to time is position.

$$\int v(t)\, dt = x(t)$$

$$\int (6t^2 - 18t + 12)\, dt = 2t^3 - 9t^2 + 12t + C = x(t)$$

If we plug in the information that at time $t = 1$, $x(1) = 9$, we can solve for C.

$$2(1)^3 - 9(1)^2 + 12(1) + C = 9$$
$$5 + C = 9$$
$$C = 4$$

so $x(t) = 2t^3 - 9t^2 + 12t + 4$

(d) Find the total distance traveled by the particle from $t = \dfrac{3}{2}$ to $t = 6$?

Step 1: Normally, all that we have to do to find the distance traveled is to integrate the velocity equation from the starting time to the ending time. But we have to watch out for whether the particle changes direction. If so, we have to break the integration into two parts—a positive integral for when it is traveling to the right, and a negative integral for when it is traveling to the left.

One way to solve this is to find two integrals and add them together. Because you can use a calculator, it is simpler to use the fnInt calculation of the absolute value for $t = \dfrac{3}{2}$ and $t = 6$, using the function of velocity $6x^2 - 18x + 12$.

Graphing Calculator (TI-83 and TI-84)

Press MATH and select 9: fnInt from the list.

Press MATH then select the NUM menu, and choose 1: abs(

Enter the function $6x^2 - 18x + 12$ and follow with the closing parentheses. List the variable and low and high values for t, separated by commas, and follow with final closing parentheses so your expression looks like the following:

fnInt(abs($6x^2 - 18x + 12$), x, 3/2, 6)

Press ENTER

The result is 176.500

2.　Let R be the region enclosed by the graphs of $y = 2 \ln x$ and $y = \dfrac{x}{2}$ and the lines $x = 2$ and $x = 8$.

(a)　Find the area of R.

Step 1: If there are two curves, $f(x)$ and $g(x)$, where $f(x)$ is always above $g(x)$, on the interval $[a, b]$, then the area of the region between the two curves is found by

$$\int_a^b (f(x) - g(x))\, dx$$

In order to determine whether one of the curves is above the other, we can graph them on the calculator.

The graph looks like the following:

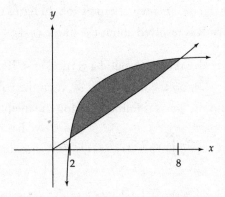

As we can see, the graph of $y = 2 \ln x$ is above $y = \dfrac{x}{2}$ on the entire interval, so all we have to do is evaluate the integral $\displaystyle\int_2^8 \left(2 \ln x - \dfrac{x}{2} \right) dx =$.

Step 2: We can do the integration one of two ways—on the calculator or analytically.

Calculator:　Evaluate **fnint** $\left(\left(2 \ln x - \left(\dfrac{x}{2} \right) \right),\ x, 2, 8 \right) = 3.498$

Analytically:　$\displaystyle\int_2^8 \left(2 \ln x - \dfrac{x}{2} \right) dx = 2\int_2^8 \ln x\, dx - \dfrac{1}{2}\int_2^8 x\, dx =$

$$2\left. (x \ln x - x) \right|_2^8 - \dfrac{1}{2}\left. \left(\dfrac{x^2}{2} \right) \right|_2^8 \approx 18.498 - 15 = 3.498$$

You can use the formula $\displaystyle\int \ln x\, dx = x \ln x - x$, or you can do it as one of the basic integration-by-parts integrals.

(b) Set up, but <u>do not integrate</u>, an integral expression, in terms of a single variable, for the volume of the solid generated when R is revolved about the *x*-axis.

Step 1: If there are two curves, $f(x)$ and $g(x)$, where $f(x)$ is always above $g(x)$, on the interval $[a, b]$, then the volume of the solid generated when the region is revolved about the *x*-axis is found by using the method of washers.

$$\pi \int_a^b \left[\left[f(x) \right]^2 - \left[g(x) \right]^2 \right] dx$$

We already know that $f(x)$ is above $g(x)$ on the interval, so the integral we need to evaluate is

$$\pi \int_2^8 \left[\left[2 \ln x \right]^2 - \left[\frac{x}{2} \right]^2 \right] dx$$

(c) Set up, but <u>do not integrate</u>, an integral expression, in terms of a single variable, for the volume of the solid generated when R is revolved about the line $x = -1$.

Step 1: Now we have to revolve the area around a <u>vertical</u> axis. If there are two curves, $f(x)$ and $g(x)$, where $f(x)$ is always above $g(x)$, on the interval $[a, b]$, then the volume of the solid generated when the region is revolved about the *y*-axis is found by using the method of shells.

$$2\pi \int_a^b x \left[f(x) - g(x) \right] dx$$

When we are rotating around a vertical axis, we use the same formula as when we rotate around the *y*-axis, but we have to account for the shift away from $x = 0$. Here we have a curve that is 1 unit farther away from the line $x = -1$ than it is from the *y*-axis, so we add 1 to the radius of the shell (for a more detailed explanation of shifting axes, see the unit on finding the volume of a solid of revolution). This gives us the equation,

$$2\pi \int_2^8 (x + 1) \left[2 \ln x - \frac{x}{2} \right] dx$$

3. Consider the equation $x^2 - 2xy + 4y^2 = 64$.

(a) Write an expression for the slope of the curve at any point (x, y).

Step 1: The slope of the curve is just the derivative. But here we have to use implicit differentiation to find the derivative. If we take the derivative of each term with respect to *x*, we get

$$2x \frac{dx}{dx} - 2 \left(x \frac{dy}{dx} + y \frac{dx}{dx} \right) + 8y \frac{dy}{dx} = 0$$

Remember that $\dfrac{dx}{dx} = 1$, which gives us

$$2x - 2\left(x\dfrac{dy}{dx} + y\right) + 8y\dfrac{dy}{dx} = 0$$

Step 2: Now just simplify and solve for $\dfrac{dy}{dx}$.

$$2x - 2x\dfrac{dy}{dx} - 2y + 8y\dfrac{dy}{dx} = 0$$

$$x - x\dfrac{dy}{dx} - y + 4y\dfrac{dy}{dx} = 0$$

$$-x\dfrac{dy}{dx} + 4y\dfrac{dy}{dx} = y - x$$

$$\left(4y - x\right)\dfrac{dy}{dx} = y - x$$

$$\dfrac{dy}{dx} = \dfrac{y - x}{4y - x}$$

(b) Find the equation of the tangent lines to the curve at the point $x = 2$.

Step 1: We are going to use the point-slope form of a line, $y - y_1 = m(x - x_1)$, where (x_1, y_1) is a point on the curve, and the derivative at that point is the slope m. First, we need to know the value of y when $x = 2$. If we plug 2 for x into the original equation, we get

$$4 - 4y + 4y^2 = 64$$

$$4y^2 - 4y - 60 = 0$$

Using the quadratic formula, we get

$$y = \dfrac{1 \pm \sqrt{61}}{2} \approx 4.41, -3.41$$

Notice that there are two values of y when $x = 2$, which is why there are two tangent lines.

Step 2: Now that we have our points, we need the slope of the tangent line at $x = 2$.

$$\frac{dy}{dx} = \frac{y - x}{4y - x}$$

At $y = 4.41$, $\dfrac{dy}{dx} = \dfrac{4.41 - 2}{4(4.41) - 2} = 0.15$

At $y = -3.41$, $\dfrac{dy}{dx} = \dfrac{-3.41 - 2}{4(-3.41) - 2} = 0.35$

Step 3: Plugging into our equation for the tangent line, we get

$$y - 4.41 = 0.15(x - 2)$$

$$y + 3.41 = 0.35(x - 2)$$

It is not necessary to simplify these equations.

(c)   Find $\dfrac{d^2 y}{dx^2}$ at $(0, 4)$.

Step 1: Once we have the first derivative, we have to differentiate again to find $\dfrac{d^2 y}{dx^2}$.

We have to use implicit differentiation again.

$$\frac{dy}{dx} = \frac{y - x}{4y - x}$$

Use the Quotient Rule.

$$\frac{d^2 y}{dx^2} = \frac{(4y - x)\left(\dfrac{dy}{dx} - \dfrac{dx}{dx}\right) - (y - x)\left(4\dfrac{dy}{dx} - \dfrac{dx}{dx}\right)}{(4y - x)^2}$$

Simplifying, we get

$$\frac{d^2 y}{dx^2} = \frac{(4y - x)\left(\dfrac{dy}{dx} - 1\right) - (y - x)\left(4\dfrac{dy}{dx} - 1\right)}{(4y - x)^2}$$

Now we plug in $\dfrac{y-x}{4y-x}$ for $\dfrac{dy}{dx}$, which gives us

$$\frac{d^2 y}{dx^2} = \frac{(4y-x)\left(\dfrac{y-x}{4y-x}-1\right)-(y-x)\left(4\dfrac{y-x}{4y-x}-1\right)}{(4y-x)^2}$$

Now we would have to use a lot of algebra to simplify this but, fortunately, we can just plug (0, 4) in immediately for x and y, and solve from there.

$$\frac{d^2 y}{dx^2} = \frac{(16)\left(\dfrac{4}{16}-1\right)-(4)\left(4\dfrac{4}{16}-1\right)}{(16)^2} = \frac{-3}{64}$$

4. Water is draining at the rate of 48π ft³/sec from the vertex at the bottom of a conical tank whose diameter at its base is 40 feet and whose height is 60 feet.

(a) Find an expression for the volume of water (in ft³/sec) in the tank in terms of its radius.

The formula for the volume of a cone is $V = \dfrac{1}{3}\pi R^2 H$, where R is the radius of the cone, and H is the height. The ratio of the height of a cone to its radius is constant at any point on the edge of the cone, so we also know that $\dfrac{h}{r} = \dfrac{60}{20} = 3$. (Remember that the radius is half the diameter.) If we solve this for H and substitute, we get

$$H = 3R$$

$$V = \frac{1}{3}\pi R^2 (3R) = \pi R^3$$

(b) At what rate (in ft/sec) is the radius of the water in the tank shrinking when the radius is 16 feet?

Step 1: This is a related rates question. We now have a formula for the volume of the cone in terms of its radius, so if we differentiate it in terms of t, we should be able to solve for the rate of change of the radius $\dfrac{dr}{dt}$.

We are given that the rate of change of the volume and the radius are, respectively

$$\frac{dV}{dt} = -48\pi \text{ and } R = 16$$

Differentiating the formula for the volume, we get $\frac{dV}{dt} = 3\pi R^2 \frac{dr}{dt}$.

Now, we plug in and get $-48\pi = 3\pi 16^2 \frac{dr}{dt}$. Finally, if we solve for $\frac{dr}{dt}$, we get

$$\frac{dr}{dt} = -\frac{1}{16} \text{ ft/sec}$$

(c)   How fast (in ft/sec) is the height of the water in the tank dropping at the instant that the radius is 16 feet?

Step 1: This is the same idea as the previous problem, except that we want to solve for $\frac{dh}{dt}$.

In order to do this, we need to go back to our ratio of height to radius and solve it for the radius.

$$\frac{h}{r} = 3 \qquad \text{or} \qquad \frac{h}{3} = r$$

Substituting for R in the original equation, we get $V = \frac{1}{3}\pi \left(\frac{h}{3}\right)^2 h = \frac{\pi h^3}{27}$.

Step 2: Now we need to know what H is when R is 16. Using our ratio,

$$H = 3\,(16) = 48.$$

Step 3: Now if we differentiate, we get

$$\frac{dV}{dt} = \frac{\pi h^2}{9} \frac{dh}{dt}$$

Now we plug in and solve.

$$48\pi = \frac{\pi (48)^2}{9} \frac{dh}{dt}$$

$$\frac{dh}{dt} = -\frac{3}{16}$$

One should also note that, because $H = 3R$, $\dfrac{dh}{dt} = 3\dfrac{dr}{dt}$. Thus, after we found $\dfrac{dr}{dt}$ in part 2, we merely had to multiply it by 3 to find the answer for part 3.

5. Let f be the function given by $y = f(x) = 2x^4 - 4x^2 + 1$.

(a) Find an equation of the line tangent to the graph at $(-2, 17)$.

In order to find the equation of a tangent line at a particular point, we need to take the derivative of the function and plug in the x- and y-values at that point to give us the slope of the line.

Step 1: The derivative is $f'(x) = 8x^3 - 8x$. If we plug in $x = -2$, we get

$$f'(-2) = 8(-2)^3 - 8(-2) = -48$$

This is the slope m.

Step 2: Now we use the slope-intercept form of the equation of a line, $y - y_1 = m(x - x_1)$, and plug in the appropriate values of x, y, and m.

$$y - 17 = -48(x + 2)$$

If we simplify this we get $y = -48x - 79$.

(b) Find the x- and y-coordinates of the relative maxima and relative minima. Verify your answer.

If we want to find the maxima/minima, we need to take the derivative and set it equal to zero. The values that we get are called critical points. We will then test each point to see if it is a maximum or a minimum.

Step 1: We already have the first derivative from part (a), so we can just set it equal to zero.

$$8x^3 - 8x = 0$$

If we now solve this for x, we get

$$8x(x^2 - 1) = 0 \quad 8x(x + 1)(x - 1) = 0 \quad x = 0, 1, -1$$

These are our critical points. In order to test if a point is a maximum or a minimum, we usually use the *second derivative test*. We plug each of the critical points into the second derivative. If we get a positive value, the point is a relative minimum. If we get a negative value, the point is a relative maximum. If we get zero, the point is a point of inflection.

Step 2: The second derivative is $f''(x) = 24x^2 - 8$. If we plug in the critical points, we get

$$f''(0) = 24(0)^2 - 8 = -8$$

$$f''(1) = 24(1)^2 - 8 = 16$$

$$f''(-1) = 24(-1)^2 - 8 = 16$$

So $x = 0$ is a relative maximum, and $x = 1, -1$ are relative minima.

Step 3: In order to find the y-coordinates, we plug the x-values back into the original equation, and solve.

$$f(0) = 1$$
$$f(1) = -1$$
$$f(-1) = -1$$

And our points are

$(0,1)$ is a relative maximum

$(1,-1)$ is a relative minimum

$(-1,-1)$ is a relative minimum

(c)   Find the x- and y- coordinates of the points of inflection. Verify your answer.

If we want to find the points of inflection, we set the second derivative equal to zero. The values that we get are the x-coordinates of the points of inflection.

Step 1: We already have the second derivative from part (b), so all we have to do is set it equal to zero and solve for x.

$$24x^2 - 8 = 0 \qquad x^2 = \frac{1}{3} \qquad x = \pm\sqrt{\frac{1}{3}}$$

Step 2: In order to find the y-coordinates, we plug the x-values back into the original equation, and solve.

$$f\left(\sqrt{\frac{1}{3}}\right) = 2\left(\sqrt{\frac{1}{3}}\right)^4 - 4\left(\sqrt{\frac{1}{3}}\right)^2 + 1 = \frac{2}{9} - \frac{4}{3} + 1 = -\frac{1}{9}$$

$$f\left(-\sqrt{\frac{1}{3}}\right) = 2\left(-\sqrt{\frac{1}{3}}\right)^4 - 4\left(-\sqrt{\frac{1}{3}}\right)^2 + 1 = \frac{2}{9} - \frac{4}{3} + 1 = -\frac{1}{9}$$

So the points of inflection are $\left(\sqrt{\frac{1}{3}}, -\frac{1}{9}\right)$ and $\left(-\sqrt{\frac{1}{3}}, -\frac{1}{9}\right)$.

6. Let $F(x) = \displaystyle\int_0^x \left[\cos\left(\frac{t}{2}\right) + \left(\frac{3}{2}\right)\right] dt$ on the closed interval $[0, 4\pi]$.

(a) Approximate $F(2\pi)$ using four inscribed rectangles.

This means that we need to find $\displaystyle\int_0^{2\pi} \left[\cos\left(\frac{t}{2}\right) + \left(\frac{3}{2}\right)\right] dt$.

Step 1: The graph of $\cos\left(\frac{t}{2}\right) + \left(\frac{3}{2}\right)$ from 0 to 2π, using four inscribed rectangles looks like the following:

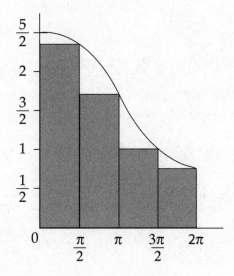

If we are cutting the interval $[0, 2\pi]$ into 4 rectangles, the width of each rectangle is $\dfrac{\pi}{2}$.

The height of each rectangle depends on the x-coordinate.

Step 2: We can now set up the calculation for the area of the rectangles.

$$\text{Area} = \frac{\pi}{2}\left[\left(\cos\frac{\pi}{4} + \frac{3}{2}\right) + \left(\cos\frac{\pi}{2} + \frac{3}{2}\right) + \left(\cos\frac{3\pi}{4} + \frac{3}{2}\right) + \left(\cos\pi + \frac{3}{2}\right)\right]$$

$$= \frac{\pi}{2}\left[\left(\frac{3}{2} + \frac{1}{\sqrt{2}}\right) + \left(\frac{3}{2}\right) + \left(\frac{3}{2} - \frac{1}{\sqrt{2}}\right) + \left(\frac{1}{2}\right)\right] = \frac{5\pi}{2} \approx 7.854$$

(b) Find $F'(2\pi)$.

Step 1: The Second Fundamental Theorem of Calculus says that if $f(x)$ is a continuous function, and a is a constant, then $\dfrac{d}{dx}\displaystyle\int_0^x f(t)\,dt = f(x)$.

So here we have: $\dfrac{d}{dx}\displaystyle\int_0^x\left[\cos\left(\frac{t}{2}\right) + \frac{3}{2}\right]dt = \cos\left(\frac{x}{2}\right) + \frac{3}{2}$.

Step 2: Now we plug in 2π for x and we get $\cos\pi + \dfrac{3}{2} = \dfrac{1}{2}$.

(c) Find the average value of $F'(x)$ on the interval $[0, 4\pi]$.

Step 1: The Mean Value Theorem for integrals says that if you want to find the average value of $f(x)$ on an interval $[a, b]$, you need to evaluate the integral $\dfrac{1}{b-a}\displaystyle\int_a^b f(x)\,dx$. So here we would evaluate the integral $\dfrac{1}{4\pi - 0}\displaystyle\int_0^{4\pi}\left(\cos\left(\frac{x}{2}\right) + \frac{3}{2}\right)dx$.

$$\frac{1}{4\pi - 0}\int_0^{4\pi}\left(\cos\left(\frac{x}{2}\right) + \frac{3}{2}\right)dx = \frac{1}{4\pi}\int_0^{4\pi}\cos\frac{x}{2}\,dx + \frac{3}{8\pi}\int_0^{4\pi}dx =$$

$$\frac{1}{4\pi}2\sin\frac{x}{2}\Big|_0^{4\pi} + \frac{3}{8\pi}x\Big|_0^{4\pi} = \frac{1}{2\pi}\sin\frac{x}{2}\Big|_0^{4\pi} + \frac{3}{8\pi}x\Big|_0^{4\pi}$$

Step 2: Now we evaluate at the limits of integration, and we get

$$\frac{1}{2\pi}\sin\frac{x}{2}\Big|_0^{4\pi} + \frac{3}{8\pi}x\Big|_0^{4\pi} = \frac{1}{2\pi}(\sin 2\pi - \sin 0) + \frac{3}{8\pi}(4\pi) = \frac{3}{2}$$

Chapter 22
Practice Test 2

AP® Calculus AB Exam

SECTION I: Multiple-Choice Questions

DO NOT OPEN THIS BOOKLET UNTIL YOU ARE TOLD TO DO SO.

At a Glance

Total Time
1 hour and 45 minutes
Number of Questions
45
Percent of Total Grade
50%
Writing Instrument
Pencil required

Instructions

Section I of this examination contains 45 multiple-choice questions. Fill in only the ovals for numbers 1 through 45 on your answer sheet.

CALCULATORS MAY NOT BE USED IN THIS PART OF THE EXAMINATION.

Indicate all of your answers to the multiple-choice questions on the answer sheet. No credit will be given for anything written in this exam booklet, but you may use the booklet for notes or scratch work. After you have decided which of the suggested answers is best, completely fill in the corresponding oval on the answer sheet. Give only one answer to each question. If you change an answer, be sure that the previous mark is erased completely. Here is a sample question and answer.

Sample Question   Sample Answer

Chicago is a
(A) state
(B) city
(C) country
(D) continent
(E) village

Use your time effectively, working as quickly as you can without losing accuracy. Do not spend too much time on any one question. Go on to other questions and come back to the ones you have not answered if you have time. It is not expected that everyone will know the answers to all the multiple-choice questions.

About Guessing

Many candidates wonder whether or not to guess the answers to questions about which they are not certain. Multiple choice scores are based on the number of questions answered correctly. Points are not deducted for incorrect answers, and no points are awarded for unanswered questions. Because points are not deducted for incorrect answers, you are encouraged to answer all multiple-choice questions. On any questions you do not know the answer to, you should eliminate as many choices as you can, and then select the best answer among the remaining choices.

CALCULUS AB

SECTION I, Part A

Time—55 Minutes

Number of questions—28

A CALCULATOR MAY NOT BE USED ON THIS PART OF THE EXAMINATION

Directions: Solve each of the following problems, using the available space for scratchwork. After examining the form of the choices, decide which is the best of the choices given and fill in the corresponding oval on the answer sheet. No credit will be given for anything written in the test book. Do not spend too much time on any one problem.

In this test: Unless otherwise specified, the domain of a function f is assumed to be the set of all real numbers x for which $f(x)$ is a real number.

1. If $g(x) = \dfrac{1}{32} x^4 - 5x^2$, find $g'(4)$.

    (A) −72
    (B) −32
    (C) −24
    (D) 24
    (E) 32

2. The domain of the function $f(x) = \sqrt{4 - x^2}$ is

    (A) $x < -2$ or $x > 2$
    (B) $x \leq -2$ or $x \geq 2$
    (C) $-2 < x < 2$
    (D) $-2 \leq x \leq 2$
    (E) $x \leq 2$

GO ON TO THE NEXT PAGE.

3. $\lim\limits_{x \to 5} \dfrac{x^2 - 25}{x - 5}$ is

(A) 0
(B) 10
(C) −10
(D) 5
(E) The limit does not exist.

---

4. If $f(x) = \dfrac{x^5 - x + 2}{x^3 + 7}$, find $f'(x)$.

(A) $\dfrac{(5x^4 - 1)}{(3x^2)}$

(B) $\dfrac{(5x^4 - 1) - (3x^2)}{(x^3 + 7)}$

(C) $\dfrac{(x^3 + 7)(5x^4 - 1) - (x^5 - x + 2)(3x^2)}{(x^3 + 7)}$

(D) $\dfrac{(x^5 - x + 2)(3x^2) - (x^3 + 7)(5x^4 - 1)}{(x^3 + 7)^2}$

(E) $\dfrac{(x^3 + 7)(5x^4 - 1) - (x^5 - x + 2)(3x^2)}{(x^3 + 7)^2}$

GO ON TO THE NEXT PAGE.

5. Evaluate $\lim\limits_{h \to 0} \dfrac{5\left(\dfrac{1}{2} + h\right)^4 - 5\left(\dfrac{1}{2}\right)^4}{h}$.

(A) $\dfrac{5}{2}$

(B) $\dfrac{5}{16}$

(C) 40

(D) 160

(E) The limit does not exist.

6. $\displaystyle\int x\sqrt{3x}\, dx =$

(A) $\dfrac{2\sqrt{3}}{5} x^{\frac{5}{2}} + C$

(B) $\dfrac{5\sqrt{3}}{2} x^{\frac{5}{2}} + C$

(C) $\dfrac{\sqrt{3}}{2} x^{\frac{1}{2}} + C$

(D) $2\sqrt{3x} + C$

(E) $\dfrac{5\sqrt{3}}{2} x^{\frac{3}{2}} + C$

GO ON TO THE NEXT PAGE.

7. Find k so that $f(x) = \begin{cases} \dfrac{x^2 - 16}{x - 4} & ; x \neq 4 \\ k & ; x = 4 \end{cases}$ is continuous for all x.

(A) All real values of k make $f(x)$ continuous for all x.
(B) 0
(C) 16
(D) 8
(E) There is no real value of k that makes $f(x)$ continuous for all x.

8. Which of the following integrals correctly gives the area of the region consisting of all points above the x-axis and below the curve $y = 8 + 2x - x^2$?

(A) $\int_{-2}^{4} (x^2 - 2x - 8)\, dx$

(B) $\int_{-4}^{2} (8 + 2x - x^2)\, dx$

(C) $\int_{-2}^{4} (8 + 2x - x^2)\, dx$

(D) $\int_{-4}^{2} (x^2 - 2x - 8)\, dx$

(E) $\int_{2}^{4} (8 + 2x - x^2)\, dx$

GO ON TO THE NEXT PAGE.

9. If $f(x) = x^2 \cos 2x$, find $f'(x)$.

   (A) $2x \sin 2x$
   (B) $-2x \cos 2x + 2x^2 \sin 2x$
   (C) $-4x \sin 2x$
   (D) $2x \cos 2x - 2x^2 \sin 2x$
   (E) $2x - 2 \sin 2x$

10. An equation of the line tangent to $y = 4x^3 - 7x^2$ at $x = 3$ is

   (A) $y + 45 = 66(x + 3)$

   (B) $y - 45 = 66(x - 3)$

   (C) $y = 66x$

   (D) $y = 66(x - 3)$

   (E) $y + 45 = \dfrac{-1}{66}(x - 3)$

GO ON TO THE NEXT PAGE.

11. $\int_0^{\frac{1}{2}} \dfrac{2}{\sqrt{1-x^2}} \, dx =$

(A) $\dfrac{\pi}{6}$

(B) $\dfrac{\pi}{3}$

(C) $-\dfrac{\pi}{3}$

(D) $\dfrac{2\pi}{3}$

(E) $-\dfrac{2\pi}{3}$

---

12. Find a positive value c, for x, that satisfies the conclusion of the Mean Value Theorem for Derivatives for $f(x) = 3x^2 - 5x + 1$ on the interval $[2, 5]$.

(A) 1

(B) $\dfrac{13}{6}$

(C) $\dfrac{11}{6}$

(D) $\dfrac{23}{6}$

(E) $\dfrac{7}{2}$

GO ON TO THE NEXT PAGE.

13. Given $f(x) = 2x^2 - 7x - 10$, find the absolute maximum of $f(x)$ on $[-1, 3]$.

    (A) -1

    (B) $\dfrac{7}{4}$

    (C) -13

    (D) $-\dfrac{129}{8}$

    (E) 0

14. Find $\dfrac{dy}{dx}$ if $x^3y + xy^3 = -10$.

    (A) $(3x^2 + 3xy^2)$

    (B) $-(3x^2 + 3xy^2)$

    (C) $\dfrac{(3x^2y + y^3)}{(3xy^2 + x^3)}$

    (D) $-\dfrac{(3x^2y + y^3)}{(3xy^2 + x^3)}$

    (E) $-\dfrac{(x^2y + y^3)}{(xy^2 + x^3)}$

GO ON TO THE NEXT PAGE.

15. If $f(x) = \sqrt{1 + \sqrt{x}}$, find $f'(x)$.

   (A) $\dfrac{-1}{4\sqrt{x}\sqrt{1 + \sqrt{x}}}$

   (B) $\dfrac{1}{2\sqrt{x}\sqrt{1 + \sqrt{x}}}$

   (C) $\dfrac{1}{4\sqrt{1 + \sqrt{x}}}$

   (D) $\dfrac{1}{4\sqrt{x}\sqrt{1 + \sqrt{x}}}$

   (E) $\dfrac{-1}{2\sqrt{x}\sqrt{1 + \sqrt{x}}}$

16. $\displaystyle\int 7xe^{3x^2}\,dx =$

   (A) $\dfrac{1}{42}e^{3x^2} + C$

   (B) $\dfrac{6}{7}e^{3x^2} + C$

   (C) $\dfrac{7}{6}e^{3x^2} + C$

   (D) $7e^{3x^2} + C$

   (E) $42e^{3x^2} + C$

GO ON TO THE NEXT PAGE.

17. Find the equation of the tangent line to $9x^2 + 16y^2 = 52$ through $(2, -1)$.

    (A) $-9x + 8y - 26 = 0$
    (B) $9x - 8y - 26 = 0$
    (C) $9x - 8y - 106 = 0$
    (D) $8x + 9y - 17 = 0$
    (E) $9x + 16y - 2 = 0$

---

18. A particle's position is given by $s = t^3 - 6t^2 + 9t$. What is its acceleration at time $t = 4$?

    (A) 0
    (B) 9
    (C) -9
    (D) -12
    (E) 12

GO ON TO THE NEXT PAGE.

19. If $f(x) = 3^{\pi x}$, then $f'(x) =$

    (A) $\dfrac{3^{\pi x}}{\pi \ln 3}$

    (B) $\dfrac{3^{\pi x}}{\ln 3}$

    (C) $\dfrac{3^{\pi x}}{\pi}$

    (D) $\pi(3^{\pi x - 1})$

    (E) $\pi \ln 3 (3^{\pi x})$

20. The average value of $f(x) = \dfrac{1}{x}$ from $x = 1$ to $x = e$ is

    (A) $\dfrac{1}{e + 1}$

    (B) $\dfrac{1}{1 - e}$

    (C) $e - 1$

    (D) $1 - \dfrac{1}{e^2}$

    (E) $\dfrac{1}{e - 1}$

GO ON TO THE NEXT PAGE.

21. If $f(x) = \sin^2 x$, find $f'''(x)$.

    (A) $-\sin^2 x$
    (B) $2 \cos 2x$
    (C) $\cos 2x$
    (D) $-4 \sin 2x$
    (E) $-\sin 2x$

22. Find the slope of the normal line to $y = x + \cos xy$ at $(0, 1)$.

    (A) 1
    (B) -1
    (C) 0
    (D) 2
    (E) Undefined

GO ON TO THE NEXT PAGE.

23. $\int e^x(e^{3x})\,dx =$

    (A) $\dfrac{1}{3}e^{3x} + C$

    (B) $\dfrac{1}{4}e^{4x} + C$

    (C) $\dfrac{1}{4}e^{5x} + C$

    (D) $4e^{4x} + C$

    (E) $4e^{5x} + C$

---

24. $\displaystyle\lim_{x \to 0} \dfrac{\tan^3(2x)}{x^3} =$

    (A) -8
    (B) -2
    (C) 2
    (D) 8
    (E) The limit does not exist.

GO ON TO THE NEXT PAGE.

25. A solid is generated when the region in the first quadrant bounded by the graph of $y = 1 + \sin^2 x$, the line $x = \dfrac{\pi}{2}$, the x-axis, and the y-axis is revolved about the x-axis. Its volume is found by evaluating which of the following integrals?

(A) $\pi \displaystyle\int_0^1 (1 + \sin^4 x)\, dx$

(B) $\pi \displaystyle\int_0^1 (1 + \sin^2 x)^2\, dx$

(C) $\pi \displaystyle\int_0^{\frac{\pi}{2}} (1 + \sin^4 x)\, dx$

(D) $\pi \displaystyle\int_0^{\frac{\pi}{2}} (1 + \sin^2 x)^2\, dx$

(E) $\pi \displaystyle\int_0^{\frac{\pi}{2}} (1 + \sin^2 x)\, dx$

26. If $y = \left(\dfrac{x^3 - 2}{2x^5 - 1}\right)^4$, find $\dfrac{dy}{dx}$ at $x = 1$.

(A) -52
(B) -28
(C) -13
(D) 13
(E) 52

GO ON TO THE NEXT PAGE.

27. $\int x\sqrt{5-x}\,dx =$

(A) $-\dfrac{10}{3}(5-x)^{\frac{3}{2}}$

(B) $\sqrt{\dfrac{5x^2}{2}-\dfrac{x^3}{3}}+C$

(C) $\dfrac{10}{3}\sqrt{\dfrac{5x^2}{2}-\dfrac{x^3}{3}}+C$

(D) $10(5-x)^{\frac{1}{2}}+\dfrac{2}{3}(5-x)^{\frac{3}{2}}+C$

(E) $-\dfrac{10}{3}(5-x)^{\frac{3}{2}}+\dfrac{2}{5}(5-x)^{\frac{5}{2}}+C$

28. If $\dfrac{dy}{dx}=\dfrac{x^3+1}{y}$ and $y=2$ when $x=1$, then, when $x=2$, $y=$

(A) $\sqrt{\dfrac{27}{2}}$

(B) $\sqrt{\dfrac{27}{8}}$

(C) $\pm\sqrt{\dfrac{27}{8}}$

(D) $\pm\dfrac{3}{2}$

(E) $\pm\sqrt{\dfrac{27}{2}}$

END OF PART A, SECTION I
IF YOU FINISH BEFORE TIME IS CALLED, YOU MAY CHECK YOUR WORK ON PART A ONLY.
DO NOT GO ON TO PART B UNTIL YOU ARE TOLD TO DO SO.

CALCULUS AB

SECTION I, Part B

Time—50 Minutes

Number of questions—17

A GRAPHING CALCULATOR IS REQUIRED FOR SOME QUESTIONS ON THIS PART OF THE EXAMINATION

Directions: Solve each of the following problems, using the available space for scratchwork. After examining the form of the choices, decide which is the best of the choices given and fill in the corresponding oval on the answer sheet. No credit will be given for anything written in the test book. Do not spend too much time on any one problem.

In this test:

1. The **exact** numerical value of the correct answer does not always appear among the choices given. When this happens, select from among the choices the number that best approximates the exact numerical value.

2. Unless otherwise specified, the domain of a function f is assumed to be the set of all real numbers x for which $f(x)$ is a real number.

29. The graph of $y = 5x^4 - x^5$ has an inflection point (or points) at

    (A) $x = 0$ only
    (B) $x = 3$ only
    (C) $x = 0, 3$
    (D) $x = -3$ only
    (E) $x = 0, -3$

GO ON TO THE NEXT PAGE.

30. The average value of $f(x) = e^{4x^2}$ on the interval $\left[-\dfrac{1}{4}, \dfrac{1}{4}\right]$ is

(A) 0.272
(B) 0.545
(C) 1.090
(D) 2.180
(E) 4.360

31. $\displaystyle\int_0^1 \tan x \, dx =$

(A) 0

(B) $\dfrac{\tan^2 1}{2}$

(C) $\ln(\cos(1))$

(D) $\ln(\sec(1))$

(E) $\ln(\sec(1)) - 1$

GO ON TO THE NEXT PAGE.

32. $\dfrac{d}{dx}\displaystyle\int_0^{x^2}\sin^2 t\,dt =$

    (A) $x^2\sin^2(x^2)$

    (B) $2x\sin^2(x^2)$

    (C) $\sin^2(x^2)$

    (D) $x^2\cos^2(x^2)$

    (E) $2x\cos^2(x^2)$

---

33. Find the value(s) of $\dfrac{dy}{dx}$ of $x^2y + y^2 = 5$ at $y = 1$.

    (A) $-\dfrac{3}{2}$ only

    (B) $-\dfrac{2}{3}$ only

    (C) $\dfrac{2}{3}$ only

    (D) $\pm\dfrac{2}{3}$

    (E) $\pm\dfrac{3}{2}$

GO ON TO THE NEXT PAGE.

34. The graph of $y = x^3 - 2x^2 - 5x + 2$ has a local maximum at

   (A) (2.120, 0)
   (B) (2.120, −8.061)
   (C) (−0.786, 0)
   (D) (−0.786, 4.209)
   (E) (0.666, −1.926)

35. Approximate $\int_0^1 \sin^2 x \, dx$ using the Trapezoid Rule with $n = 4$, to three decimal places.

   (A) 0.277
   (B) 0.273
   (C) 0.555
   (D) 1.109
   (E) 2.219

GO ON TO THE NEXT PAGE.

36. The volume generated by revolving about the x-axis the region above the curve $y = x^3$, below the line $y = 1$, and between $x = 0$ and $x = 1$ is

(A) $\dfrac{\pi}{42}$

(B) 0.143π

(C) $\dfrac{\pi}{7}$

(D) 0.643π

(E) $\dfrac{6\pi}{7}$

37. A 20-foot ladder slides down a wall at 5 ft/sec. At what speed is the bottom sliding out when the top is 10 feet from the floor (in ft/sec)?

(A) 0.346
(B) 2.887
(C) 0.224
(D) 5.774
(E) 4.472

GO ON TO THE NEXT PAGE.

38. $\int \dfrac{\ln x}{3x}\, dx =$

    (A) $6 \ln^2|x| + C$

    (B) $\dfrac{1}{6} \ln(\ln|x|) + C$

    (C) $\dfrac{1}{3} \ln^2|x| + C$

    (D) $\dfrac{1}{6} \ln^2|x| + C$

    (E) $\dfrac{1}{3} \ln|x| + C$

---

39. Find two non-negative numbers x and y whose sum is 100 and for which x^2y is a maximum.

    (A) $x = 33.333$ and $y = 33.333$
    (B) $x = 50$ and $y = 50$
    (C) $x = 33.333$ and $y = 66.667$
    (D) $x = 100$ and $y = 0$
    (E) $x = 66.667$ and $y = 33.333$

GO ON TO THE NEXT PAGE.

40. Find the distance traveled (to three decimal places) from $t = 1$ to $t = 5$ seconds, for a particle whose velocity is given by $v(t) = t + \ln t$.

(A) 6.000

(B) 1.609

(C) 16.047

(D) 0.800

(E) 148.413

---

41. $\int \sin^5(2x)\cos(2x)\,dx =$

(A) $\dfrac{\sin^6 2x}{12} + C$

(B) $\dfrac{\sin^6 2x}{6} + C$

(C) $\dfrac{\sin^6 2x}{3} + C$

(D) $\dfrac{\cos^5 2x}{3} + C$

(E) $\dfrac{\cos^5 2x}{6} + C$

GO ON TO THE NEXT PAGE.

42. The volume of a cube is increasing at a rate proportional to its volume at any time t. If the volume is 8 ft³ originally, and 12 ft³ after 5 seconds, what is its volume at $t = 12$ seconds?

   (A) 21.169
   (B) 22.941
   (C) 16.000
   (D) 28.800
   (E) 17.600

43. If $f(x) = \left(1 + \dfrac{x}{20}\right)^5$, find $f''(40)$.

   (A) 0.068
   (B) 1.350
   (C) 5.400
   (D) 6.750
   (E) 540.000

GO ON TO THE NEXT PAGE.

44. A particle's height at a time $t \geq 0$ is given by $h(t) = 100t - 16t^2$. What is its maximum height?

    (A) 312.500
    (B) 156.250
    (C) 78.125
    (D) 6.250
    (E) 3.125

45. If $f(x)$ is continuous and differentiable and $f(x) = \begin{cases} ax^4 + 5x; \ x \leq 2 \\ bx^2 - 3x; \ x > 2 \end{cases}$, then $b =$

    (A) 0.5
    (B) 0
    (C) 2
    (D) 6
    (E) There is no value of b.

STOP

END OF PART B, SECTION I

IF YOU FINISH BEFORE TIME IS CALLED, YOU MAY CHECK YOUR WORK ON PART B ONLY.

DO NOT GO ON TO SECTION II UNTIL YOU ARE TOLD TO DO SO.

SECTION II
GENERAL INSTRUCTIONS

You may wish to look over the problems before starting to work on them, since it is not expected that everyone will be able to complete all parts of all problems. All problems are given equal weight, but the parts of a particular problem are not necessarily given equal weight.

A GRAPHING CALCULATOR IS REQUIRED FOR SOME PROBLEMS OR PARTS OF PROBLEMS ON THIS SECTION OF THE EXAMINATION.

- You should write all work for each part of each problem in the space provided for that part in the booklet. Be sure to write clearly and legibly. If you make an error, you may save time by crossing it out rather than trying to erase it. Erased or crossed-out work will not be graded.

- Show all your work. You will be graded on the correctness and completeness of your methods as well as your answers. Correct answers without supporting work may not receive credit.

- Justifications require that you give mathematical (noncalculator) reasons and that you clearly identify functions, graphs, tables, or other objects you use.

- You are permitted to use your calculator to solve an equation, find the derivative of a function at a point, or calculate the value of a definite integral. However, you must clearly indicate the setup of your problem, namely the equation, function, or integral you are using. If you use other built-in features or programs, you must show the mathematical steps necessary to produce your results.

- Your work must be expressed in standard mathematical notation rather than calculator syntax. For example, $\int_1^5 x^2\,dx$ may not be written as fnInt (X², X, 1, 5).

- Unless otherwise specified, answers (numeric or algebraic) need not be simplified. If your answer is given as a decimal approximation, it should be correct to three places after the decimal point.

- Unless otherwise specified, the domain of a function f is assumed to be the set of all real numbers x for which $f(x)$ is a real number.

SECTION II, PART A
Time—30 minutes
Number of problems—2

A graphing calculator is required for some problems or parts of problems.

During the timed portion for Part A, you may work only on the problems in Part A.

On Part A, you are permitted to use your calculator to solve an equation, find the derivative of a function at a point, or calculate the value of a definite integral. However, you must clearly indicate the setup of your problem, namely the equation, function, or integral you are using. If you use other built-in features or programs, you must show the mathematical steps necessary to produce your results.

GO ON TO THE NEXT PAGE.

1. The temperature on New Year's Day in Hinterland was given by $T(H) = -A - B \cos\left(\dfrac{\pi H}{12}\right)$, where T is the temperature in degrees Fahrenheit and H is the number of hours from midnight ($0 \leq H < 24$).

   (a) The initial temperature at midnight was $-15°\,F$ and at noon of New Year's Day was $5°\,F$. Find A and B.

   (b) Find the average temperature for the first 10 hours.

   (c) Use the Trapezoid Rule with 4 equal subdivisions to estimate $\displaystyle\int_6^8 T(H)\,dH$.

   (d) Find an expression for the rate that the temperature is changing with respect to H.

---

2. Sea grass grows on a lake. The rate of growth of the grass is $\dfrac{dG}{dt} = kG$, where k is a constant.

   (a) Find an expression for G, the amount of grass in the lake (in tons), in terms of t, the number of years, if the amount of grass is 100 tons initially and 120 tons after one year.

   (b) In how many years will the amount of grass available be 300 tons?

   (c) If fish are now introduced into the lake and consume a consistent 80 tons/year of sea grass, how long will it take for the lake to be completely free of sea grass?

GO ON TO THE NEXT PAGE.

SECTION II, PART B
Time—1 hour
Number of problems—4

No calculator is allowed for these problems.

During the timed portion for Part B, you may continue to work on the problems in Part A without the use of any calculator.

3. Consider the curve defined by $y = x^4 + 4x^3$.

    (a) Find the equation of the tangent line to the curve at $x = -1$.

    (b) Find the coordinates of the absolute minimum.

    (c) Find the coordinates of the point(s) of inflection.

4. Water is being poured into a hemispherical bowl of radius 6 inches at the rate of 4 in³/sec.

    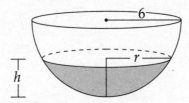

    (a) Given that the volume of the water in the spherical segment shown above is $V = \pi h^2\left(R - \dfrac{h}{3}\right)$, where R is the radius of

    the *sphere,* find the rate that the water level is rising when the water is 2 inches deep.

    (b) Find an expression for r, the radius of the *surface of the spherical segment* of water, in terms of h.

    (c) How fast is the circular area of the surface of the spherical segment of water growing (in in²/sec) when the water is 2 inches deep?

GO ON TO THE NEXT PAGE.

5. Let R be the region in the first quadrant bounded by $y^2 = x$ and $x^2 = y$.

    (a) Find the area of region R.

    (b) Find the volume of the solid generated when R is revolved about the x-axis.

    (c) The section of a certain solid cut by any plane perpendicular to the x-axis is a circle with the endpoints of its diameter lying on the parabolas $y^2 = x$ and $x^2 = y$. Find the volume of the solid.

---

6. An object moves with velocity $v(t) = t^2 - 8t + 7$.

    (a) Write a polynomial expression for the position of the particle at any time $t \geq 0$.

    (b) At what time(s) is the particle changing direction?

    (c) Find the total distance traveled by the particle from time $t = 0$ to $t = 4$.

STOP

END OF EXAM

Chapter 23
Practice Test 2
Answers and
Explanations

ANSWER KEY TO SECTION I

1.	B	11.	B	21.	D	31.	D	41.	A
2.	D	12.	E	22.	B	32.	B	42.	A
3.	B	13.	A	23.	B	33.	D	43.	B
4.	E	14.	D	24.	D	34.	D	44.	B
5.	A	15.	D	25.	D	35.	A	45.	D
6.	A	16.	C	26.	A	36.	E		
7.	D	17.	B	27.	E	37.	B		
8.	C	18.	E	28.	E	38.	D		
9.	D	19.	E	29.	B	39.	E		
10.	B	20.	E	30.	C	40.	C		

ANSWERS AND EXPLANATIONS TO SECTION I

1. **B** If $g(x) = \dfrac{1}{32}x^4 - 5x^2$, find $g'(4)$.

First, take the derivative.

$$g'(x) = \frac{1}{32}\left(4x^3\right) - 5(2x) = \frac{x^3}{8} - 10x$$

Now, plug in 4 for x.

$$\frac{(4)^3}{8} - 10(4) = 8 - 40 = -32$$

2. **D** The domain of the function $f(x) = \sqrt{4 - x^2}$ is

When you have a square root in a function, the domain will require that the expression under the radical (the "radicand") not be negative. Thus, the domain will be those values where $4 - x^2$ is not negative.

In other words, $4 - x^2 \geq 0$.

We solve this by, first, factoring the expression on the left: $(2 + x)(2 - x) \geq 0$.

Next, we take the roots of the left side, which are –2 and 2, and put them on a number line.

Now, we pick a value in each of the three regions on the number line $x < -2$, $-2 < x < 2$, and $x > 2$. We plug the value into the expression $4 - x^2$ to see if we get a positive or negative value. If it's positive, then we include that region in the domain. If it's negative, then we exclude that region from the domain.

Let's try –3 for a value in the region $x < -2$.

We get $4 - (-3)^2 = -5$, so we exclude the region $x < -2$ from the domain.

Now, we try 0 for a value in the region $-2 < x < 2$.

We get $4 - (0)^2 = 4$, so we include the region $-2 < x < 2$ in the domain.

Finally, we try 3 for a value in the region $x > 2$.

We get $4 - (3)^2 = -5$, so we exclude the region $x > 2$ from the domain.

Because the radicand is allowed to be zero, we include the endpoints in the domain. Therefore, the domain is $-2 \leq x \leq 2$.

3. **B** $\lim\limits_{x \to 5} \dfrac{x^2 - 25}{x - 5}$ is

Notice that if we plug 5 into the expressions in the numerator and the denominator, we get $\dfrac{0}{0}$, which is undefined. Before we give up, we need to see if we can simplify the limit so that it can be evaluated. If we factor the expression in the numerator, we get $\dfrac{(x + 5)(x - 5)}{(x - 5)}$, which can be simplified to $x + 5$.

Now, if we take the limit (by plugging in 5 for x), we get 10.

4. **E** If $f(x) = \dfrac{x^5 - x + 2}{x^3 + 7}$, find $f'(x)$.

We need to use the Quotient Rule, which is

Given $f(x) = \dfrac{g(x)}{h(x)}$, then $f'(x) = \dfrac{h(x)g'(x) - g(x)h'(x)}{\left[h(x)\right]^2}$

Here we have

$$f'(x) = \frac{(x^3 + 7)(5x^4 - 1) - (x^5 - x + 2)(3x^2)}{(x^3 + 7)^2}$$

5. **A** Evaluate $\lim\limits_{h \to 0} \dfrac{5\left(\dfrac{1}{2} + h\right)^4 - 5\left(\dfrac{1}{2}\right)^4}{h}$.

Notice how this limit takes the form of the definition of the derivative, which is

$$f'(x) = \lim_{h \to 0} \frac{f(x + h) - f(x)}{h}$$

Here, if we think of $f(x)$ as $5x^4$, then this expression gives the derivative of $5x^4$ at the point $x = \dfrac{1}{2}$.

The derivative of $5x^4$ is $f'(x) = 20x^3$.

At $x = \dfrac{1}{2}$, we get $f'\left(\dfrac{1}{2}\right) = 20\left(\dfrac{1}{2}\right)^3 = \dfrac{5}{2}$.

6. **A** $\int x\sqrt{3x}\, dx =$

First, rewrite the integral as $\int x\left(\sqrt{3}\right)x^{\frac{1}{2}}\, dx$.

Now, we can simplify the integral to $\sqrt{3} \int x^{\frac{3}{2}}\, dx$.

Next, use the power rule for integrals, which is $\int x^n \, dx = \frac{x^{n+1}}{n+1} + C$.

Then, we get $\sqrt{3}\int x^{\frac{3}{2}} \, dx = \sqrt{3}\,\frac{x^{\frac{5}{2}}}{\frac{5}{2}} + C = \frac{2\sqrt{3}}{5}x^{\frac{5}{2}} + C$.

7. **D** Find k so that $f(x) = \begin{cases} \dfrac{x^2 - 16}{x - 4}; & x \neq 4 \\ k & ; x = 4 \end{cases}$ is continuous for all x.

In order for $f(x)$ to be continuous at a point c, there are three conditions that need to be fulfilled.

(1) $f(c)$ exists.

(2) $\lim\limits_{x \to c} f(x)$ exists.

(3) $\lim\limits_{x \to c} f(x) = f(c)$.

First, let's check condition (1): $f(4)$ exists; it's equal to k.

Next, let's check condition (2).

From the left side, we get

$$\lim_{x \to 4^-} \frac{x^2 - 16}{x - 4} = \lim_{x \to 4^-} \frac{(x-4)(x+4)}{x-4} = \lim_{x \to 4^-} (x+4) = 8$$

From the right side, we get

$$\lim_{x \to 4^+} \frac{x^2 - 16}{x - 4} = \lim_{x \to 4^+} \frac{(x-4)(x+4)}{x-4} = \lim_{x \to 4^+} (x+4) = 8$$

Therefore, the limit exists, and $\lim\limits_{x \to 4} \dfrac{x^2 - 16}{x - 4} = 8$.

Now, let's check condition (3). In order for this condition to be fulfilled, k must equal 8.

8. **C** Which of the following integrals correctly gives the area of the region consisting of all points above the x-axis and below the curve $y = 8 + 2x - x^2$?

The curve $y = 8 + 2x - x^2$ is an upside-down parabola and looks like the following:

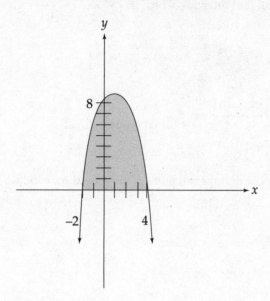

Notice that it crosses the x-axis at $x = -2$ and at $x = 4$.

The formula for the area of the region under the curve $f(x)$ and above the x-axis from $x = a$ to $x = b$ is $\int_a^b f(x)\,dx$.

Thus, in order to find the area of the desired region, we need to evaluate the integral $\int_{-2}^{4} (8 + 2x - x^2)\,dx$.

9. **D** If $f(x) = x^2 \cos 2x$, find $f'(x)$.

Here we need to use the Product Rule, which is

If $f(x) = uv$, where u and v are both functions of x,

$$\text{then } f'(x) = u\frac{dv}{dx} + v\frac{du}{dx}$$

Here we get

$$f'(x) = x^2(-2 \sin 2x) + 2x(\cos 2x)$$

10. **B**   An equation of the line tangent to $y = 4x^3 - 7x^2$ at $x = 3$ is

If we want to find the equation of the tangent line, first we need to find the y-coordinate that corresponds to $x = 3$.

$$y = 4(3)^3 - 7(3)^2 = 108 - 63 = 45$$

Next, we need to find the derivative of the curve at $x = 3$.

$$\frac{dy}{dx} = 12x^2 - 14x \text{ and at } x = 3, \left.\frac{dy}{dx}\right|_{x=3} = 12(3)^2 - 14(3) = 66$$

Now, we have the slope of the tangent line and a point that it goes through. We can use the point-slope formula for the equation of a line, $(y - y_1) = m(x - x_1)$, and plug in what we have just found. We get

$$(y - 45) = 66(x - 3)$$

11. **B** $\int_0^{\frac{1}{2}} \frac{2}{\sqrt{1 - x^2}}\, dx =$

This integral is of the form $\int \frac{dx}{\sqrt{a^2 - x^2}} = \sin^{-1}\left(\frac{x}{a}\right) + C$, where $a = 1$.

Thus, we get

$$\int_0^{\frac{1}{2}} \frac{2\,dx}{\sqrt{1 - x^2}} = 2\sin^{-1}(x)\Big|_0^{\frac{1}{2}} = 2\left[\sin^{-1}\left(\frac{1}{2}\right) - \sin^{-1}(0)\right] = 2\left(\frac{\pi}{6} - 0\right) = \frac{\pi}{3}$$

12. **E**   Find a positive value c, for x, that satisfies the conclusion of the Mean Value Theorem for derivatives for $f(x) = 3x^2 - 5x + 1$ on the interval $[2, 5]$.

The Mean Value Theorem for derivatives says that, given a function $f(x)$ which is continuous and differentiable on $[a, b]$, there exists some value c on (a, b) where $\frac{f(b) - f(a)}{b - a} = f'(c)$.

Here, we have $\frac{f(b) - f(a)}{b - a} = \frac{f(5) - f(2)}{5 - 2} = \frac{51 - 3}{3} = 16$.

Plus, we have $f'(c) = 6c - 5$, so we simply set $6c - 5 = 16$.

If we solve for c, we get $c = \frac{7}{2}$.

13.   **A**   Given $f(x) = 2x^2 - 7x - 10$ find the absolute maximum of $f(x)$ on $[-1, 3]$.

First, let's take the derivative and then set it equal to zero to determine any critical points of the function.

$$f'(x) = 4x - 7$$

$$4x - 7 = 0$$

$$x = \frac{7}{4}$$

Now, we can use the second derivative test to determine if this is a local minimum or maximum.

$$f''(x) = 4$$

Because the second derivative is always positive, the function is concave up everywhere, and thus $x = \frac{7}{4}$ must be a local minimum.

How, then, do we find the absolute maximum? Anytime we are given a function that is defined on an interval, the endpoints of the interval are also critical points. Thus, all that we have to do now is to plug the endpoints into the function and see which one gives us the bigger value. That will be the absolute maximum.

$$f(-1) = 2(-1)^2 - 7(-1) - 10 = -1$$
$$f(3) = 2(3)^2 - 7(3) - 10 = -13$$

Therefore, the absolute maximum of $f(x)$ on the interval $[-1, 3]$ is -1.

14.   **D**   Find $\dfrac{dy}{dx}$ if $x^3y + xy^3 = -10$.

We need to use implicit differentiation to find $\dfrac{dy}{dx}$.

$$3x^2y + x^3\frac{dy}{dx} + y^3 + 3xy^2\frac{dy}{dx} = 0$$

Now, in order to isolate $\dfrac{dy}{dx}$, we move all of the terms that do not contain $\dfrac{dy}{dx}$ to the right side of the equals sign.

$$x^3\frac{dy}{dx} + 3xy^2\frac{dy}{dx} = -3x^2y - y^3$$

Factor out $\dfrac{dy}{dx}$.

$$\frac{dy}{dx}(x^3 + 3xy^2) = -3x^2y - y^3$$

Divide both sides by $(x^3 + 3xy^2)$ to isolate $\dfrac{dy}{dx}$.

$$\frac{dy}{dx} = -\frac{3x^2y + y^3}{x^3 + 3xy^2}$$

15.   **D**   If $f(x) = \sqrt{1 + \sqrt{x}}$, find $f'(x)$.

First, rewrite the equation using fractional powers instead of radical signs.

$$f(x) = \left(1 + x^{\frac{1}{2}}\right)^{\frac{1}{2}}$$

Now, take the derivative.

$$f'(x) = \frac{1}{2}\left(1 + x^{\frac{1}{2}}\right)^{-\frac{1}{2}}\left(\frac{1}{2}x^{-\frac{1}{2}}\right)$$

This can be rewritten as

$$f'(x) = \frac{1}{4}\,\frac{1}{\sqrt{x}}\,\frac{1}{\sqrt{1 + \sqrt{x}}}$$

16.   **C** $\displaystyle\int 7xe^{3x^2}\,dx =$

We can use u-substitution to evaluate the integral.

Let $u = 3x^2$ and $du = 6x\,dx$. If we solve the second term for $x\,dx$, we get

$$\frac{1}{6}\,du = x\,dx$$

Now we can rewrite the integral as

$$\frac{7}{6}\int e^u\,du$$

Evaluate the integral to get

$$\frac{7}{6}e^u + C$$

Now substitute back to get

$$\frac{7}{6}e^{3x^2} + C$$

17.   **B**   Find the equation of the tangent line to $9x^2 + 16y^2 = 52$ through $(2, -1)$.

First, we need to find $\dfrac{dy}{dx}$. It's simplest to find it implicitly.

$$18x + 32y\,\frac{dy}{dx} = 0$$

Now, solve for $\dfrac{dy}{dx}$.

$$\frac{dy}{dx} = -\frac{18x}{32y} = -\frac{9x}{16y}$$

Next, plug in $x = 2$ and $y = -1$ to get the slope of the tangent line at the point.

$$\frac{dy}{dx} = \frac{-18}{-16} = -\frac{9}{8}$$

Now, use the point-slope formula to find the equation of the tangent line.

$$(y + 1) = \frac{9}{8}\,(x - 2)$$

If we multiply through by 8, we get $8y + 8 = 9x - 18$ or $9x - 8y - 26 = 0$.

18.   **E**   A particle's position is given by $s = t^3 - 6t^2 + 9t$. What is its acceleration at time $t = 4$?

Acceleration is the second derivative of position with respect to time (velocity is the first derivative).

The first derivative is $v(t) = 3t^2 - 12t + 9$.

The second derivative is $a(t) = 6t - 12$.

Now we simply plug in $t = 4$ and we get $a(4) = 24 - 12 = 12$.

19.   **E**   If $f(x) = 3^{\pi x}$, then $f'(x) =$

The derivative of an expression of the form a^u, where u is a function of x, is

$$\frac{d}{dx}\,a^u = a^u\,(\ln a)\,\frac{du}{dx}$$

Here we get

$$\frac{d}{dx}\,3^{\pi x} = 3^{\pi x}\,(\ln 3)\,\pi$$

20. **E** The average value of $f(x) = \dfrac{1}{x}$ from $x = 1$ to $x = e$ is

In order to find the average value, we use the Mean Value Theorem for integrals, which says that the average value of $f(x)$ on the interval $[a, b]$ is $\dfrac{1}{b-a}\displaystyle\int_a^b f(x)\, dx$.

Here we have $\dfrac{1}{e-1}\displaystyle\int_1^e \dfrac{1}{x}\, dx$.

Evaluating the integral, we get $\ln x\big|_1^e = \ln e - \ln 1 = 1$. Therefore, the answer is $\dfrac{1}{e-1}$.

21. **D** If $f(x) = \sin^2 x$, find $f'''(x)$.

We just use the Chain Rule three times.

$$f'(x) = 2\sin x \cos x = \sin 2x$$

$$f''(x) = 2\cos 2x$$

$$f'''(x) = -4\sin 2x$$

22. **B** Find the slope of the normal line to $y = x + \cos xy$ at $(0, 1)$.

First, we need to find $\dfrac{dy}{dx}$ using implicit differentiation.

$$\frac{dy}{dx} = 1 - \left(x\frac{dy}{dx} + y\right)\sin xy$$

Rather than simplifying this, simply plug in $(0, 1)$ to find $\dfrac{dy}{dx}$.

We get $\dfrac{dy}{dx} = 1$.

This means that the slope of the tangent line at $(0, 1)$ is 1, so the slope of the normal line at $(0, 1)$ is the negative reciprocal, which is -1.

23. **B** $\displaystyle\int e^x (e^{3x})\, dx =$

First, add the exponents to get $\displaystyle\int e^{4x}\, dx$.

Evaluating the integral, we get $\dfrac{1}{4}e^{4x} + C$.

24. **D** $\displaystyle\lim_{x \to 0} \frac{\tan^3(2x)}{x^3} =$

We will need to use the fact that $\displaystyle\lim_{x \to 0}\frac{\sin x}{x} = 1$ to find the limit.

First, rewrite the limit as

$$\lim_{x \to 0} \frac{\sin^3(2x)}{x^3 \cos^3(2x)}$$

Next, break the fraction into

$$\lim_{x \to 0} \left(\frac{\sin^3(2x)}{x^3} \frac{1}{\cos^3(2x)} \right)$$

Now, if we multiply the top and bottom of the first fraction by 8, we get

$$\lim_{x \to 0} \frac{8\sin^3(2x)}{(2x)^3} \frac{1}{\cos^3(2x)}$$

Now, we can take the limit, which gives us 8(1)(1) = 8.

25. **D** A solid is generated when the region in the first quadrant bounded by the graph of $y = 1 + \sin^2 x$, the line $x = \dfrac{\pi}{2}$, the x-axis, and the y-axis is revolved about the x-axis. Its volume is found by evaluating which of the following integrals?

First, let's graph the curve.

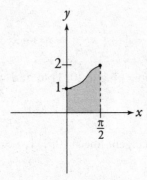

We can find the volume by taking a vertical slice of the region. The formula for the volume of a solid of revolution around the x-axis, using a vertical slice bounded from above by the curve $f(x)$ and from below by $g(x)$, on the interval $[a, b]$ is

$$\pi \int_a^b \left[f(x)^2 - g(x)^2 \right] dx$$

Here we get

$$\pi \int_0^{\frac{\pi}{2}} (1 + \sin^2 x)^2 \, dx$$

26.   **A**   If $y = \left(\dfrac{x^3 - 2}{2x^5 - 1}\right)^4$, find $\dfrac{dy}{dx}$ at $x = 1$.

We use the Chain Rule and the Quotient Rule.

$$\frac{dy}{dx} = 4\left(\frac{x^3 - 2}{2x^5 - 1}\right)^3 \left[\frac{(2x^5 - 1)(3x^2) - (x^3 - 2)(10x^4)}{(2x^5 - 1)^2}\right]$$

If we plug in 1 for x, we get

$$\frac{dy}{dx} = 4(-1)^3 \left[\frac{3 + 10}{1^2}\right] = -52$$

27.   **E** $\displaystyle\int x\sqrt{5 - x}\, dx =$

We can evaluate this integral using u-substitution.

Let $u = 5 - x$ and $5 - u = x$. Then $-du = dx$.

Substituting, we get

$$-\int (5 - u)u^{\frac{1}{2}}\, du$$

The integral can be rewritten as

$$-\int \left(5u^{\frac{1}{2}} - u^{\frac{3}{2}}\right) du$$

Evaluating the integral, we get

$$-5\frac{u^{\frac{3}{2}}}{\frac{3}{2}} + \frac{u^{\frac{5}{2}}}{\frac{5}{2}} + C$$

This can be simplified to

$$-\frac{10}{3}u^{\frac{3}{2}} + \frac{2}{5}u^{\frac{5}{2}} + C$$

Finally, substituting back, we get

$$-\frac{10}{3}(5 - x)^{\frac{3}{2}} + \frac{2}{5}(5 - x)^{\frac{5}{2}} + C$$

28.   **E**   If $\dfrac{dy}{dx} = \dfrac{x^3 + 1}{y}$ and $y = 2$ when $x = 1$, then, when $x = 2$, $y =$

This is a differential equation that can be solved using separation of variables. Put all of the terms containing y on the left and all of the terms containing x on the right.

$$y \, dy = (x^3 + 1) \, dx$$

Next we integrate both sides.

$$\int y \, dy = \int (x^3 + 1) \, dx$$

Evaluating the integrals, we get

$$\frac{y^2}{2} = \frac{x^4}{4} + x + C$$

Next we plug in $y = 2$ and $x = 1$ to solve for C. We get $2 = \dfrac{1}{4} + 1 + C$ and so $C = \dfrac{3}{4}$. This gives us

$$\frac{y^2}{2} = \frac{x^4}{4} + x + \frac{3}{4}$$

Now if we substitute $x = 2$, we get

$$\frac{y^2}{2} = 4 + 2 + \frac{3}{4} = \frac{27}{4}$$

Solving for y, we get

$$y = \pm \sqrt{\frac{27}{2}}$$

29.   **B**   The graph of $y = 5x^4 - x^5$ has an inflection point (or points) at

In order to find the inflection point(s) of a polynomial, we need to find the values of x where its second derivative is zero.

First, we find the second derivative.

$$\frac{dy}{dx} = 20x^3 - 5x^4$$

$$\frac{d^2 y}{dx^2} = 60x^2 - 20x^3$$

Now, let's set the second derivative equal to zero and solve for x.

$$60x^2 - 20x^3 = 0$$

$$20x^2(3 - x) = 0$$

$$x = 3$$

This is the point of inflection. $x = 0$ is not a point of inflection because $\dfrac{d^2y}{dx^2}$ does not change sign there.

You can use a calculator on this part of the exam, and you can use it to find the inflection point(s) of this graph.

Graphing Calculator (TI-83 and TI-84)

Press the Y= button, and enter the following values to the list:

$Y_1 = 5X^4 - X^5$

$Y_2 = nDeriv(Y_1,X,X)$

$Y_3 = nDeriv(Y_2,X,X)$

Graph Y_3 and find its zero.

Make sure that only the equation sign in Y_3 is darkened.

Press 2nd and TRACE to access the CALC menu.

Select 2:zero and press ENTER, and a graph will appear.

Move the cursor anywhere to the left of where the second derivative graph crosses the x-axis, and press ENTER to mark the left bound.

Move the cursor anywhere to the right of where the second derivative graph crosses the x-axis, and press ENTER to mark the right bound.

Press ENTER when you are prompted for a guess.

The bottom of the graph will now show "Zero" and the value of x when $y = 0$.

$x = 2.9999996$ when $y = 0$

30.　**C**　The average value of $f(x) = e^{4x^2}$ on the interval $\left[-\dfrac{1}{4}, \dfrac{1}{4}\right]$ is

In order to find the average value, we use the Mean Value Theorem for Integrals, which says that the average value of $f(x)$ on the interval $[a, b]$ is $\dfrac{1}{b-a} \displaystyle\int_a^b f(x)\, dx$.

Here we have

$$\frac{1}{\dfrac{1}{4} + \dfrac{1}{4}} \int_{-\frac{1}{4}}^{\frac{1}{4}} e^{4x^2}\, dx = 2\int_{-\frac{1}{4}}^{\frac{1}{4}} e^{4x^2}\, dx$$

You can't evaluate this integral using any of the techniques that you have studied so far, so use the calculator to evaluate the integral numerically.

Remember this: Use your calculator to evaluate integrals whenever you can.

You should get approximately 1.090. **The AP exam always expects you to round to three decimal places.**

31.　**D**　$\displaystyle\int_0^1 \tan x\, dx =$

First, rewrite the integral as $\displaystyle\int_0^1 \frac{\sin x}{\cos x}\, dx$.

Now, we can use u-substitution to evaluate the integral. Let $u = \cos x$. Then $du = -\sin x$. We can also change the limits of integration. The lower limit becomes $\cos 0 = 1$ and the upper limit becomes $\cos 1$, which we leave alone. Now we perform the substitution, and we get

$$-\int_1^{\cos 1} \frac{du}{u}$$

Evaluating the integral, we get $-\ln u \Big|_1^{\cos 1} = -\ln(\cos 1) + \ln 1 = -\ln(\cos 1)$. This log is also equal to $\ln(\sec 1)$.

32.　**B**　$\dfrac{d}{dx}\displaystyle\int_0^{x^2} \sin^2 t\, dt =$

The Second Fundamental Theorem of Calculus tells us how to find the derivative of an integral. It says that $\dfrac{d}{dx}\displaystyle\int_c^u f(t)\, dt = f(u)\dfrac{du}{dx}$, where c is a constant and u is a function of x.

Here we can use the theorem to get

$$\frac{d}{dx}\int_0^{x^2} \sin^2 t \ dt = (\sin^2(x^2))(2x) \text{ or } 2x\sin^2(x^2)$$

33. **D** Find the value(s) of $\dfrac{dy}{dx}$ of $x^2y + y^2 = 5$ at $y = 1$.

Here we use implicit differentiation to find $\dfrac{dy}{dx}$.

$$2xy + x^2\frac{dy}{dx} + 2y\frac{dy}{dx} = 0$$

Now we plug $y = 1$ into the original equation to find its corresponding x-values.

$$x^2 + 1 = 5$$

$$x^2 = 4$$

$$x = \pm 2$$

Now plug in the x- and y-values to find the value of $\dfrac{dy}{dx}$.

For $y = 1$ and $x = 2$, we get

$$2(2)(1) + (2)^2\frac{dy}{dx} + 2(1)\frac{dy}{dx} = 0$$

Solving for $\dfrac{dy}{dx}$, we get

$$4 + 6\frac{dy}{dx} = 0 \text{ and } \frac{dy}{dx} = -\frac{2}{3}$$

For $y = 1$ and $x = -2$, we get

$$2(-2)(1) + (-2)^2\frac{dy}{dx} + 2(1)\frac{dy}{dx} = 0$$

Solving for $\dfrac{dy}{dx}$, we get

$$-4 + 6\frac{dy}{dx} = 0 \text{ and } \frac{dy}{dx} = \frac{2}{3}$$

34. **D** The graph of $y = x^3 - 2x^2 - 5x + 2$ has a local maximum at

First, let's find $\dfrac{dy}{dx}$.

$$\frac{dy}{dx} = 3x^2 - 4x - 5$$

Next, set the derivative equal to zero and solve for x.

$$3x^2 - 4x - 5 = 0$$

Using the quadratic formula (or your calculator), we get

$$x = \frac{4 \pm \sqrt{16 + 60}}{6} \approx 2.120, -0.786$$

Let's use the second derivative test to determine which is the maximum. We take the second derivative, and then plug in the critical values that we found when we set the first derivative equal to zero. **If the sign of the second derivative at a critical value is positive, then the curve has a local minimum there. If the sign of the second derivative is negative, then the curve has a local maximum there.**

The second derivative is: $\dfrac{d^2y}{dx^2} = 6x - 4$. This is negative at $x = -0.786$, so the curve has a local maximum there. Now we plug $x = -0.786$ into the original equation to find the y-coordinate of the maximum. We get approximately 4.209. Therefore, the curve has a local maximum at $(-0.786, 4.209)$.

Graphing Calculator (TI-83 and TI-84)

Press the Y= button, and enter the following values to the list:

$Y_1 = X^3 - 2X^2 - 5X + 2$

$Y_2 = nDeriv(Y_1, X, X)$

$Y_3 = nDeriv(Y_2, X, X)$

Graph Y_2 and find its zeros.

First use 0 as the left bound: When $y = 0$, $x = 2.1196329$.

Now use 0 as the right bound: When $y = 0$, $x = -0.7862995$

Graph Y_3 and use TRACE to determine that the sign of the second derivative is negative at $x = -0.7862995$, and positive at $x = 2.1196329$, and therefore the x-coordinate of the local maximum will be -0.786.

Plug this value for x into the original equation or use TABLE to find the y-value for the local maximum, which is about 4.209.

35. **A**   Approximate $\displaystyle\int_0^1 \sin^2 x\, dx$ using the Trapezoid Rule with $n = 4$, to three decimal places.

The Trapezoid Rule enables us to approximate the area under a curve with a fair degree of accuracy. The rule says that the area between the x-axis and the curve $y = f(x)$, on the interval $[a, b]$, with n trapezoids, is

$$\frac{1}{2}\frac{b-a}{n}\left[y_0 + 2y_1 + 2y_2 + 2y_3 + \ldots + 2y_{n-1} + y_n\right]$$

Using the rule here, with $n = 4$, $a = 0$, and $b = 1$, we get

$$\frac{1}{2}\left(\frac{1}{4}\right)\left[\sin^2 0 + 2\sin^2\frac{1}{4} + 2\sin^2\frac{1}{2} + 2\sin^2\frac{3}{4} + \sin^2 1\right]$$

This is approximately 0.277.

36. **E**   The volume generated by revolving about the x-axis the region above the curve $y = x^3$, below the line $y = 1$, and between $x = 0$ and $x = 1$ is

First, make a quick sketch of the region.

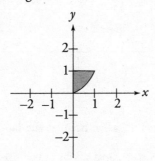

We can find the volume by taking a vertical slice of the region. The formula for the volume of a solid of revolution around the x-axis, using a vertical slice bounded from above by the curve $f(x)$ and from below by $g(x)$, on the interval $[a, b]$, is

$$\pi\int_a^b \left[f(x)^2 - g(x)^2\right] dx$$

Here we get

$$\pi\int_0^1 [(1)^2 - (x^3)^2]\ dx$$

Now we have to evaluate the integral. First, expand the integrand to get

$$\pi\int_0^1 (1 - x^6)\ dx$$

Next integrate to get

$$\pi\left(x - \frac{x^7}{7}\right)\Big|_0^1 = \pi\left(1 - \frac{1}{7}\right) = \frac{6\pi}{7}$$

37. **B**   A 20-foot ladder slides down a wall at 5 ft/sec. At what speed is the bottom sliding out when the top is 10 feet from the floor (in ft/sec)?

First, let's make a sketch of the situation.

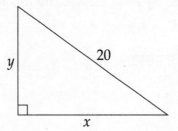

We are given that $\dfrac{dy}{dt}$ = −5 (it's negative because the ladder is sliding down and it's customary to make the upward direction positive), and we want to find $\dfrac{dx}{dt}$ when y = 10.

We can find a relationship between x and y using the Pythagorean theorem. We get $x^2 + y^2 = 400$.

Now, taking the derivative with respect to t, we get

$$2x\frac{dx}{dt} + 2y\frac{dy}{dt} = 0,$$ which can be simplified to $x\dfrac{dx}{dt} = -y\dfrac{dy}{dt}$

Next, we need to find x when y = 10.

Using the Pythagorean theorem,

$$x^2 + 10^2 = 400, \text{ so } x = \sqrt{300} \approx 17.321$$

Now, plug into the equation above to get

$$17.321 \, \frac{dx}{dt} = -10(-5) \text{ and } \frac{dx}{dt} \approx 2.887$$

38. **D** $\displaystyle\int \frac{\ln x}{3x} \, dx =$

We can evaluate the integral with *u*-substitution.

Let $u = \ln x$. Then $du = \dfrac{dx}{x}$.

Substituting, we get $\dfrac{1}{3} \displaystyle\int u \, du$

Now, we can evaluate the integral: $\left(\dfrac{1}{3}\right)\left(\dfrac{u^2}{2}\right) + C$

Substituting back, we get $\dfrac{\ln^2 |x|}{6} + C$

39. **E** Find two nonnegative numbers x and y whose sum is 100 and for which $x^2 y$ is a maximum.

Let's set $P = x^2 y$. We want to maximize P, so we need to eliminate one of the variables. We are also given that $x + y = 100$, so we can solve this for y and substitute: $y = 100 - x$, so $P = x^2(100 - x) = 100x^2 - x^3$.

Now we can take the derivative.

$$\frac{dP}{dx} = 200x - 3x^2$$

Set the derivative equal to zero and solve for x.

$$200x - 3x^2 = 0$$
$$x(200 - 3x) = 0$$

$$x = 0 \text{ or } x = \frac{200}{3} \approx 66.667$$

Now we can use the second derivative to find the maximum: $\dfrac{d^2 P}{dx^2} = 200 - 6x$.

If we plug in $x = 66.667$, the second derivative is negative, so P is a maximum at $x = 66.667$. Solving for y, we get $y \approx 33.333$.

40. **C**   Find the distance traveled (to three decimal places) from $t = 1$ to $t = 5$ seconds, for a particle whose velocity is given by $v(t) = t + \ln t$.

The function $t + \ln t$ is always positive on the interval, so we can find the distance traveled by evaluating the integral.

$$\int_1^5 (t + \ln t)\, dt$$

We can evaluate the integral numerically using the calculator.

You should get approximately 16.047. **The AP exam always expects you to round to three decimal places.**

41. **A** $\int \sin^5 (2x) \cos (2x)\, dx =$

We can evaluate this integral using u-substitution.

Let $u = \sin(2x)$. Then $du = 2 \cos(2x)\, dx$, which we can rewrite as $\dfrac{1}{2}\, du = \cos(2x)\, dx$.

Substituting into the integrand, we get

$$\frac{1}{2} \int u^5\, du$$

Evaluating the integral gives us

$$\frac{1}{2} \frac{u^6}{6} + C = \frac{u^6}{12} + C$$

Substituting back, we get

$$\frac{\sin^6 (2x)}{12} + C$$

42. **A**   The volume of a cube is increasing at a rate proportional to its volume at any time t. If the volume is 8 ft^3 originally, and 12 ft^3 after 5 seconds, what is its volume at $t = 12$ seconds?

When we see a phrase where something is increasing at a rate "proportional to itself at any time t," this means that we set up the differential equation.

$$\frac{dV}{dt} = kV$$

(or whatever the appropriate variable is)

We solve this differential equation using separation of variables.

First, move the V to the left side and the dt to the right side, to get

$$\frac{dV}{V} = k \, dt$$

Now, integrate both sides.

$$\int \frac{dV}{V} = k \int dt$$

$$\ln V = kt + C$$

Next, it's traditional to put the equation in terms of V. We do this by exponentiating both sides to the base e. We get

$$V = e^{kt+C}$$

Using the rules of exponents, we can rewrite this as

$$V = e^{kt} e^{C}$$

Finally, because e^{C} is a constant, we can rewrite the equation as

$$V = Ce^{kt}$$

Now, we use the initial condition that $V = 8$ at time $t = 0$ to solve for C.

$$8 = Ce^{0} = C(1) = C$$

This gives us

$$V = 8e^{kt}$$

Next, we use the condition that $V = 12$ at time $t = 5$ to solve for k.

$$12 = 8e^{5k}$$

$$\frac{3}{2} = e^{5k}$$

$$\ln \frac{3}{2} = 5k$$

$$k = \frac{1}{5}\ln\frac{3}{2}$$

This gives us

$$V = 8e^{\left(\frac{1}{5}\ln\frac{3}{2}\right)t}$$

Finally, we plug in $t = 12$, and solve for V.

$$V = 8e^{\left(\frac{1}{5}\ln\frac{3}{2}\right)(12)} \approx 21.169$$

43. **B** If $f(x) = \left(1 + \dfrac{x}{20}\right)^5$, find $f''(40)$.

The first derivative is

$$f'(x) = 5\left(1 + \frac{x}{20}\right)^4\left(\frac{1}{20}\right) = \frac{1}{4}\left(1 + \frac{x}{20}\right)^4$$

The second derivative is

$$f''(x) = \frac{1}{4}(4)\left(1 + \frac{x}{20}\right)^3\left(\frac{1}{20}\right) = \left(\frac{1}{20}\right)\left(1 + \frac{x}{20}\right)^3$$

Evaluating this at $x = 40$, we get

$$f''(40) = \frac{1}{20}\left(1 + \frac{40}{20}\right)^3 = \frac{27}{20} = 1.350$$

44. **B** A particle's height at a time $t \geq 0$ is given by $h(t) = 100t - 16t^2$. What is its maximum height?

First, let's take the derivative: $h'(t) = 100 - 32t$.

Now, we set it equal to zero, and solve for t: $100 - 32t = 0$.

$$t = \frac{100}{32}$$

Now, to solve for the maximum height, we simply plug $t = \dfrac{100}{32}$ back into the original equation for height.

$$h\left(\frac{100}{32}\right) = 100\left(\frac{100}{32}\right) - 16\left(\frac{100}{32}\right)^2 = 156.250$$

By the way, we know that this is a maximum not a minimum because the second derivative is −32, which means that the critical value will give us a maximum not a minimum.

45. **D** If $f(x)$ is continuous and differentiable and $f(x) = \begin{cases} ax^4 + 5x; \ x \le 2 \\ bx^2 - 3x; \ x > 2 \end{cases}$, then $b =$

In order to solve this for b, we need $f(x)$ to be continuous at $x = 2$.

If we plug $x = 2$ into both pieces of this piecewise function, we get

$$f(x) = \begin{cases} 16a + 10; \ x \le 2 \\ 4b - 6; \ x > 2' \end{cases}$$

So, we need $16a + 10 = 4b - 6$.

Now, if we take the derivative of both pieces of this function and plug in $x = 2$, we get

$$f'(x) = \begin{cases} 32a + 5; \ x \le 2 \\ 4b - 3; \ x > 2 \end{cases}, \text{ so we need } 32a + 5 = 4b - 3$$

Solving the simultaneous equations, we get $a = \dfrac{1}{2}$ and $b = 6$.

ANSWERS AND EXPLANATIONS TO SECTION II

1. The temperature on New Year's Day in Hinterland was given by $T(H) = -A - B\cos\left(\dfrac{\pi H}{12}\right)$, where T is the temperature in degrees Fahrenheit and H is the number of hours from midnight ($0 \le H < 24$).

(a) The initial temperature at midnight was $-15°$ F, and at noon of New Year's Day was $5°$ F. Find A and B.

Simply plug in the temperature, -15, for T and the time, midnight ($H = 0$), for H into the equation. We get $-15 = -A - B \cos 0$, which simplifies to $-15 = -A - B$.

Now plug the temperature, 5, for T and the time, noon ($H = 12$), for H into the equation. We get $5 = -A - B \cos(\pi)$, which simplifies to $5 = -A + B$.

Now we can solve the pair of simultaneous equations for A and B, and we get $A = 5°$ F and $B = 10°$ F.

(b) Find the average temperature for the first 10 hours.

In order to find the average value, we use the Mean Value Theorem for integrals, which says that the average value of $f(x)$ on the interval $[a, b]$ is

$$\frac{1}{b - a} \int_a^b f(x) \, dx$$

Here, we have: $\dfrac{1}{10 - 0} \int_0^{10} \left(-5 - 10\cos\left(\dfrac{\pi H}{12}\right) \right) dH$

Evaluating the integral, we get

$$\frac{1}{10}\left[\left(-5H - \frac{120}{\pi}\sin\left(\frac{\pi H}{12}\right) \right) \right]_0^{10} = \frac{1}{10}\left[\left(-50 - \frac{120}{\pi}\sin\left(\frac{5\pi}{6}\right) \right) \right] = \frac{1}{10}\left[\left(-50 - \frac{60}{\pi} \right) \right] \approx -6.910\degree F$$

(c) Use the Trapezoid Rule with 4 equal subdivisions to estimate $\int_6^8 T(H) \, dH$.

The Trapezoid Rule enables us to approximate the area under a curve with a fair degree of accuracy. The rule says that the area between the x-axis and the curve $y = f(x)$ on the interval $[a, b]$, with n trapezoids, is

$$\frac{1}{2}\frac{b - a}{n}\left[y_0 + 2y_1 + 2y_2 + 2y_3 + \ldots + 2y_{n-1} + y_n \right]$$

Using the rule here, with $n = 4$, $a = 6$, and $b = 8$, we get

$$\frac{1}{2} \cdot \frac{1}{2}\left[\left(-5 - 10\cos\frac{6\pi}{12} \right) + 2\left(-5 - 10\cos\frac{13\pi}{24} \right) + 2\left(-5 - 10\cos\frac{7\pi}{12} \right) + 2\left(-5 - 10\cos\frac{15\pi}{24} \right) + \left(-5 - 10\cos\frac{8\pi}{12} \right) \right]$$

This is approximately $-4.890\degree F$.

(d) Find an expression for the rate that the temperature is changing with respect to H.

We simply take the derivative with respect to H.

$$\frac{dT}{dH} = -10\left(\frac{\pi}{12}\right)\left(-\sin\frac{\pi H}{12} \right) = \frac{5\pi}{6}\sin\frac{\pi H}{12}$$

2.   Sea grass grows on a lake. The rate of growth of the grass is $\frac{dG}{dt} = kG$, where k is a constant.

(a)   Find an expression for G, the amount of grass in the lake (in tons), in terms of t, the number of years, if the amount of grass is 100 tons initially, and 120 tons after one year.

We solve this differential equation using separation of variables.

First, move the G to the left side and the dt to the right side, to get: $\frac{dG}{G} = k\,dt$.

Now, integrate both sides.

$$\int \frac{dG}{G} = k \int dt$$

$$\ln G = kt + C$$

Next, solve for G by exponentiating both sides to the base e. We get $G = e^{kt+C}$.

Using the rules of exponents, we can rewrite this as: $G = e^{kt}\, e^{C}$. Finally, because e^{C} is a constant, we can rewrite the equation as $G = Ce^{kt}$.

Now, we use the initial condition that $G = 100$ at time $t = 0$ to solve for C.

$$100 = Ce^{0} = C(1) = C$$

This gives us $G = 100e^{kt}$.

Next, we use the condition that $G = 120$ at time $t = 1$ to solve for k.

$$120 = 100e^{k}$$
$$1.2 = e^{k}$$
$$\ln 1.2 = k \approx 0.1823$$

This gives us $G = 100e^{0.1823t}$.

(b)   In how many years will the amount of grass available be 300 tons?

All we need to do is set G equal to 300, and solve for t.

$$300 = 100e^{0.1823t}$$
$$3 = e^{0.1823t}$$
$$\ln 3 = 0.1823t$$
$$t \approx 6.026 \text{ years}$$

(c)   If fish are now introduced into the lake and consume a consistent 80 tons/year of sea grass, how long will it take for the lake to be completely free of sea grass?

Now we have to account for the fish's consumption of the sea grass. So we have to evaluate the differential equation $\dfrac{dG}{dt} = kG - 80$.

First, separate the variables, to get

$$\frac{dG}{kG - 80} = dt$$

Now, integrate both sides.

$$\int \frac{dG}{kG - 80} = \int dt \text{ or } \int \frac{dG}{G - \dfrac{80}{k}} = k \int dt$$

$$\ln\left(G - \frac{80}{k} \right) = kt + C$$

Next, exponentiate both sides to the base e. We get

$$G - \frac{80}{k} = Ce^{kt}$$

Solving for G, we get

$$G = \left(G_0 - \frac{80}{k} \right) e^{kt} + \frac{80}{k}$$

Now, set $G = 0$. We get

$$0 = \left(G_0 - \frac{80}{k} \right) e^{kt} + \frac{80}{k}$$

Now, set $G_0 = 300$, and solve for e^{kt}.

$$e^{kt} = \frac{-\dfrac{80}{k}}{300 - \dfrac{80}{k}} = \frac{80}{80 - 300k}$$

Take the log of both sides.

$$kt = \ln\left(\frac{80}{80 - 300k} \right)$$

$$\text{and } t = \frac{1}{k} \ln\left(\frac{80}{80 - 300k} \right)$$

Now, we plug in the value for k that we got in part (a) above, and we get $t \approx 6.313$ years.

3.   Consider the curve defined by $y = x^4 + 4x^3$.

(a)   Find the equation of the tangent line to the curve at $x = -1$.

If we want to find the equation of the tangent line, first we need to find the y-coordinate that corresponds to $x = -1$.

$$y = (-1)^4 + 4(-1)^3 = 1 - 4 = -3$$

Next, we need to find the derivative of the curve at $x = -1$.

It is $\dfrac{dy}{dx} = 4x^3 + 12x^2$ and, at $x = -1$, $\left.\dfrac{dy}{dx}\right|_{x=-1} = 4(-1)^3 + 12(-1)^2 = 8$.

Now, we have the slope of the tangent line and a point that it goes through. We can use the point-slope formula for the equation of a line, $(y - y_1) = m(x - x_1)$, and plug in what we have just found.

We get

$$(y + 3) = 8(x + 1), \text{ which can be rewritten as } y = 8x + 5$$

(b)   Find the coordinates of the absolute minimum.

First, we set the derivative equal to zero and solve for x.

$$\frac{dy}{dx} = 4x^3 + 12x^2 = 0$$

$$4x^2(x + 3) = 0$$

$$x = 0 \text{ or } x = -3$$

Now, we can use the second derivative test to determine whether a critical value is the x-coordinate of a minimum or a maximum. The second derivative test is the following:

If c is a critical point, then

c is the x-coordinate of a maximum if $f''(c) < 0$, and

c is the x-coordinate of a minimum if $f''(c) > 0$.

By the way, c is the x-coordinate of a point of inflection if $f''(c) = 0$, and the second derivative changes sign at that point.

So now we need to find the second derivative.

$$\frac{d^2 y}{dx^2} = 12x^2 + 24x$$

If we plug in $x = -3$, we get

$$\frac{d^2 y}{dx^2} = 12(-3)^2 + 24(-3) = 36$$

So, the curve has a minimum at $x = -3$. Finally, to get the y-coordinate of the minimum, we plug $x = -3$ into the original equation, and we get

$$y = (-3)^4 + 4(-3)^3 = 81 - 108 = -27$$

Thus, the curve has an absolute minimum at $(-3, -27)$.

(c)   Find the coordinates of the point(s) of inflection.

In order to find points of inflection, we need to set the second derivative equal to zero. We have the second derivative from part (b) above.

$$12x^2 + 24x = 0$$
$$12x(x + 2) = 0$$
$$x = 0 \text{ or } x = -2$$

Next, we need to check if the second derivative changes sign at both of these points. We can

do this by trying points on the number line in the different intervals created by these points. If

we try a point to the left of $x = -2$, for example $x = -3$, and plug it into the second derivative,

we get $\frac{d^2 y}{dx^2} = 36$. If we then try a point between $x = 0$ and $x = -2$, for example $x = -1$, we get

$\frac{d^2 y}{dx^2} = -12$. Finally, if we try a point to the right of $x = 0$, for example $x = 1$, we get $\frac{d^2 y}{dx^2} = 36$.

The second derivative changes sign at both $x = 0$ and $x = -2$, so these are both the x-coordinates

of points of inflection. To get the y-coordinates, simply plug $x = 0$ and $x = -2$ into the original

equation: $y = x^4 + 4x^3$. We find that the points of inflection are $(0, 0)$ and $(-2, -16)$.

4.   Water is being poured into a hemispherical bowl of radius 6 inches at the rate of 4 in³/sec.

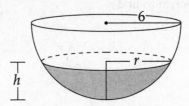

(a)   Given that the volume of the water in the spherical segment shown above is $V = \pi h^2 \left(R - \dfrac{h}{3} \right)$, where R is the radius of the *sphere*, find the rate that the water level is rising when the water is 2 inches deep.

First, rewrite the equation as

$$V = \pi R h^2 - \frac{\pi}{3} h^3$$

Now, take the derivative of the equation with respect to t.

$$\frac{dV}{dt} = 2\pi R h \frac{dh}{dt} - \pi h^2 \frac{dh}{dt}$$

If we plug in $\dfrac{dV}{dt} = 4$, $R = 6$, and $h = 2$, we get

$$4 = 20\pi \frac{dh}{dt} \quad \text{or} \quad \frac{dh}{dt} = \frac{4}{20\pi} = \frac{1}{5\pi}$$

(b)   Find an expression for r, the radius of the *surface of the spherical segment* of water, in terms of h.

Notice that we can construct a right triangle using the radius of the sphere and the radius of the surface of the water.

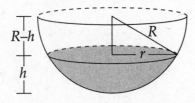

Notice that the distance from the center of the sphere to the surface of the water is $R - h$. Now, we can use the Pythagorean theorem to find r.

$$R^2 = (R - h)^2 + r^2$$

We can rearrange this to get

$$r = \sqrt{R^2 - (R - h)^2} = \sqrt{2Rh - h^2}$$

Because $R = 6$, we get

$$r = \sqrt{12h - h^2}$$

(c)   How fast is the circular area of the surface of the spherical segment of water growing (in in^2/sec) when the water is 2 inches deep?

The area of the surface of the water is $A = \pi r^2$, where $r = \sqrt{12h - h^2}$. Thus, $A = \pi(12h - h^2)$.

Taking the derivative of the equation with respect to t, we get

$$\frac{dA}{dt} = \pi\left(12\frac{dh}{dt} - 2h\frac{dh}{dt}\right)$$

We found in part (a) above that

$$\frac{dh}{dt} = \frac{1}{5\pi}, \text{ so } \frac{dA}{dt} = \pi\left(\frac{12}{5\pi} - \frac{4}{5\pi}\right) = \frac{8}{5} \text{ in}^2 / \text{sec}$$

5.   Let R be the region in the first quadrant bounded by $y^2 = x$ and $x^2 = y$.

(a)   Find the area of region R.

First, let's sketch the region.

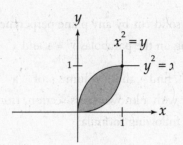

In order to find the area, we "slice" the region vertically and add up all of the slices. Now, we use the formula for the area of the region between $y = f(x)$ and $y = g(x)$, from $x = a$ to $x = b$.

$$\int_a^b \left[f(x) - g(x) \right] dx$$

We need to rewrite the equation $y^2 = x$ as $y = \sqrt{x}$ so that we can integrate with respect to x. Our integral for the area is

$$\int_0^1 \left(\sqrt{x} - x^2 \right) dx$$

Evaluating the integral, we get

$$\left(\frac{2x^{\frac{3}{2}}}{3} - \frac{x^3}{3} \right)\Bigg|_0^1 = \frac{2}{3} - \frac{1}{3} = \frac{1}{3}$$

(b)   Find the volume of the solid generated when R is revolved about the x-axis.

In order to find the volume of a region between $y = f(x)$ and $y = g(x)$, from $x = a$ to $x = b$, when it is revolved around the x-axis, we use the following formula:

$$\pi \int_a^b \left[f(x)^2 - g(x)^2 \right] dx$$

Here, our integral for the area is

$$\pi \int_0^1 \left(x - x^4 \right) dx$$

Evaluating the integral, we get

$$\pi\left(\frac{x^2}{2} - \frac{x^5}{5}\right)\Bigg|_0^1 = \pi\left(\frac{1}{2} - \frac{1}{5}\right) = \frac{3\pi}{10}$$

(c) The section of a certain solid cut by any plane perpendicular to the x-axis is a circle with the endpoints of its diameter lying on the parabolas $y^2 = x$ and $x^2 = y$. Find the volume of the solid.

Whenever we want to find the volume of a solid, formed by the region between $y = f(x)$ and $y = g(x)$, with a known cross-section, from $x = a$ to $x = b$, when it is revolved around the x-axis, we use the following formula:

$$\int_a^b A(x)\ dx$$

(Note: $A(x)$ is the area of the cross section.) We find the area of the cross-section by using the vertical slice formed by $f(x) - g(x)$, and then plugging it into the appropriate area formula. In the case of a circle, $f(x) - g(x)$ gives us the length of the diameter and we use the following formula:

$$A(x) = \frac{\pi(diameter)^2}{4}$$

This gives us the integral,

$$\int_0^1 \frac{\pi}{4}\left(\sqrt{x} - x^2\right)^2 dx$$

Expand the integrand.

$$\int_0^1 \frac{\pi}{4}\left(\sqrt{x} - x^2\right)^2 dx = \frac{\pi}{4}\int_0^1\left(x - 2x^{\frac{5}{2}} + x^4\right) dx =$$

Evaluate the integral.

$$\frac{\pi}{4}\int_0^1\left(x - 2x^{\frac{5}{2}} + x^4\right) dx = \frac{\pi}{4}\left(\frac{x^2}{2} - \frac{4x^{\frac{7}{2}}}{7} + \frac{x^5}{5}\right)\Bigg|_0^1 = \frac{\pi}{4}\left(\frac{1}{2} - \frac{4}{7} + \frac{1}{5}\right) = \frac{9\pi}{280}$$

6.   An object moves with velocity $v(t) = t^2 - 8t + 7$.

(a) Write a polynomial expression for the position of the particle at any time $t \geq 0$.

The velocity of an object is the derivative of its position with respect to time. Thus, if we want to find the position, we take the integral of velocity with respect to time.

$$s(t) = \int (t^2 - 8t + 7)\, dt = \frac{t^3}{3} - \frac{8t^2}{2} + 7t + C = \frac{t^3}{3} - 4t^2 + 7t + C$$

(b) At what time(s) is the particle changing direction?

If we want to find when the particle is changing direction, we need to find where the velocity of the particle is zero.

$$v(t) = t^2 - 8t + 7 = (t - 1)(t - 7) = 0$$

Thus, at $t = 1$ or $t = 7$, the particle could be changing direction. To make sure, we need to check that the acceleration of the particle is <u>not</u> zero at those times. The acceleration of a particle is the derivative of the velocity with respect to time.

$$a(t) = 2t - 8$$

At $t = 1$,

$$a(1) = 2 - 8 = -6$$

It does not equal zero, so the particle is changing direction at $t = 1$.

At $t = 7$,

$$a(7) = 14 - 8 = 6$$

It does not equal zero, so the particle is changing direction at $t = 7$.

(c) Find the total distance traveled by the particle from time $t = 0$ to $t = 4$.

If we want to find the total distance that a particle travels from time a to time b, we need to evaluate

$$\int_a^b |v(t)|\, dt$$

This means that, over an interval where the particle's velocity is negative, we multiply the integral by –1. So, we need to find where the velocity is negative and where it is positive.

We know that the velocity is zero at $t = 1$ and at $t = 7$.

We can find that the velocity is positive when $t < 1$ and when $t > 7$, and that the velocity is negative when $1 < t < 7$.

Thus, the distance that the particle travels from $t = 0$ to $t = 4$ is

$$\int_0^1 (t^2 - 8t + 7)\, dt - \int_1^4 (t^2 - 8t + 7)\, dt$$

Evaluating the integrals, we get

$$\left(\frac{t^3}{3} - 4t^2 + 7t \right)\Bigg|_0^1 - \left(\frac{t^3}{3} - 4t^2 + 7t \right)\Bigg|_1^4 = \left(\frac{10}{3} - 0 \right) - \left(-\frac{44}{3} - \frac{10}{3} \right) = \frac{64}{3}$$

Chapter 24
Practice Test 3

AP® Calculus AB Exam

SECTION I: Multiple-Choice Questions

DO NOT OPEN THIS BOOKLET UNTIL YOU ARE TOLD TO DO SO.

At a Glance

Total Time
1 hour and 45 minutes
Number of Questions
45
Percent of Total Grade
50%
Writing Instrument
Pencil required

Instructions

Section I of this examination contains 45 multiple-choice questions. Fill in only the ovals for numbers 1 through 45 on your answer sheet.

CALCULATORS MAY NOT BE USED IN THIS PART OF THE EXAMINATION.

Indicate all of your answers to the multiple-choice questions on the answer sheet. No credit will be given for anything written in this exam booklet, but you may use the booklet for notes or scratch work. After you have decided which of the suggested answers is best, completely fill in the corresponding oval on the answer sheet. Give only one answer to each question. If you change an answer, be sure that the previous mark is erased completely. Here is a sample question and answer.

Sample Question   Sample Answer

Chicago is a
(A) state
(B) city
(C) country
(D) continent
(E) village

Use your time effectively, working as quickly as you can without losing accuracy. Do not spend too much time on any one question. Go on to other questions and come back to the ones you have not answered if you have time. It is not expected that everyone will know the answers to all the multiple-choice questions.

About Guessing

Many candidates wonder whether or not to guess the answers to questions about which they are not certain. Multiple choice scores are based on the number of questions answered correctly. Points are not deducted for incorrect answers, and no points are awarded for unanswered questions. Because points are not deducted for incorrect answers, you are encouraged to answer all multiple-choice questions. On any questions you do not know the answer to, you should eliminate as many choices as you can, and then select the best answer among the remaining choices.

CALCULUS AB

SECTION I, Part A

Time—55 Minutes

Number of questions—28

A CALCULATOR MAY NOT BE USED ON THIS PART OF THE EXAMINATION

Directions: Solve each of the following problems, using the available space for scratchwork. After examining the form of the choices, decide which is the best of the choices given and fill in the corresponding oval on the answer sheet. No credit will be given for anything written in the test book. Do not spend too much time on any one problem.

In this test: Unless otherwise specified, the domain of a function f is assumed to be the set of all real numbers x for which $f(x)$ is a real number.

1. $\int_{\frac{\pi}{4}}^{x} \cos(2t)\, dt =$

    (A) $\cos(2x)$

    (B) $\dfrac{\sin(2x) - 1}{2}$

    (C) $\cos(2x) - 1$

    (D) $\sin(2x)$

    (E) $\dfrac{\sin 2(x)}{2}$

2. What are the coordinates of the point of inflection on the graph of $y = x^3 - 15x^2 + 33x + 100$?

    (A) $(9, 0)$
    (B) $(5, -48)$
    (C) $(1, 119)$
    (D) $(9, -89)$
    (E) $(5, 15)$

GO ON TO THE NEXT PAGE.

3. If $3x^2 - 2xy + 3y = 1$, then when $x = 2$, $\dfrac{dy}{dx} =$

   (A) -12

   (B) -10

   (C) $-\dfrac{10}{7}$

   (D) 12

   (E) 32

4. $\displaystyle\int_1^3 \dfrac{8}{x^3}\, dx =$

   (A) $\dfrac{32}{9}$

   (B) $\dfrac{40}{9}$

   (C) 0

   (D) $-\dfrac{40}{9}$

   (E) $-\dfrac{32}{9}$

GO ON TO THE NEXT PAGE.

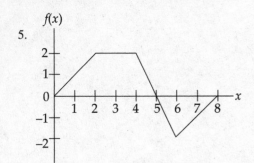

5.

The graph of a piecewise linear function f, for $0 \leq x \leq 8$, is shown above. What is the value of $\int_0^8 f(x)\, dx$?

(A) 1
(B) 4
(C) 8
(D) 10
(E) 13

6. If f is continuous for $a \leq x \leq b$, then at any point $x = c$, where $a < c < b$, which of the following must be true?

(A) $f(c) = \dfrac{f(b) - f(a)}{b - a}$

(B) $f(a) = f(b)$

(C) $f(c) = 0$

(D) $\int_a^b f(x)\, dx = f(c)$

(E) $\lim\limits_{x \to c} f(x) = f(c)$

GO ON TO THE NEXT PAGE.

7. If $f(x) = x^2\sqrt{3x+1}$, then $f'(x) =$

(A) $\dfrac{-3x^2 - 2x}{\sqrt{3x+1}}$

(B) $\dfrac{9x^2 + 2x}{\sqrt{3x+1}}$

(C) $\dfrac{-9x^2 + 4x}{2\sqrt{3x+1}}$

(D) $\dfrac{15x^2 + 4x}{2\sqrt{3x+1}}$

(E) $\dfrac{-9x^2 - 4x}{2\sqrt{3x+1}}$

GO ON TO THE NEXT PAGE.

8. What is the instantaneous rate of change at $t = -1$ of the function f, if $f(t) = \dfrac{t^3 + t}{4t + 1}$?

(A) $\dfrac{12}{9}$

(B) $\dfrac{4}{9}$

(C) $-\dfrac{20}{9}$

(D) $-\dfrac{4}{9}$

(E) $-\dfrac{12}{9}$

9. $\displaystyle\int_{2}^{e+1} \left(\dfrac{4}{x-1} \right) dx =$

(A) 4
(B) $4e$
(C) 0
(D) $-4e$
(E) -4

GO ON TO THE NEXT PAGE.

10.

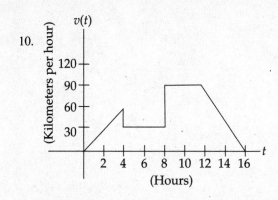

A car's velocity is shown on the graph above. Which of the following gives the total distance traveled from $t = 0$ to $t = 16$ (in kilometers)?

(A) 360
(B) 390
(C) 780
(D) 1,000
(E) 1,360

11. $\dfrac{d}{dx} \tan^2(4x) =$

(A) $8 \tan(4x)$
(B) $4 \sec^4(4x)$
(C) $8 \tan(4x) \sec^2(4x)$
(D) $4 \tan(4x) \sec^2(4x)$
(E) $8 \sec^4(4x)$

GO ON TO THE NEXT PAGE.

12. What is the equation of the line tangent to the graph of $y = \sin^2 x$ at $x = \dfrac{\pi}{4}$?

    (A) $y - \dfrac{1}{2} = -\left(x - \dfrac{\pi}{4}\right)$

    (B) $y - \dfrac{1}{2} = \left(x - \dfrac{\pi}{4}\right)$

    (C) $y - \dfrac{1}{\sqrt{2}} = \left(x - \dfrac{\pi}{4}\right)$

    (D) $y - \dfrac{1}{\sqrt{2}} = \dfrac{1}{2}\left(x - \dfrac{\pi}{4}\right)$

    (E) $y - \dfrac{1}{2} = \dfrac{1}{2}\left(x - \dfrac{\pi}{4}\right)$

13. If the function $f(x) = \begin{cases} 3ax^2 + 2bx + 1; & x \le 1 \\ ax^4 - 4bx^2 - 3x; & x > 1 \end{cases}$ is differentiable for all real values of x, then $b =$

    (A) $-\dfrac{11}{4}$

    (B) $\dfrac{1}{4}$

    (C) $-\dfrac{7}{16}$

    (D) 0

    (E) $-\dfrac{1}{4}$

GO ON TO THE NEXT PAGE.

14. The graph of $y = x^4 + 8x^3 - 72x^2 + 4$ is concave down for

(A) $-6 < x < 2$

(B) $x > 2$

(C) $x < -6$

(D) $x < -3 - 3\sqrt{5}$ or $x > -3 + 3\sqrt{5}$

(E) $-3 - 3\sqrt{5} < x < -3 + 3\sqrt{5}$

15. If $f(x) = \dfrac{x^2 + 5x - 24}{x^2 + 10x + 16}$, then $\lim\limits_{x \to -8} f(x)$ is

(A) 0

(B) 1

(C) $-\dfrac{3}{2}$

(D) $\dfrac{11}{6}$

(E) Nonexistent

GO ON TO THE NEXT PAGE.

16.

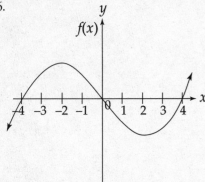

The graph of $f(x)$ is shown in the figure above. Which of the following could be the graph of $f'(x)$?

(A)

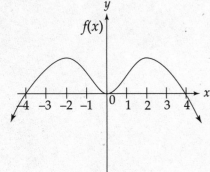

(D)

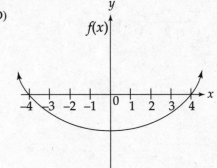

(B)

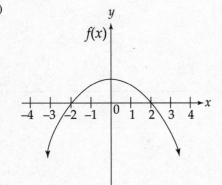

(E)

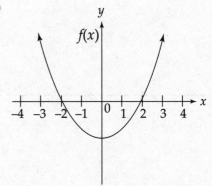

(C)

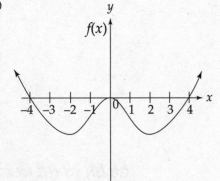

GO ON TO THE NEXT PAGE.

17. If $f(x) = \ln(\cos(3x))$, then $f'(x) =$

    (A) $-3 \csc(3x)$
    (B) $3 \sec(3x)$
    (C) $3 \tan(3x)$
    (D) $-3 \tan(3x)$
    (E) $-3 \cot(3x)$

18. If $f(x) = \int_0^{x+1} \sqrt[3]{t^2 - 1} \, dt$, then $f'(-4) =$

    (A) $\sqrt[3]{-9}$

    (B) -2

    (C) 2

    (D) $\sqrt[3]{15}$

    (E) 0

GO ON TO THE NEXT PAGE.

19. A particle moves along the x-axis so that its position at time t, in seconds, is given by $x(t) = t^2 - 7t + 6$. For what value(s) of t is the velocity of the particle zero?

(A) 1
(B) 6
(C) 1 or 6
(D) 3.5
(E) 1 or 3.5 or 6

---

20. $\int_0^{\frac{\pi}{2}} \sin(2x) e^{\sin^2 x}\, dx =$

(A) e
(B) $e - 1$
(C) $1 - e$
(D) $e + 1$
(E) 1

GO ON TO THE NEXT PAGE.

21. The average value of $\sec^2 x$ on the interval $\left[\dfrac{\pi}{6}, \dfrac{\pi}{4}\right]$ is

(A) $\dfrac{8}{\pi}$

(B) $\dfrac{12\sqrt{3} - 12}{\pi}$

(C) $\dfrac{12 - 4\sqrt{3}}{\pi}$

(D) $\dfrac{6\sqrt{2} - 6}{\pi}$

(E) $\dfrac{6 - 6\sqrt{2}}{\pi}$

22. Find the area of the region bounded by the parabolas $y = x^2$ and $y = 6x - x^2$.

(A) 9
(B) 27
(C) 6
(D) −9
(E) −18

GO ON TO THE NEXT PAGE.

23. The function f is given by $f(x) = x^4 + 4x^3$. On which of the following intervals is f decreasing?

    (A) $(-3, 0)$

    (B) $\left(0, \infty\right)$

    (C) $\left(-3, \infty\right)$

    (D) $\left(-\infty, -3\right)$

    (E) $\left(-\infty, 0\right)$

---

24. $\displaystyle\lim_{x \to 0} \frac{\tan(3x) + 3x}{\sin(5x)} =$

    (A) 0

    (B) $\dfrac{3}{5}$

    (C) 1

    (D) $\dfrac{6}{5}$

    (E) Nonexistent

GO ON TO THE NEXT PAGE.

25. If the region enclosed by the y-axis, the curve $y = 4\sqrt{x}$, and the line $y = 8$ is revolved about the x-axis, the volume of the solid generated is

    (A) $\dfrac{32\pi}{3}$

    (B) 128π

    (C) $\dfrac{128}{3}$

    (D) 128

    (E) $\dfrac{128\pi}{3}$

26. The maximum velocity attained on the interval $0 \leq t \leq 5$, by the particle whose displacement is given by $s(t) = 2t^3 - 12t^2 + 16t + 2$ is

    (A) 286
    (B) 46
    (C) 16
    (D) 0
    (E) −8

GO ON TO THE NEXT PAGE.

27. The value of c that satisfies the Mean Value Theorem for derivatives on the interval $[0, 5]$ for the function $f(x) = x^3 - 6x$ is

   (A) $-\dfrac{5}{\sqrt{3}}$

   (B) 0

   (C) 1

   (D) $\dfrac{5}{3}$

   (E) $\dfrac{5}{\sqrt{3}}$

28. If $f(x) = \sec(4x)$, then $f'\left(\dfrac{\pi}{16}\right)$ is

   (A) $4\sqrt{2}$

   (B) $\sqrt{2}$

   (C) 0

   (D) $\dfrac{1}{\sqrt{2}}$

   (E) $\dfrac{4}{\sqrt{2}}$

END OF PART A, SECTION I

IF YOU FINISH BEFORE TIME IS CALLED, YOU MAY CHECK YOUR WORK ON PART A ONLY.

DO NOT GO ON TO PART B UNTIL YOU ARE TOLD TO DO SO.

CALCULUS AB

SECTION I, Part B

Time—50 Minutes

Number of questions—17

A GRAPHING CALCULATOR IS REQUIRED FOR SOME QUESTIONS ON THIS PART OF THE EXAMINATION

Directions: Solve each of the following problems, using the available space for scratchwork. After examining the form of the choices, decide which is the best of the choices given and fill in the corresponding oval on the answer sheet. No credit will be given for anything written in the test book. Do not spend too much time on any one problem.

In this test:

1. The **exact** numerical value of the correct answer does not always appear among the choices given. When this happens, select from among the choices the number that best approximates the exact numerical value.

2. Unless otherwise specified, the domain of a function f is assumed to be the set of all real numbers x for which $f(x)$ is a real number.

29. If $f(x)$ is the function given by $f(x) = e^{3x} + 1$, at what value of x is the slope of the tangent line to $f(x)$ equal to 2 ?

(A) −0.135
(B) 0
(C) 0.231
(D) −0.366
(E) 0.693

GO ON TO THE NEXT PAGE.

30. The graph of the function $y = x^3 + 12x^2 + 15x + 3$ has a relative maximum at $x =$

    (A) -10.613
    (B) -0.248
    (C) -7.317
    (D) -1.138
    (E) -0.683

31. The side of a square is increasing at a constant rate of 0.4 cm/sec. In terms of the perimeter, P, what is the rate of change of the area of the square, in cm^2 / sec?

    (A) $0.05P$
    (B) $0.2P$
    (C) $0.4P$
    (D) $6.4P$
    (E) $51.2P$

GO ON TO THE NEXT PAGE.

32. Let f be the function given by $f(x) = 3^x$. For what value of x is the slope of the line tangent to the curve at $(x, f(x))$ equal to 1 ?

    (A) 1.099
    (B) 0.086
    (C) 0
    (D) −0.086
    (E) −1.099

---

33. Given f and g are differentiable functions and

$$f(a) = -4, \ g(a) = c, \ g(c) = 10, \ f(c) = 15$$

$$f'(a) = 8, \ g'(a) = b, \ g'(c) = 5, \ f'(c) = 6$$

    If $h(x) = f(g(x))$, find $h'(a)$

    (A) $6b$
    (B) $8b$
    (C) $-4b$
    (D) 80
    (E) $15b$

GO ON TO THE NEXT PAGE.

34. What is the area of the region in the first quadrant enclosed by the graph of $y = e^{-\frac{x^2}{4}}$ and the line $y = 0.5$?

(A) 0.240
(B) 0.516
(C) 0.480
(D) 1.032
(E) 1.349

35. What is the trapezoidal approximation of $\int_0^3 e^x \, dx$ using $n = 4$ subintervals?

(A) 6.407
(B) 13.565
(C) 19.972
(D) 27.879
(E) 34.944

GO ON TO THE NEXT PAGE.

36. The second derivative of a function f is given by $f''(x) = x \sin x - 2$. How many points of inflection does f have on the interval $(-10, 10)$?

    (A) Zero
    (B) Two
    (C) Four
    (D) Six
    (E) Eight

---

37. $\displaystyle \lim_{h \to 0} \frac{\sin\left(\dfrac{5\pi}{6} + h\right) - \dfrac{1}{2}}{h} =$

    (A) $\dfrac{\sqrt{3}}{2}$

    (B) $\dfrac{1}{2}$

    (C) 0

    (D) $-\dfrac{1}{2}$

    (E) $-\dfrac{\sqrt{3}}{2}$

GO ON TO THE NEXT PAGE.

38. $\dfrac{d}{dx} \displaystyle\int_{2x}^{5x} \cos t \; dt =$

    (A) $5\cos 5x - 2\cos 2x$

    (B) $5\sin 5x - 2\sin 2x$

    (C) $\cos 5x - \cos 2x$

    (D) $\sin 5x - \sin 2x$

    (E) $\dfrac{1}{5}\cos 5x - \dfrac{1}{2}\sin 2x$

39. The base of a solid S is the region enclosed by the graph of $4x + 5y = 20$, the x-axis, and the y-axis. If the cross-sections of S perpendicular to the x-axis are semicircles, then the volume of S is

    (A) $\dfrac{5\pi}{3}$

    (B) $\dfrac{10\pi}{3}$

    (C) $\dfrac{50\pi}{3}$

    (D) $\dfrac{225\pi}{3}$

    (E) $\dfrac{425\pi}{3}$

GO ON TO THE NEXT PAGE.

40. Which of the following is an equation of the line tangent to the graph of $y = x^3 + x^2$ at $y = 3$?

    (A) $y = 33x - 63$
    (B) $y = 33x - 135$
    (C) $y = 6.488x - 1.175$
    (D) $y = 6.488x - 4.620$
    (E) $y = 6.488x - 10.620$

41. If $f'(x) = \ln x - x + 2$, at which of the following values of x does f have a relative minimum value?

    (A) 5.146
    (B) 3.146
    (C) 1.000
    (D) 0.159
    (E) 0

GO ON TO THE NEXT PAGE.

42. Find the total area of the region between the curve $y = \cos x$ and the x-axis from $x = 1$ to $x = 2$ in radians.

   (A) 0
   (B) 0.068
   (C) 0.249
   (D) 1.751
   (E) 2.592

43. Let $f(x) = \int \cot x \, dx; \, 0 < x < \neq$. If $f\left(\dfrac{\pi}{6}\right) = 1$, then $f(1) =$

   (A) −1.861
   (B) −0.480
   (C) 0.134
   (D) 0.524
   (E) 1.521

GO ON TO THE NEXT PAGE.

44. A radioactive isotope, y, decays according to the equation $\dfrac{dy}{dt} = ky$, where k is a constant and t is measured in seconds. If the half-life of y is 1 minute, then the value of k is

(A) −41.589
(B) −0.012
(C) 0.027
(D) 0.693
(E) 98.923

45.

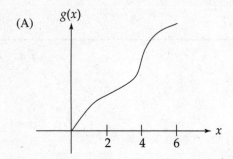

Let $g(x) = \int_0^x f(t)\, dt$, where $f(t)$ has the graph shown above. Which of the following could be the graph of g ?

(A)

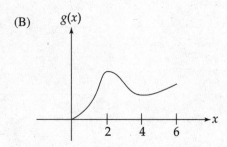

(D)

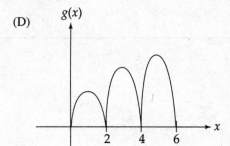

(B)

(E)

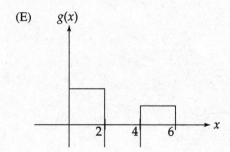

(C)

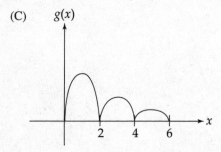

STOP

END OF PART B, SECTION I

IF YOU FINISH BEFORE TIME IS CALLED, YOU MAY CHECK YOUR WORK ON PART B ONLY.

DO NOT GO ON TO SECTION II UNTIL YOU ARE TOLD TO DO SO.

SECTION II
GENERAL INSTRUCTIONS

You may wish to look over the problems before starting to work on them, since it is not expected that everyone will be able to complete all parts of all problems. All problems are given equal weight, but the parts of a particular problem are not necessarily given equal weight.

A GRAPHING CALCULATOR IS REQUIRED FOR SOME PROBLEMS OR PARTS OF PROBLEMS ON THIS SECTION OF THE EXAMINATION.

- You should write all work for each part of each problem in the space provided for that part in the booklet. Be sure to write clearly and legibly. If you make an error, you may save time by crossing it out rather than trying to erase it. Erased or crossed-out work will not be graded.

- Show all your work. You will be graded on the correctness and completeness of your methods as well as your answers. Correct answers without supporting work may not receive credit.

- Justifications require that you give mathematical (noncalculator) reasons and that you clearly identify functions, graphs, tables, or other objects you use.

- You are permitted to use your calculator to solve an equation, find the derivative of a function at a point, or calculate the value of a definite integral. However, you must clearly indicate the setup of your problem, namely the equation, function, or integral you are using. If you use other built-in features or programs, you must show the mathematical steps necessary to produce your results.

- Your work must be expressed in standard mathematical notation rather than calculator syntax. For example, $\int_1^5 x^2 \, dx$ may not be written as fnInt (X^2, X, 1, 5).

- Unless otherwise specified, answers (numeric or algebraic) need not be simplified. If your answer is given as a decimal approximation, it should be correct to three places after the decimal point.

- Unless otherwise specified, the domain of a function f is assumed to be the set of all real numbers x for which $f(x)$ is a real number.

SECTION II, PART A
Time—30 minutes
Number of problems—2

A graphing calculator is required for some problems or parts of problems.

During the timed portion for Part A, you may work only on the problems in Part A.

On Part A, you are permitted to use your calculator to solve an equation, find the derivative of a function at a point, or calculate the value of a definite integral. However, you must clearly indicate the setup of your problem, namely the equation, function, or integral you are using. If you use other built-in features or programs, you must show the mathematical steps necessary to produce your results.

GO ON TO THE NEXT PAGE.

1.

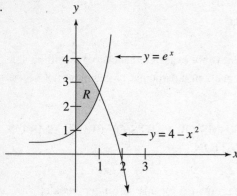

Let R be the region in the first quadrant shown in the figure above.

(a) Find the area of R.

(b) Find the volume of the solid generated when R is revolved about the x-axis.

(c) Find the volume of the solid generated when R is revolved about the line $x = -1$.

GO ON TO THE NEXT PAGE.

2. A body is coasting to a stop and the only force acting on it is a resistance proportional to its speed, according to the equation $\frac{ds}{dt} = v_f = v_0 e^{-\left(\frac{k}{m}\right)t}$; $s(0) = 0$, where v_0 is the body's initial velocity (in m/s), v_f is its final velocity, m is its mass, k is a constant, and t is time.

   (a) If a body with mass $m = 50$ kg and $k = 1.5$ kg/sec initially has a velocity of 30 m/s, how long, to the nearest second, will it take to slow to 1 m/s?

   (b) How far, to the 10 nearest meters, will the body coast during the time it takes to slow from 30 m/s to 1 m/s?

   (c) If the body coasts from 30 m/s to a stop, how far will it coast?

GO ON TO THE NEXT PAGE.

SECTION II, PART B
Time—1 hour
Number of problems—4

No calculator is allowed for these problems.

During the timed portion for Part B, you may continue to work on the problems in Part A without the use of any calculator.

3. An object moves with velocity $v(t) = 9t^2 + 18t - 7$ for $t \geq 0$ from an initial position of $s(0) = 3$.

    (a) Write an equation for the position of the particle.

    (b) When is the particle changing direction?

    (c) What is the total distance covered from $t = 2$ to $t = 5$?

GO ON TO THE NEXT PAGE.

4.

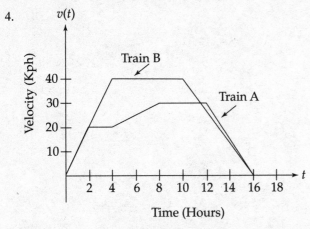

Three trains, *A, B,* and *C* each travel on a straight track for $0 \leq t \leq 16$ hours. The graphs above, which consist of line segments, show the velocities, in kilometers per hour, of trains *A* and *B*. The velocity of *C* is given by

$$v(t) = 8t - 0.25t^2$$

(Indicate units of measure for all answers.)

(a) Find the velocities of *A* and *C* at time *t* = 6 hours.

(b) Find the accelerations of *B* and *C* at time *t* = 6 hours.

(c) Find the positive difference between the total distance that *A* traveled and the total distance that *B* traveled in 16 hours.

(d) Find the total distance that *C* traveled in 16 hours.

GO ON TO THE NEXT PAGE.

5.

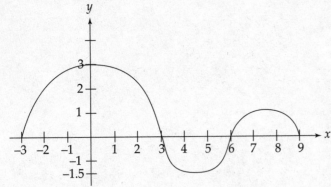

The figure above shows the graph of $g(x)$, where g is the derivative of the function f, for $-3 \le x \le 9$. The graph consists of three semicircular regions and has horizontal tangent lines at $x = 0$, $x = 4.5$, and $x = 7.5$.

(a) Find all values of x, for $-3 < x \le 9$, at which f attains a relative minimum. Justify your answer.

(b) Find all values of x, for $-3 < x \le 9$, at which f attains a relative maximum. Justify your answer.

(c) If $f(x) = \int_{-3}^{x} g(t)\, dt$, find $f(6)$.

(d) Find all points where $f''(x) = 0$.

GO ON TO THE NEXT PAGE.

6. Consider the curve given by $x^2y - 4x + y^2 = 2$.

(a) Find $\dfrac{dy}{dx}$.

(b) Find $\dfrac{d^2y}{dx^2}$.

(c) Find the equation of the tangent lines at each of the two points on the curve whose x-coordinate is 1.

STOP

END OF EXAM

Chapter 25
Practice Test 3
Answers and
Explanations

ANSWER KEY TO SECTION I

1.	B	11.	C	21.	C	31.	B	41.	D
2.	E	12.	B	22.	A	32.	D	42.	C
3.	B	13.	B	23.	D	33.	A	43.	E
4.	A	14.	A	24.	D	34.	B	44.	B
5.	B	15.	D	25.	B	35.	C	45.	B
6.	E	16.	E	26.	B	36.	C		
7.	D	17.	D	27.	E	37.	E		
8.	D	18.	C	28.	A	38.	A		
9.	A	19.	D	29.	A	39.	B		
10.	C	20.	B	30.	C	40.	D		

ANSWERS AND EXPLANATIONS TO SECTION I

1. **B** $\int_{\frac{\pi}{4}}^{x} \cos(2t)\, dt =$

    First, take the antiderivative: $\int \cos(2t)\, dt = \frac{1}{2}\sin(2t)$

    Next, plug in x and $\frac{\pi}{4}$ for t and take the difference: $\frac{1}{2}\sin(2x) - \frac{1}{2}\sin\left(2\left(\frac{\pi}{4}\right)\right)$

    This can be simplified to $\dfrac{\sin(2x) - 1}{2}$.

2. **E**   What are the coordinates of the point of inflection on the graph of $y = x^3 - 15x^2 + 33x + 100$?

    In order to find the inflection point(s) of a polynomial, we need to find the values of x where its second derivative is zero.

    First, we find the first and second derivative.

    $$\frac{dy}{dx} = 3x^2 - 30x + 33$$

    $$\frac{d^2 y}{dx^2} = 6x - 30$$

    Now, let's set the second derivative equal to zero and solve for x.

    $$6x - 30 = 0;\ x = 5$$

    In order to find the y-coordinate, we plug in 5 for x in the original equation.

    $$y = 5^3 - 15(5^2) + 33(5) + 100 = 15$$

    Therefore, the coordinates of the point of inflection are (5, 15).

3. **B**   If $3x^2 - 2xy + 3y = 1$, then when $x = 2$, $\dfrac{dy}{dx} =$

    We need to use implicit differentiation to find $\dfrac{dy}{dx}$.

    $$6x - 2\left(x\frac{dy}{dx} + y\right) + 3\frac{dy}{dx} = 0$$

    $$6x - 2x\frac{dy}{dx} - 2y + 3\frac{dy}{dx} = 0$$

Now, if we wanted to solve for $\frac{dy}{dx}$ in terms of x and y, we would have to do some algebra to isolate $\frac{dy}{dx}$. But because we are asked to solve for $\frac{dy}{dx}$ at a specific value of x, we don't need to simplify.

We need to find the y-coordinate that corresponds to the x-coordinate $x = 2$. We plug $x = 2$ into the original equation and solve for y.

$$3(2)^2 - 2(2)y + 3y = 12 - y = 1$$
$$y = 11$$

Finally, we plug $x = 2$ and $y = 11$ into the derivative, and we get

$$6(2) - 2(2)\frac{dy}{dx} - 2(11) + 3\frac{dy}{dx} = 0$$

$$12 - 4\frac{dy}{dx} - 22 + 3\frac{dy}{dx} = 0$$

$$\frac{dy}{dx} = -10$$

4. **A** $\int_1^3 \frac{8}{x^3}\, dx =$

First, rewrite the integral as $\int_1^3 8x^{-3}\, dx =$

Using the power rule for integrals, which is $\int x^n\, dx = \frac{x^{n+1}}{n+1} + C$, we get

$$\int 8x^{-3}\, dx = \frac{8}{-2}x^{-2} = -\frac{4}{x^2}$$

Next, plug in 3 and 1 for x and take the difference: $-\frac{4}{3^2} + \frac{4}{1^2} = -\frac{4}{9} + 4 = \frac{32}{9}$

5. **B**

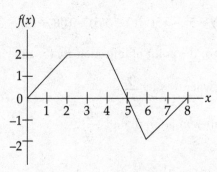

The graph of a piecewise linear function f, for $0 \le x \le 8$, is shown above. What is the value of $\int_0^8 f(x)\, dx$?

We need to add the areas of the regions between the graph and the x-axis. Note that the area of the region between 0 and 5 has a positive value and the area of the region between 5 and 8 has a negative value. The area of the former region can be found by calculating the area of a trapezoid with bases of 2 and 5 and a height of 2. The area is $\frac{1}{2}(2 + 5)(2) = 7$. The area of the latter region can be found by calculating the area of a triangle with a base of 3 and a height of 2. The area is $\frac{1}{2}(3)(2) = 3$. Thus the value of the integral is $7 - 3 = 4$.

6. **E** If f is continuous for $a \leq x \leq b$, then at any point $x = c$, where $a < c < b$, which of the following must be true?

In order for $f(x)$ to be continuous at a point c, there are three conditions that need to be fulfilled.

(1) $f(c)$ exists

(2) $\lim\limits_{x \to c} f(x)$ exists

(3) $\lim\limits_{x \to c} f(x) = f(c)$

Answer choices (A), (B), (C), and (D) are not necessarily true.

7. **D** If $f(x) = x^2 \sqrt{3x + 1}$, then $f'(x) =$

Here we need to use the Product Rule, which is: If $f(x) = uv$, where u and v are both functions of x, then $f'(x) = u\dfrac{dv}{dx} + v\dfrac{du}{dx}$.

We get $f'(x) = x^2 \left[\dfrac{1}{2}(3x + 1)^{-\frac{1}{2}}(3) \right] + 2x\sqrt{3x + 1}$.

This can be simplified to $\dfrac{3x^2}{2\sqrt{3x + 1}} + 2x\sqrt{3x + 1}$.

Multiply the numerator and denominator of the second expression by $2\sqrt{3x + 1}$ to get a common denominator: $\dfrac{3x^2}{2\sqrt{3x + 1}} + 2x\sqrt{3x + 1}\left(\dfrac{2\sqrt{3x + 1}}{2\sqrt{3x + 1}} \right)$

This simplifies to $\dfrac{3x^2}{2\sqrt{3x + 1}} + 4x\left(\dfrac{3x + 1}{2\sqrt{3x + 1}} \right) = \dfrac{3x^2}{2\sqrt{3x + 1}} + \dfrac{12x^2 + 4x}{2\sqrt{3x + 1}} = \dfrac{15x^2 + 4x}{2\sqrt{3x + 1}}$.

8. **D** What is the instantaneous rate of change at $t = -1$ of the function f, if $f(t) = \dfrac{t^3 + t}{4t + 1}$?

We find the instantaneous rate of change of the function by taking the derivative and plugging in $t = -1$.

We need to use the Quotient Rule, which is

$$\text{Given } f(x) = \frac{g(x)}{h(x)}, \text{ then } f'(x) = \frac{h(x)g'(x) - g(x)h'(x)}{\left[h(x)\right]^2}$$

Here we have $f'(t) = \dfrac{(4t + 1)(3t^2 + 1) - (t^3 + t)(4)}{(4t + 1)^2}$.

Next plug in $t = -1$ and solve.

$$f'(-1) = \frac{\left(4(-1) + 1\right)\left(3(-1)^2 + 1\right) - \left((-1)^3 + (-1)\right)(4)}{(4(-1) + 1)^2} = \frac{(-3)(4) - (-2)(4)}{(-3)^2} = -\frac{4}{9}$$

9. **A** $\displaystyle\int_2^{e+1} \left(\frac{4}{x - 1}\right) dx =$

You should know that $\displaystyle\int \frac{dx}{x} = \ln|x| + C$.

We take the antiderivative and we get $\displaystyle\int \left(\frac{4}{x - 1}\right) dx = 4\ln|x - 1| + C$.

Next, plug in $e + 1$ and 2 for x, and take the difference: $4\ln(e) - 4\ln(1)$.

You should know that $\ln e = 1$ and $\ln 1 = 0$. Thus, we get $4\ln(e) - 4\ln(1) = 4$.

10. **C**

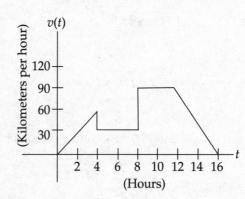

A car's velocity is shown on the graph above. Which of the following gives the total distance traveled from $t = 0$ to $t = 16$ (in kilometers)?

We find the total distance traveled by finding the area of the region between the curve and the x-axis. Normally, we would have to integrate but here we can find the area of the region easily because it consists of geometric objects whose areas are simple to calculate.

The area of the region between $t = 0$ and $t = 4$ can be found by calculating the area of a triangle with a base of 4 and a height of 60. The area is $\frac{1}{2}(4)(60) = 120$.

The area of the region between $t = 4$ and $t = 8$ can be found by calculating the area of a rectangle with a base of 4 and a height of 30.

The area is $(4)(30) = 120$.

The area of the region between $t = 8$ and $t = 16$ can be found by calculating the area of a trapezoid with bases of 4 and 8, and a height of 90 (or you could break it up into a rectangle and a triangle). The area is $\frac{1}{2}(4 + 8)(90) = 540$.

Thus, the total distance traveled is $120 + 120 + 540 = 780$ kilometers.

11. **C** $\dfrac{d}{dx}\tan^2(4x) =$

The derivative of $\tan(u) = \sec^2 u \dfrac{du}{dx}$. Here we need to use the Chain Rule.

$$\frac{d}{dx}\tan^2(4x) = 2[\tan(4(x)][\sec^2(4x)](4) = 8[\tan(4x)][\sec^2(4x)]$$

12. **B** What is the equation of the line tangent to the graph of $y = \sin^2 x$ at $x = \dfrac{\pi}{4}$?

If we want to find the equation of the tangent line, first we need to find the y-coordinate that corresponds to $x = \dfrac{\pi}{4}$. It is $y = \sin^2\left(\dfrac{\pi}{4}\right) = \left(\dfrac{1}{\sqrt{2}}\right)^2 = \dfrac{1}{2}$.

Next, we need to find the derivative of the curve at $x = \dfrac{\pi}{4}$, using the Chain Rule.

We get $\dfrac{dy}{dx} = 2\sin x \cos x$. At $x = \dfrac{\pi}{4}$, $\left.\dfrac{dy}{dx}\right|_{x=\frac{\pi}{4}} = 2\sin\left(\dfrac{\pi}{4}\right)\cos\left(\dfrac{\pi}{4}\right) = 2\left(\dfrac{1}{\sqrt{2}}\right)\left(\dfrac{1}{\sqrt{2}}\right) = 1$.

Now we have the slope of the tangent line and a point that it goes through. We can use the point-slope formula for the equation of a line, $(y - y_1) = m(x - x_1)$, and plug in what we have just found.

We get $\left(y - \dfrac{1}{2}\right) = (1)\left(x - \dfrac{\pi}{4}\right)$.

13. **B** If the function $f(x) = \begin{cases} 3ax^2 + 2bx + 1; \ x \leq 1 \\ ax^4 - 4bx^2 - 3x; \ x > 1 \end{cases}$ is differentiable for all real values of x, then $b =$

In order to solve this for b, we need $f(x)$ to be differentiable at $x = 1$, which means that it must

be continuous at $x = 1$. If we plug $x = 1$ into both pieces of this piecewise function, we get

$f(x) = \begin{cases} 3a + 2b + 1; \ x \leq 1 \\ a - 4b - 3; \ x > 1 \end{cases}$, so we need $3a + 2b + 1 = a - 4b - 3$, which can be simplified to

$2a + 6b = -4$.

Now we take the derivative of both pieces of this function.

$$f'(x) = \begin{cases} 6ax + 2b; \ x < 1 \\ 4ax^3 - 8bx - 3; \ x > 1 \end{cases}$$

Plug in $x = 1$ to get $f'(x) = \begin{cases} 6a + 2b; \ x < 1 \\ 4a - 8b - 3; \ x > 1 \end{cases}$.

From there, we can simplify $6a + 2b = 4a - 8b - 3$ to get $2a + 10b = -3$.

Solving the simultaneous equations, we get $a = -\dfrac{11}{4}$ and $b = \dfrac{1}{4}$.

14. **A** The graph of $y = x^4 + 8x^3 - 72x^2 + 4$ is concave down for

A graph is concave down where the second derivative is negative.

First, we find the first and second derivative.

$$\frac{dy}{dx} = 4x^3 + 24x^2 - 144x$$

$$\frac{d^2y}{dx^2} = 12x^2 + 48x - 144$$

Next, we want to determine on which intervals the second derivative of the function is positive and on which it is negative. We do this by finding where the second derivative is zero.

$$12x^2 + 48x - 144 = 0$$
$$x^2 + 4x - 12 = 0$$
$$(x + 6)(x - 2) = 0$$
$$x = -6 \text{ or } x = 2$$

We can test where the second derivative is positive and negative by picking a point in each of the three regions $-\infty < x < -6$, $-6 < x < 2$, and $2 < x < \infty$, plugging the point into the second derivative and seeing what the sign of the answer is. You should find that the second derivative is negative on the interval $-6 < x < 2$.

15. **D**   If $f(x) = \dfrac{x^2 + 5x - 24}{x^2 + 10x + 16}$, then $\lim\limits_{x \to -8} f(x)$ is

First, try plugging $x = -8$ into $f(x) = \dfrac{x^2 + 5x - 24}{x^2 + 10x + 16}$.

We get $f(x) = \dfrac{(-8)^2 + 5(-8) - 24}{(-8)^2 + 10(-8) + 16} = \dfrac{0}{0}$. This does NOT necessarily mean that the limit does not

exist. When we get a limit of the form $\dfrac{0}{0}$, we first try to simplify the function by factoring and

canceling like terms. We get

$$f(x) = \dfrac{x^2 + 5x - 24}{x^2 + 10x + 16} = \dfrac{(x + 8)(x - 3)}{(x + 8)(x + 2)} = \dfrac{(x - 3)}{(x + 2)}$$

Now, if we plug in $x = -8$, we get $f(x) = \dfrac{(-8 - 3)}{(-8 + 2)} = \dfrac{-11}{-6} = \dfrac{11}{6}$.

16. **E**

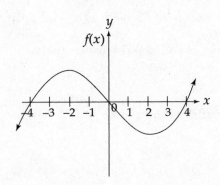

The graph of $f(x)$ is shown in the figure above. Which of the following could be the graph of $f'(x)$?

Here we want to examine the slopes of various pieces of the graph of $f(x)$. Notice that the graph has a positive slope from $x = -\infty$ to $x = -2$, where the slope is zero. Thus, we are looking for a graph of $f'(x)$ that is positive from $x = -\infty$ to $x = -2$ and zero at $x = -2$. Notice that the graph of $f(x)$ has a negative slope from $x = -2$ to $x = 2$, where the slope is zero. Thus, we are looking for a graph of $f'(x)$ that is negative from $x = -2$ to $x = 2$ and zero at $x = 2$. Finally, notice that the graph of $f(x)$ has a positive slope from $x = 2$ to $x = \infty$. Thus, we are looking for a graph of $f'(x)$ that is positive from $x = 2$ to $x = \infty$. Graph (E) satisfies all of these requirements.

17. **D** If $f(x) = \ln(\cos(3x))$, then $f'(x) =$

Remember that $\dfrac{d}{dx} \ln(u(x)) = \dfrac{u'(x)}{u(x)}$.

We will need to use the Chain Rule to find the derivative.

$$f'(x) = \left(\frac{-\sin(3x)}{\cos(3x)} \right)(3) = -3\tan(3x)$$

18. **C** If $f(x) = \displaystyle\int_0^{x+1} \sqrt[3]{t^2 - 1}\, dt$, then $f'(-4)$

The Second Fundamental Theorem of Calculus tells us how to find the derivative of an integral. It

says that $\dfrac{d}{dx}\displaystyle\int_c^u f(t)\, dt = f(u)\dfrac{du}{dx}$, where c is a constant and u is a function of x.

Here we can use the theorem to get $\dfrac{d}{dx}\displaystyle\int_0^{x+1} \sqrt[3]{t^2 - 1}\, dt = \sqrt[3]{(x+1)^2 - 1}$.

Now we evaluate the expression at $x = -4$. We get $\sqrt[3]{(-4+1)^2 - 1} = 2$.

19. **D** A particle moves along the x-axis so that its position at time t, in seconds, is given by $x(t) = t^2 - 7t + 6$. For what value(s) of t is the velocity of the particle zero?

Velocity is the first derivative of position with respect to time.

The first derivative is: $v(t) = 2t - 7$.

Thus, the velocity of the particle is zero at time $t = 3.5$ seconds.

20. **B** $\displaystyle\int_0^{\frac{\pi}{2}} \sin(2x) e^{\sin^2 x}\, dx =$

We can use u-substitution to evaluate the integral.

Let $u = \sin^2 x$ and $du = 2 \sin x \cos x\, dx$. Next, recall from trigonometry that $2 \sin x \cos x = \sin(2x)$. Now we can substitute into the integral $\displaystyle\int e^u\, du$, leaving out the limits of integration for the moment.

Evaluate the integral to get $\displaystyle\int e^u\, du = e^u$.

Now, we substitute back to get $e^{\sin^2 x}$.

Finally, we evaluate at the limits of integration, and we get

$$e^{\sin^2 x}\Big|_0^{\frac{\pi}{2}} = e^{\sin^2 \frac{\pi}{2}} - e^{\sin^2 0} = e - 1$$

21.   **C**   The average value of $\sec^2 x$ on the interval $\left[\dfrac{\pi}{6}, \dfrac{\pi}{4}\right]$ is

In order to find the average value, we use the Mean Value Theorem for integrals, which says that the average value of $f(x)$ on the interval $[a, b]$ is $\dfrac{1}{b-a}\displaystyle\int_a^b f(x)\,dx$.

Here we have $\dfrac{1}{\dfrac{\pi}{4}-\dfrac{\pi}{6}}\displaystyle\int_{\frac{\pi}{6}}^{\frac{\pi}{4}} \sec^2 x\,dx$.

Next, recall that $\dfrac{d}{dx}\tan x = \sec^2 x$.

We evaluate the integral.

$$\dfrac{1}{\dfrac{\pi}{4}-\dfrac{\pi}{6}}(\tan x)_{\frac{\pi}{6}}^{\frac{\pi}{4}} = \dfrac{1}{\dfrac{\pi}{4}-\dfrac{\pi}{6}}\left[\tan\dfrac{\pi}{4}-\tan\dfrac{\pi}{6}\right] = \dfrac{1}{\dfrac{\pi}{4}-\dfrac{\pi}{6}}\left(1-\dfrac{\sqrt{3}}{3}\right)$$

Get a common denominator for each of the two expressions.

$$\dfrac{1}{\dfrac{\pi}{4}-\dfrac{\pi}{6}}\left(1-\dfrac{\sqrt{3}}{3}\right) = \dfrac{1}{\dfrac{6\pi}{24}-\dfrac{4\pi}{24}}\left(\dfrac{3}{3}-\dfrac{\sqrt{3}}{3}\right)$$

We can simplify this to $\dfrac{1}{\dfrac{2\pi}{24}}\left(\dfrac{3-\sqrt{3}}{3}\right) = \dfrac{12}{\pi}\left(\dfrac{3-\sqrt{3}}{3}\right) = \dfrac{12-4\sqrt{3}}{\pi}$.

22.   **A**   Find the area of the region bounded by the parabolas $y = x^2$ and $y = 6x - x^2$.

First, we should graph the two curves.

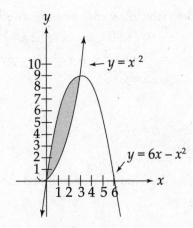

Next, we need to find the points of intersection of the two curves, which we do by setting them equal to each other and solving for x.

$$x^2 = 6x - x^2$$

$$2x^2 = 6x$$

$$2x^2 - 6x = 0$$

$$2x(x - 3) = 0$$

$$x = 0 \text{ or } x = 3$$

We can find the area between the two curves by integrating the top curve minus the bottom curve, using the points of intersection as the limits of integration. We get

$$\int_0^3 \left[(6x - x^2) - (x^2)\right] dx$$

We evaluate the integral and we get $\int_0^3 (6x - 2x^2)\, dx = \left(3x^2 - \frac{2}{3}x^3\right)\Big|_0^3 = 9$.

23. **D** The function f is given by $f(x) = x^4 + 4x^3$. On which of the following intervals is f decreasing?

A function is decreasing on an interval where the derivative is negative.
The derivative is $f'(x) = 4x^3 + 12x^2$.

We want to determine on which intervals the derivative of the function is positive and on which it is negative. We do this by finding where the derivative is zero.

$$4x^3 + 12x^2 = 0$$

$$4x^2(x + 3) = 0$$

$$x = -3 \text{ or } x = 0$$

We can test where the derivative is positive and negative by picking a point in each of the three regions $-\infty < x < -3$, $-3 < x < 0$, and $0 < x < \infty$, plugging the point into the derivative and seeing what the sign of the answer is. Because x^2 is never negative, you should find that the derivative is negative on the interval $-\infty < x < -3$.

24. **D** $\displaystyle \lim_{x \to 0} \frac{\tan(3x) + 3x}{\sin(5x)} =$

We will need to use the fact that $\displaystyle \lim_{x \to 0} \frac{\sin x}{x} = 1$ to find the limit.

First, rewrite the limit as $\displaystyle\lim_{x\to 0}\dfrac{\dfrac{\sin(3x)}{\cos(3x)}+3x}{\sin(5x)} =$

Next, break the expression into two rational expressions.

$$\lim_{x\to 0}\dfrac{\sin(3x)}{\sin(5x)\cos(3x)}+\dfrac{3x}{\sin(5x)} =$$

This can be broken up further into

$$\lim_{x\to 0}\dfrac{\sin(3x)}{\sin(5x)}\dfrac{1}{\cos(3x)}+\dfrac{3x}{\sin(5x)} =$$

We will evaluate the limit of each separately.

First expression

Divide the top and bottom by x: $\displaystyle\lim_{x\to 0}\dfrac{\dfrac{\sin(3x)}{x}}{\dfrac{\sin(5x)}{x}}.$

Then, multiply the top and bottom of the upper expression by 3, and the top and bottom of the

lower expression by 5: $\displaystyle\lim_{x\to 0}\dfrac{\dfrac{3\sin(3x)}{3x}}{\dfrac{5\sin(5x)}{5x}}.$

Now, if we take the limit, we get $\displaystyle\lim_{x\to 0}\dfrac{\dfrac{3\sin(3x)}{3x}}{\dfrac{5\sin(5x)}{5x}}=\dfrac{3(1)}{5(1)}=\dfrac{3}{5}.$

Second expression

This limit is straightforward: $\displaystyle\lim_{x\to 0}\dfrac{1}{\cos(3x)}=\dfrac{1}{\cos(0)}=1.$

Third expression

First, pull the constant, 3, out of the limit: $\displaystyle\lim_{x\to 0}\dfrac{3x}{\sin(5x)}=3\lim_{x\to 0}\dfrac{x}{\sin(5x)}.$

Now, if we multiply the top and bottom of the expression by 5, we get $3\lim\limits_{x\to 0}\dfrac{5x}{5\sin(5x)}$.

Now, if we take the limit, we get $3\lim\limits_{x\to 0}\dfrac{5x}{5\sin(5x)} = 3\left(\dfrac{1}{5}\right) = \dfrac{3}{5}$.

Combine the three numbers, and we get $\dfrac{3}{5}(1) + \dfrac{3}{5} = \dfrac{6}{5}$.

25. **B** If the region enclosed by the y-axis, the curve $y = 4\sqrt{x}$, and the line $y = 8$ is revolved about the x-axis, the volume of the solid generated is

First, we graph the curves.

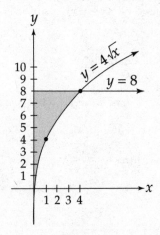

We can find the volume by taking a vertical slice of the region. The formula for the volume of a solid of revolution around the x-axis, using a vertical slice bounded from above by the curve $f(x)$ and from below by $g(x)$, on the interval $[a, b]$, is

$$\pi\int_a^b \left[f(x)^2 - g(x)^2 \right] dx$$

The upper curve is $y = 8$, and the lower curve is $y = 4\sqrt{x}$.

Next, we need to find the point(s) of intersection of the two curves, which we do by setting them equal to each other and solving for x.

$$8 = 4\sqrt{x}$$

$$2 = \sqrt{x}$$

$$x = 4$$

Thus, the limits of integration are $x = 0$ and $x = 4$.

Now, we evaluate the integral.

$$\pi \int_0^4 \left[(8)^2 - \left[4\sqrt{x} \right]^2 \right] dx = \pi \int_0^4 \left(64 - 16x \right) dx = \pi \left(64x - 8x^2 \right)\Big|_0^4 = 128\pi$$

26. **B** The maximum velocity attained on the interval $0 \le t \le 5$ by the particle whose displacement is given by $s(t) = 2t^3$ $12t^2 + 16t + 2$ is

Velocity is the first derivative of position with respect to time.

The first derivative is
$$v(t) = 6t^2 - 24t + 16$$

If we want to find the maximum velocity, we take the derivative of velocity (which is acceleration) and find where the derivative is zero.

$$v'(t) = 12t - 24$$

Next, we set the derivative equal to zero and solve for t, in order to find the critical value.

$$12t - 24 = 0$$

$$t = 2$$

Note that the second derivative of velocity is 12, which is positive. Remember the second derivative test: If the sign of the second derivative at a critical value is positive, then the curve has a local minimum there. If the sign of the second derivative is negative, then the curve has a local maximum there.

Thus, the velocity is a *minimum* at $t = 2$. In order to find where it has an absolute *maximum*, we plug the endpoints of the interval into the original equation for velocity, and the larger value will be the answer.

At $t = 0$ the velocity is 16. At $t = 5$, the velocity is 46.

27. **E** The value of c that satisfies the Mean Value Theorem for derivatives on the interval $[0, 5]$ for the function $f(x) = x^3 - 6x$ is

The Mean Value Theorem for derivatives says that, given a function $f(x)$ which is continuous and differentiable on $[a, b]$, then there exists some value c on (a, b) where $\dfrac{f(b) - f(a)}{b - a} = f'(c)$.

Here we have $\dfrac{f(b) - f(a)}{b - a} = \dfrac{f(5) - f(0)}{5 - 0} = \dfrac{95 - 0}{5} = 19$.

Plus, $f'(c) = 3c^2 - 6$, so we simply set $3c^2 - 6 = 19$. If we solve for c, we get $c = \pm\dfrac{5}{\sqrt{3}}$. Both of these values satisfy the Mean Value Theorem for derivatives, but only the positive value, $c = \dfrac{5}{\sqrt{3}}$, is in the interval.

28. **A** If $f(x) = \sec(4x)$, then $f'\left(\dfrac{\pi}{16}\right)$ is

Recall that $\dfrac{d}{dx}\sec x = \sec x \tan x$.

Therefore, using the Chain Rule, we get $f'(x) = 4\sec(4x)\tan(4x)$.

If we plug in $x = \dfrac{\pi}{16}$, we get $f'(x) = 4\sec\left(\dfrac{\pi}{4}\right)\tan\left(\dfrac{\pi}{4}\right) = 4\sqrt{2}$.

29. **A** If $f(x)$ is the function given by $f(x) = e^{3x} + 1$, at what value of x is the slope of the tangent line to $f(x)$ equal to 2?

The slope of the tangent line is the derivative of the function. We get $f'(x) = 3e^{3x}$. Now we set the derivative equal to 2 and solve for x.

$$3e^{3x} = 2$$

$$e^{3x} = \frac{2}{3}$$

$$3x = \ln\frac{2}{3}$$

$$x = \frac{1}{3}\ln\frac{2}{3} \approx -.135$$

Remember to round all answers to three decimal places on the AP exam.

30. **C** The graph of the function $y = x^3 + 12x^2 + 15x + 3$ has a relative maximum at $x =$

First, let's find the derivative: $\dfrac{dy}{dx} = 3x^2 + 24x + 15$.

Next, set the derivative equal to zero and solve for x.

$$3x^2 + 24x + 15 = 0$$
$$x^2 + 8x + 5 = 0$$

Using the quadratic formula (or your calculator), we get

$$x = \frac{-8 \pm \sqrt{64 - 20}}{2} \approx -0.683, -7.317$$

Let's use the second derivative test to determine which is the maximum. We take the second derivative and then plug in the critical values that we found when we set the first derivative equal to zero. If the sign of the second derivative at a critical value is positive, then the curve has a local minimum there. If the sign of the second derivative is negative, then the curve has a local maximum there.

The second derivative is $\dfrac{d^2 y}{dx^2} = 6x + 24$. The second derivative is negative at $x = -7.317$, so the curve has a local maximum there.

31. **B** The side of a square is increasing at a constant rate of 0.4 cm/sec. In terms of the perimeter, P, what is the rate of change of the area of the square, in cm²/sec?

The formula for the perimeter of a square is $P = 4s$, where s is the length of a side of the square.

If we differentiate this with respect to t, we get $\dfrac{dP}{dt} = 4\dfrac{ds}{dt}$. We plug in $\dfrac{ds}{dt} = 0.4$, and we get $\dfrac{dP}{dt} = 4(0.4) = 1.6$.

The formula for the area of a square is $A = s^2$. If we solve the perimeter equation for s in terms of P and substitute it into the area equation, we get

$$s = \frac{P}{4}, \text{ so } A = \left(\frac{P}{4}\right)^2 = \frac{P^2}{16}$$

If we differentiate this with respect to t, we get $\dfrac{dA}{dt} = \dfrac{P}{8}\dfrac{dP}{dt}$.

Now we plug in $\dfrac{dP}{dt} = 1.6$, and we get $\dfrac{dA}{dt} = \dfrac{P}{8}(1.6) = 0.2P.$

32. **D** Let f be the function given by $f(x) = 3^x$. For what value of x is the slope of the line tangent to the curve at $(x, f(x))$ equal to 1?

The slope of the tangent line is the derivative of the function.

Recall that $\dfrac{d}{dx}a^x = a^x \ln a$. Here we get $f'(x) = 3^x \ln 3$.

Now we set the derivative equal to 1 and solve for x.

Using the calculator, we get $3^x \ln 3 = 1$, so $x \approx -.086$

33.   A   Given f and g are differentiable functions and

$$f(a) = -4, g(a) = c, g(c) = 10, f(c) = 15$$

$$f'(a) = 8, g'(a) = b, g'(c) = 5, f'(c) = 6$$

If $h(x) = f(g(x))$, find $h'(a)$.

Use the Chain Rule to find $h'(a)$: $h'(x) = f'(g(x))(g'(x))$.

We substitute a for x, and because $g(a) = c$, we get $h'(a) = f'(c)(g'(a)) = 6b$.

34.   B   What is the area of the region in the first quadrant enclosed by the graph of $y = e^{-\frac{x^2}{4}}$ and the line $y = 0.5$?

First, we should graph the two curves.

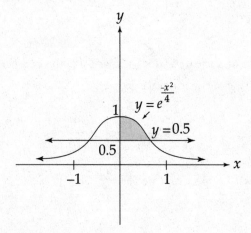

Next, we need to find the points of intersection of the two curves, which we do by setting them equal to each other and solving for x.

$$e^{-\frac{x^2}{4}} = 0.5$$

You will need to use a calculator to solve for x. The answers are (to three decimal places): $x = -1.665$ and $x = +1.665$.

We can find the area between the two curves by integrating the top curve minus the bottom curve, using the points of intersection as the limits of integration. Because we want to find the area in the first quadrant, we use 0 as the lower limit of integration. We get

$$\int_0^{1.665} \left(e^{-\frac{x^2}{4}} - 0.5 \right) dx$$

We will need a calculator to evaluate this integral: $\int_0^{1.665}\left(e^{-\frac{x^2}{4}}-0.5\right)dx \approx 0.516$

35.   C   What is the trapezoidal approximation of $\int_0^3 e^x\,dx$ using $n = 4$ subintervals?

The Trapezoid Rule enables us to approximate the area under a curve with a fair degree of accuracy. The rule says that the area between the x-axis and the curve $y = f(x)$, on the interval $[a, b]$, with n trapezoids, is

$$\frac{1}{2}\frac{b-a}{n}\left[y_0 + 2y_1 + 2y_2 + 2y_3 + \ldots + 2y_{n-1} + y_n\right]$$

Using the rule here, with $n = 4$, $a = 0$, and $b = 3$, we get

$$\frac{1}{2}\left(\frac{3}{4}\right)\left[e^0 + 2e^{\frac{3}{4}} + 2e^{\frac{6}{4}} + 2e^{\frac{9}{4}} + e^3\right] \approx 19.972$$

36.   C   The second derivative of a function f is given by $f''(x) = x \sin x - 2$. How many points of inflection does f have on the interval $(-10, 10)$?

Use your calculator to graph the second derivative and count the number of times that it crosses the x-axis on the interval $(-10, 10)$.

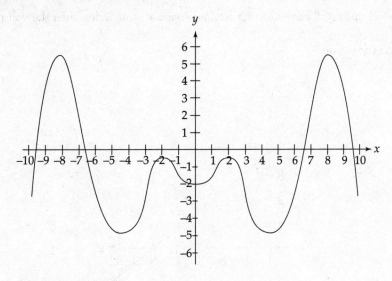

It crosses four times, so there are four points of inflection.

37.　E　$\lim\limits_{h \to 0} \dfrac{\sin\left(\dfrac{5\pi}{6} + h\right) - \dfrac{1}{2}}{h}$

Notice how this limit takes the form of the definition of the derivative, which is

$$f'(x) = \lim_{h \to 0} \frac{f(x+h) - f(x)}{h}$$

If we think of $f(x)$ as sin x, then this expression gives the derivative of sin x at the point $x = \dfrac{5\pi}{6}$.

The derivative of sin x is $f'(x) = \cos x$. At $x = \dfrac{5\pi}{6}$, we get $f'\left(\dfrac{5\pi}{6}\right) = \cos\left(\dfrac{5\pi}{6}\right) = -\dfrac{\sqrt{3}}{2}$.

38.　A　$\dfrac{d}{dx}\displaystyle\int_{2x}^{5x} \cos t\ dt =$

The Second Fundamental Theorem of Calculus tells us how to find the derivative of an integral:

$$\frac{dF}{dx} = \frac{d}{dx}\int_{a}^{x} f(t)\,dt.$$

Here we can use the theorem to get $\dfrac{d}{dx}\displaystyle\int_{2x}^{5x} \cos t\ dt = 5\cos 5x - 2\cos 2x$.

39.　B　The base of a solid S is the region enclosed by the graph of $4x + 5y = 20$, the x-axis, and the y-axis. If the cross-sections of S perpendicular to the x-axis are semicircles, then the volume of S is

First, sketch the region.

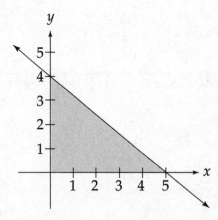

The rule for finding the volume of a solid with known cross-sections is $V = \displaystyle\int_{a}^{b} A(x)\,dx$, where A is the formula for the area of the cross-section. So x represents the diameter of a semi-circular cross-section.

The area of a semi-circle in terms of its diameter is $A = \pi \dfrac{d^2}{8}$. We find the length of the diameter by solving the equation $4x + 5y = 20$ for y: $y = \dfrac{20 - 4x}{5}$. Next, we need to find where the graph intersects the x-axis. You should get $x = 5$. Thus, we find the volume by evaluating the integral.

$$\int_0^5 \pi \dfrac{\left(\dfrac{20 - 4x}{5}\right)^2}{8} \, dx$$

This integral can be simplified to

$$\dfrac{\pi}{200} \int_0^5 \left(20 - 4x\right)^2 dx = \dfrac{\pi}{200} \int_0^5 \left(400 - 160x + 16x^2\right) dx$$

You can evaluate the integral by hand or with a calculator. You should get

$$\dfrac{\pi}{200} \int_0^5 \left(400 - 160x + 16x^2\right) dx = \dfrac{10\pi}{3}$$

40. **D** Which of the following is an equation of the line tangent to the graph of $y = x^3 + x^2$ at $y = 3$?

If we want to find the equation of the tangent line, first we need to find the x-coordinate that corresponds to $y = 3$. If you use your calculator to solve $x^3 + x^2 = 3$, you should get $x = 1.1746$.

Next, we need to find the derivative of the curve at $x = 1.1746$.

We get

$$\dfrac{dy}{dx} = 3x^2 + 2x. \text{ At } x = 1.1746, \left. \dfrac{dy}{dx} \right|_{x=1.1746} = 3(1.1746)^2 + 2(1.1746) = 6.488$$

(It is rounded to three decimal places.)

Now we have the slope of the tangent line and a point that it goes through. We can use the point-slope formula for the equation of a line, $(y - y_1) = m(x - x_1)$ and plug in what we have just found. We get $(y - 3) = (6.488)(x - 1.1746)$. This simplifies to $y = 6.488x - 4.620$.

41. **D** If $f'(x) = \ln x - x + 2$, at which of the following values of x does f have a relative minimum value?

Set the derivative equal to zero and solve for x. Using your calculator, you should get $\ln x - x + 2 = 0$.

$$x = 3.146 \text{ or } x = 0.159 \text{ (rounded to three decimal places)}$$

Let's use the second derivative test to determine which is the minimum. We take the second derivative and then plug in the critical values that we found when we set the first derivative equal to zero. If the sign of the second derivative at a critical value is positive, then the curve has a local minimum there. If the sign of the second derivative is negative, then the curve has a local maximum there.

The second derivative is $f''(x) = \dfrac{1}{x} - 1$. The second derivative is positive at $x = 0.159$, so the curve has a local minimum there.

42. C   Find the area of the region between the curve $y = \cos x$ and the x-axis from $x = 1$ to $x = 2$ radians.

First, we should graph the curve.

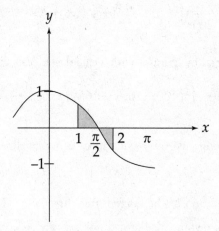

Note that the curve is above the x-axis from $x = 1$ to $x = \dfrac{\pi}{2}$ and below the x-axis from $x = \dfrac{\pi}{2}$ to

$x = 2$. Thus, we need to evaluate two integrals to find the area.

$$\int_1^{\frac{\pi}{2}} \cos x \, dx + \int_{\frac{\pi}{2}}^2 \left(-\cos x\right) dx$$

We will need a calculator to evaluate these integrals.

$$\int_1^{\frac{\pi}{2}} \cos x \, dx + \int_{\frac{\pi}{2}}^2 \left(-\cos x\right) dx \approx 0.249$$

43. E   Let $f(x) = \displaystyle\int \cot x \, dx$; $0 < x < \pi$. If $f\left(\dfrac{\pi}{6}\right) = 1$, then $f(1) =$

First, we find $\displaystyle\int \cot x \, dx$ by rewriting the integral as $\displaystyle\int \dfrac{\cos x}{\sin x} \, dx$.

Then, we use u-substitution. Let $u = \sin x$ and $du = \cos x$.

Substituting, we can get: $\int \dfrac{\cos x}{\sin x}\, dx = \int \dfrac{du}{u} = \ln|u| + C$. Then substituting back, we get $\ln(\sin x) + C$. (We can get rid of the absolute value bars because sine is always positive on the interval.)

Next, we use $f\left(\dfrac{\pi}{6}\right) = 1$ to solve for C. We get $1 = \ln\left(\sin \dfrac{\pi}{6}\right) + C$.

$$1 = \ln\left(\dfrac{1}{2}\right) + C$$

$$1 = \ln\left(\dfrac{1}{2}\right) + C = 1.693147$$

Thus, $f(x) = \ln(\sin x) + 1.693147$.

At $x = 1$, we get $f(1) = \ln(\sin 1) + 1.693147 = 1.521$ (rounded to three decimal places).

44. **B**   A radioactive isotope, y, decays according to the equation $\dfrac{dy}{dt} = ky$, where k is a constant and t is measured in seconds. If the half-life of y is 1 minute, then the value of k is

We solve this differential equation using separation of variables.
First, move the y to the left side and the dt to the right side, to get $\dfrac{dy}{y} = k\, dt$.

Now, integrate both sides.

$$\int \dfrac{dy}{y} = k \int dt$$

$\ln y = kt + C$

Next, it's traditional to put the equation in terms of y. We do this by exponentiating both sides to the base e. We get $y = e^{kt + C}$.

Using the rules of exponents, we can rewrite this as: $y = e^{kt} e^{C}$. Finally, because e^{C} is a constant, we can rewrite the equation as: $y = Ce^{kt}$.

Now, we use the initial condition to solve for k. At time $t = 60$ (seconds), $y = \dfrac{1}{2}$.

We are assuming a starting amount of $y = 1$, which will make $C = 1$. Actually, we could assume any starting amount. The half-life tells us that there will be half that amount after 1 minute. Therefore,

$$\dfrac{1}{2} = e^{60k}$$

Solve for k: $k = \dfrac{1}{60} \ln\!\left(\dfrac{1}{2}\right)$.

This gives us $k = -0.012$ (rounded to three decimal places).

45.   **B**

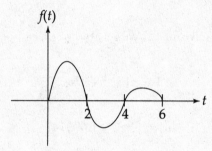

Let $g(x) = \displaystyle\int_{0}^{x} f(t)\, dt$, where $f(t)$ has the graph shown above. Which of the following could be the graph of g?

The function $g(x) = \displaystyle\int_{0}^{x} f(t)\, dt$ is called an accumulation function and stands for the area between the curve and the x-axis to the point x. At $x = 0$, the area is 0, so $g(0) = 0$. From $x = 0$ to $x = 2$ the area grows, so $g(x)$ has a positive slope. Then, from $x = 2$ to $x = 4$ the area shrinks (because we subtract the area of the region under the x-axis from the area of the region above it), so $g(x)$ has a negative slope. Finally, from $x = 4$ to $x = 6$ the area again grows, so $g(x)$ has a positive slope. The curve that best represents this is shown in (B).

ANSWERS AND EXPLANATIONS TO SECTION II

1.

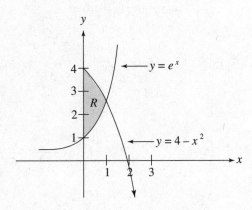

Let R be the region in the first quadrant shown in the figure above.

(a) Find the area of R.

In order to find the area, we "slice" the region vertically and add up all of the slices. We use the formula for the area of the region between $y = f(x)$ and $y = g(x)$, from $x = a$ to $x = b$,

$$\int_a^b \left[f(x) - g(x) \right] dx$$

We have

$$f(x) = 4 - x^2 \text{ and } g(x) = e^x$$

Next, we need to find the point of intersection in the first quadrant. Use your calculator to find that the point of intersection is $x = 1.058$ (rounded to three decimal places). Plugging into the formula, we get

$$\int_0^{1.058} \left[(4 - x^2) - e^x \right] dx$$

Evaluating the integral, we get $\displaystyle\int_0^{1.058} \left[(4 - x^2) - e^x \right] dx = \left(4x - \frac{x^3}{3} - e^x \right)\Bigg|_0^{1.058} = 1.957 .$

(b) Find the volume of the solid generated when R is revolved about the x-axis.

In order to find the volume of a region between $y = f(x)$ and $y = g(x)$, from $x = a$ to $x = b$, when it is revolved about the x-axis, we use the following formula:

$$\pi \int_a^b \left[f(x)^2 - g(x)^2 \right] dx$$

Here our integral is $\pi \int_0^{1.058} \left[\left(4 - x^2 \right)^2 - \left(e^x \right)^2 \right] dx$.

Evaluating the integral, we get

$$\pi \int_0^{1.058} \left[\left(4 - x^2 \right) \right]^2 - \left[e^x \right]^2 dx = \pi \int_0^{1.058} \left(16 - 8x^2 + x^4 - e^{2x} \right) dx = \pi \left[16x - \frac{8x^3}{3} + \frac{x^5}{5} - \frac{e^{2x}}{2} \right]_0^{1.058} = 32.629$$

(c) Find the volume of the solid generated when R is revolved about the line $x = -1$.

In order to find the volume of this region, if we want to use vertical slices, we will use the method of cylindrical shells. Also, because we are revolving about the line $x = -1$, we will need to add 1 to the radius of the cylindrical shell. We will use the formula

$$2\pi \int_a^b (x + 1) \left[f(x) - g(x) \right] dx$$

We get

$$2\pi \int_0^{1.058} (x + 1) \left[(4 - x^2) - e^x \right] dx$$

We suggest that you use your calculator to evaluate the integral.

$$2\pi \int_0^{1.058} (x + 1) \left[(4 - x^2) - e^x \right] dx = 2\pi \int_0^{1.058} \left[4x - x^3 - xe^x + 4 - x^2 - e^x \right] dx = 17.059$$

2. A body is coasting to a stop and the only force acting on it is a resistance proportional to its speed, according to the equation $\frac{ds}{dt} = v_f = v_0\, e^{-\left(\frac{k}{m}\right)t}$; $s(0) = 0$, where v_0 is the body's initial velocity (in m/s), v_f is its final velocity, m is its mass, k is a constant, and t is time.

(a) If a body, with mass $m = 50$ kg and $k = 1.5$ kg/sec, initially has a velocity of 30 m/s, how long, to the nearest second, will it take to slow to 1 m/s?

We simply plug into the formula and solve for t.

We get

$$v_f = v_0\, e^{-\left(\frac{k}{m}\right)t}$$

$$1 = 30\, e^{-\left(\frac{1.5}{50}\right)t}$$

Divide both sides by 30: $\frac{1}{30} = e^{-\left(\frac{1.5}{50}\right)t}$.

Take the log of both sides: $\ln\frac{1}{30} = -\left(\frac{1.5}{50}\right)t$.

Multiply both sides by $-\frac{50}{1.5}$: $-\frac{50}{1.5}\ln\frac{1}{30} = t \approx 113$ seconds.

(b) How far, to the nearest 10 meters, will the body coast during the time it takes to slow from 30 m/s to 1 m/s?

Now we need to solve the differential equation $\frac{ds}{dt} = v_0\, e^{-\left(\frac{k}{m}\right)t}$, which we can do with separation of variables.

First, multiply both sides by dt: $ds = v_0\, e^{-\left(\frac{k}{m}\right)t}\, dt$.

Integrate both sides: $\int ds = \int v_0\, e^{-\left(\frac{k}{m}\right)t}\, dt$.

Evaluate the integrals: $s = -\frac{mv_0}{k}\, e^{-\left(\frac{k}{m}\right)t} + C$.

Now, plug in the initial conditions to solve for C: $0 = -\frac{(50)(30)}{1.5}\, e^{-\left(\frac{1.5}{50}\right)(0)} + C$.

$$C = \frac{(30)(50)}{1.5} = 1{,}000$$

Therefore, $s = -\dfrac{mv_0}{k} e^{-\left(\frac{k}{m}\right)t} + 1{,}000$. Now, we plug in the time $t = 113$ that we found in part (a) as well as the initial conditions to solve for s, which yields the following:

$$s = -\dfrac{(50)(30)}{1.5} e^{-\left(\frac{1.5}{50}\right)113} + 1{,}000 \approx 970 \text{ meters}$$

(c) If the body coasts from 30 m/s to a stop, how far will it coast?

Here, because the braking force is an exponential function, the object will coast to a stop after an infinite amount of time. In other words, we need to find

$$\lim_{t \to \infty} s(t) = \lim_{t \to \infty}\left[1{,}000 - 1{,}000 e^{-\left(\frac{k}{m}\right)t} \right] = 1{,}000 \text{ meters}$$

3. An object moves with velocity $v(t) = 9t^2 + 18t - 7$ for $t \geq 0$ from an initial position of $s(0) = 3$.

(a) Write an equation for the position of the particle.

The position function of the particle can be determined by integrating the velocity with respect to time, thus $s(t) = \int v(t)\,dt$. For this problem, $s(t) = \int (9t^2 + 18t - 7)\,dt = 3t^3 + 9t^2 - 7t + C$. Because we are given the initial position, $s(0) = 3$, plug that in to solve for C. Thus, $C = 3$ and the equation for the position of the particle is $s(t) = 3t^3 + 9t^2 - 7t + 3$.

(b) When is the particle changing direction?

The particle changes direction when the velocity is zero, but the acceleration is not. In order to determine when those times are, set the velocity equal to zero and solve for t: $v(t) = 9t^2 + 18t - 7 = 0$ when $t = \dfrac{1}{3}$ and $t = -\dfrac{7}{3}$. Because the time range in question is $t \geq 0$, we can ignore $t = -\dfrac{7}{3}$. Then, take the derivative of the velocity function to find the acceleration function, as $\dfrac{d}{dt}(v(t)) = a(t)$. For the given $v(t)$, $a(t) = 18t + 18$. Check that the acceleration at time $t = \dfrac{1}{3}$ is not zero by plugging into the acceleration function: $a(t) = 24$. Therefore, the particle is changing direction at $t = \dfrac{1}{3}$ because $v(t) = 0$ and $a(t) \neq 0$.

(c) What is the total distance covered from $t = 2$ to $t = 5$?

The distance covered is found by using the position function found in part (a). Determine the position at $t = 2$ and subtract it from the position at $t = 5$. From part (b), we know that the object does not change direction over this time interval, so we do not need to find the time piecewise. Thus, $s(5) - s(2) = 568 - 49 = 519$.

4.

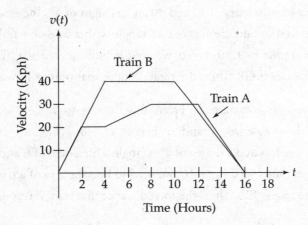

Time (Hours)

Three trains, *A, B,* and *C*, each travel on a straight track for $0 \leq t \leq 16$ hours. The graphs above, which consist of line segments, show the velocities, in kilometers per hour, of trains *A* and *B*. The velocity of *C* is given by $v(t) = 8t - 0.25t^2$.

(Indicate units of measure for all answers.)

(a) Find the velocities of *A* and *C* at time $t = 6$ hours.

We can find the velocity of train *A* at time $t = 6$ simply by reading the graph. We get $v_A(6) = 25$ kph. We find the velocity of train *C* at time $t = 6$ by plugging $t = 6$ into the formula. We get $v_C(6) = 8(6) - 0.25(6^2) = 39$ kilometers per hour.

(b) Find the accelerations of *B* and *C* at time $t = 6$ hours.

Acceleration is the derivative of velocity with respect to time. For train *B*, we look at the *slope* of the graph at time $t = 6$. We get $a_B(6) = 0$ km/hr². For train *C*, we take the derivative of *v*. We get $a(t) = 8 - 0.5t$. At time $t = 6$, we get $a_C(6) = 5$ km/hr².

(c) Find the positive difference between the total distance that A traveled and the total distance that B traveled in 16 hours.

In order to find the total distance that train A traveled in 16 hours, we need to find the area under the graph. We can find this area by adding up the areas of the different geometric objects that are under the graph. From time $t = 0$ to $t = 2$, we need to find the area of a triangle with a base of 2 and a height of 20. The area is 20. Next, from time $t = 2$ to $t = 4$, we need to find the area of a rectangle with a base of 2 and a height of 20. The area is 40. Next, from time $t = 4$ to $t = 8$, we need to find the area of a trapezoid with bases of 20 and 30 and a height of 4. The area is 100. Next, from time $t = 8$ to $t = 12$, we need to find the area of a rectangle with a base of 4 and a height of 30. The area is 120. Finally, from time $t = 12$ to $t = 16$, we need to find the area of a triangle with a base of 4 and a height of 30. The area is 60. Thus, the total distance that train A traveled is 340 km.

Let's repeat the process for train B. From time $t = 0$ to $t = 4$, we need to find the area of a triangle with a base of 4 and a height of 40. The area is 80. Next, from time $t = 4$ to $t = 10$, we need to find the area of a rectangle with a base of 6 and a height of 40. The area is 240. Finally, from time $t = 10$ to $t = 16$, we need to find the area of a triangle with a base of 6 and a height of 40. The area is 120. Thus, the total distance that train B traveled is 440 km.

Therefore, the positive difference between their distances is 100 km.

(d) Find the total distance that C traveled in 16 hours.

First, note that the graph of train C's velocity, $v(t) = 8t - 0.25t^2$, is above the x-axis on the entire interval. Therefore, in order to find the total distance traveled, we integrate $v(t)$ over the interval.

We get

$$\int_0^{16} (8t - 0.25t^2)\, dt$$

Evaluate the integral: $\int_0^{16} (8t - 0.25t^2)\, dt = \left(4t^2 - \dfrac{t^3}{12} \right)_0^{16} = \dfrac{2{,}048}{3} \text{ km}.$

5.

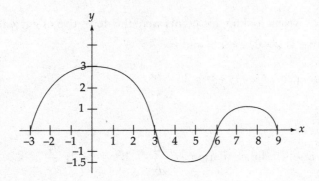

The figure above shows the graph of $g(x)$, where g is the derivative of the function f, for $-3 \le x \le 9$. The graph consists of three semicircular regions and has horizontal tangent lines at $x = 0$, $x = 4.5$, and $x = 7.5$.

(a) Find all values of x, for $-3 < x \le 9$, at which f attains a relative minimum. Justify your answer.

Because g is the derivative of the function f, f will attain a relative minimum at a point where $g = 0$ and where g is negative to the left of that point and positive to the right of it. This occurs at $x = 6$.

(b) Find all values of x, for $-3 < x \le 9$, at which f attains a relative maximum. Justify your answer.

Because g is the derivative of the function f, f will attain a relative maximum at a point where $g = 0$ and where g is positive to the left of that point and negative to the right of it. This occurs at $x = 3$.

(c) If $f(x) = \int_{-3}^{x} g(t)\, dt$, find $f(6)$.

We are trying to find the area between the graph and the x-axis from $x = -3$ to $x = 6$. From $x = -3$ to $x = 3$, the region is a semicircle of radius 3, so the area is $\dfrac{9\pi}{2}$.

From $x = 3$ to $x = 6$, the region is a semicircle of radius $\dfrac{3}{2}$, so the area is $\dfrac{9\pi}{8}$. We subtract the latter region from the former to obtain: $\dfrac{9\pi}{2} - \dfrac{9\pi}{8} = \dfrac{27\pi}{8}$.

(d) Find all points where $f''(x) = 0$.

Because $f''(x) = g'(x)$, we are looking for points were the derivative of g is zero. This occurs at the horizontal tangent lines at $x = 0$, $x = 4.5$, and $x = 7.5$.

6.   Consider the curve given by $x^2y - 4x + y^2 = 2$.

(a) Find $\dfrac{dy}{dx}$.

We can find $\dfrac{dy}{dx}$ by implicit differentiation: $x^2\dfrac{dy}{dx} + 2xy - 4 + 2y\dfrac{dy}{dx} = 0$.

Now we need to do some algebra to isolate $\dfrac{dy}{dx}$. First, we move all of the terms that do not contain $\dfrac{dy}{dx}$ to the right side of the equals sign.

$$x^2\frac{dy}{dx} + 2y\frac{dy}{dx} = 4 - 2xy$$

Next, we factor out $\dfrac{dy}{dx}$: $\dfrac{dy}{dx}(x^2 + 2y) = 4 - 2xy$.

Finally, we divide through by $(x^2 + 2y)$ to isolate $\dfrac{dy}{dx}$.

$$\frac{dy}{dx} = \frac{4 - 2xy}{x^2 + 2y}$$

(b) Find $\dfrac{d^2 y}{dx^2}$.

We need to use the Quotient Rule and implicit differentiation.

$$\frac{d^2y}{dx^2} = \frac{(x^2 + 2y)\left(-2x\dfrac{dy}{dx} - 2y\right) - (4 - 2xy)\left(2x + 2\dfrac{dy}{dx}\right)}{\left(x^2 + 2y\right)^2}$$

Next, substitute $\dfrac{dy}{dx} = \dfrac{4 - 2xy}{x^2 + 2y}$ into the derivative.

$$\frac{d^2y}{dx^2} = \frac{(x^2 + 2y)\left(-2x\left(\dfrac{4 - 2y}{x^2 + 2y}\right) - 2y\right) - (4 - 2xy)\left(2x + 2\left(\dfrac{4 - 2y}{x^2 + 2y}\right)\right)}{\left(x^2 + 2y\right)^2}$$

There is no need to simplify this.

(c) Find the equation of the tangent lines at each of the two points on the curve whose x-coordinate is 1.

First, we need to find the y-coordinates that correspond to $x = 1$. We plug $x = 1$ into $x^2y - 4x + y^2 = 2$, and rearrange a little, and we get $y^2 + y - 6 = 0$.

Next, we factor the quadratic to get $(y + 3)(y - 2) = 0$, so we will be finding tangent lines at the coordinates $(1, -3)$ and $(1, 2)$.

At $(1, -3)$, we get $\dfrac{dy}{dx} = \dfrac{4 - 2(1)(-3)}{(1)^2 + 2(-3)} = \dfrac{10}{-5} = -2$.

Therefore, the equation of the tangent line is $y + 3 = -2(x - 1)$.

At $(1, 2)$, we get $\dfrac{dy}{dx} = \dfrac{4 - 2(1)(2)}{(1)^2 + 2(2)} = \dfrac{0}{5} = 0$.

Therefore, the equation of the tangent line is $y = 2$.

Appendix

DERIVATIVES AND INTEGRALS THAT YOU SHOULD KNOW

1. $\dfrac{d}{dx}\left[ku\right] = k\dfrac{du}{dx}$

2. $\dfrac{d}{dx}\left[k\right] = 0$

3. $\dfrac{d}{dx}\left[uv\right] = u\dfrac{dv}{dx} + v\dfrac{du}{dx}$

4. $\dfrac{d}{dx}\left[\dfrac{u}{v}\right] = \dfrac{v\dfrac{du}{dx} - u\dfrac{dv}{dx}}{v^2}$

5. $\dfrac{d}{dx}\left[e^u\right] = e^u\dfrac{du}{dx}$

6. $\dfrac{d}{dx}\left[\ln u\right] = \dfrac{1}{u}\dfrac{du}{dx}$

7. $\dfrac{d}{dx}\left[\sin u\right] = \cos u\dfrac{du}{dx}$

8. $\dfrac{d}{dx}\left[\cos u\right] = -\sin u\dfrac{du}{dx}$

9. $\dfrac{d}{dx}\left[\tan u\right] = \sec^2 u\dfrac{du}{dx}$

10. $\dfrac{d}{dx}\left[\cot u\right] = -\csc^2 u\dfrac{du}{dx}$

11. $\dfrac{d}{dx}\left[\sec u\right] = \sec u \tan u\dfrac{du}{dx}$

12. $\dfrac{d}{dx}\left[\csc u\right] = -\csc u \cot u\dfrac{du}{dx}$

13. $\dfrac{d}{dx}\left[\sin^{-1} u\right] = \dfrac{1}{\sqrt{1-u^2}}\dfrac{du}{dx}$

14. $\dfrac{d}{dx}\left[\tan^{-1} u\right] = \dfrac{1}{1+u^2}\dfrac{du}{dx}$

15. $\dfrac{d}{dx}\left[\sec^{-1} u\right] = \dfrac{1}{|u|\sqrt{u^2-1}}\dfrac{du}{dx}$

1. $\displaystyle\int k\,du = ku + C$

2. $\displaystyle\int u^n\,du = \frac{u^{n+1}}{n+1} + C;\ n \neq -1$

3. $\displaystyle\int \frac{du}{u} = \ln|u| + C$

4. $\displaystyle\int e^u\,du = e^u + C$

5. $\displaystyle\int \sin u\,du = -\cos u + C$

6. $\displaystyle\int \cos u\,du = \sin u + C$

7. $\displaystyle\int \tan u\,du = -\ln|\cos u| + C$

8. $\displaystyle\int \cot u\,du = \ln|\sin u| + C$

9. $\displaystyle\int \sec u\,du = \ln|\sec u + \tan u| + C$

10. $\displaystyle\int \csc u\,du = -\ln|\csc u + \cot u| + C$

11. $\displaystyle\int \sec^2 u\,du = \tan u + C$

12. $\displaystyle\int \csc^2 u\,du = -\cot u + C$

13. $\displaystyle\int \sec u \tan u\,du = \sec u + C$

14. $\displaystyle\int \csc u \cot u\,du = -\csc u + C$

15. $\displaystyle\int \frac{du}{\sqrt{a^2 - u^2}} = \sin^{-1}\frac{|u|}{a} + C;\ |u| < a$

16. $\displaystyle\int \frac{du}{a^2 + u^2}\,du = \frac{1}{a}\tan^{-1}\frac{u}{a} + C$

17. $\displaystyle\int \frac{du}{u\sqrt{u^2 - a^2}} = \frac{1}{a}\sec^{-1}\frac{u}{a} + C;\ |u| > a$

PREREQUISITE MATHEMATICS

One of the biggest problems that students have with calculus is that their algebra, geometry, and trigonometry are not solid enough. In calculus, you'll be expected to do a lot of graphing. This requires more than just graphing equations with your calculator. You'll be expected to look at an equation and have a "feel" for what the graph looks like. You'll be expected to factor, combine, simplify, and otherwise rearrange algebraic expressions. You'll be expected to know your formulas for the volume and area of various shapes. You'll be expected to remember trigonometric ratios, their values at special angles, and various identities. You'll be expected to be comfortable with logarithms. And so on. Throughout this book, we spend a lot of time reminding you of these things as they come up, but we thought we should summarize them here at the end.

Powers

When you multiply exponential expressions with like bases, you add the powers.

$$x^a \cdot x^b = x^{a+b}$$

When you divide exponentiated expressions with like bases, you subtract the powers.

$$\frac{x^a}{x^b} = x^{a-b}$$

When you raise an exponentiated expression to a power, you multiply the powers.

$$\left(x^a\right)^b = x^{ab}$$

When you raise an expression to a fractional power, the denominator of the fraction is the root of the expression, and the numerator is the power.

$$x^{\frac{a}{b}} = \sqrt[b]{x^a}$$

When you raise an expression to the power of zero, you get one.

$$x^0 = 1$$

When you raise an expression to the power of one, you get the expression.

$$x^1 = x$$

When you raise an expression to a negative power, you get the reciprocal of the expression to the absolute value of the power.

$$x^{-a} = \frac{1}{x^a}$$

Logarithms

A logarithm is the power to which you raise a base, in order to get a value. In other words, $\log_b x = a$ means that $b^a = x$. There are several rules of logarithms that you should be familiar with.

When you take the logarithm of the product of two expressions, you add the logarithms.

$$\log(ab) = \log a + \log b$$

When you take the logarithm of the quotient of two expressions, you subtract the logarithms.

$$\log\left(\frac{a}{b}\right) = \log a - \log b$$

When you take the logarithm of an expression to a power, you multiply the logarithm by the power.

$$\log(a^b) = b \log a$$

The logarithm of 1 is zero.

$$\log 1 = 0$$

The logarithm of its base is 1.

$$\log_b b = 1$$

You cannot take the logarithm of zero or of a negative number.

In calculus, and virtually all mathematics beyond calculus, you will work with natural logarithms. These are logs with base e and are denoted by *ln*. Thus, you should know the following:

$$\ln 1 = 0$$

$$\ln e = 1$$

$$\ln e^x = x$$

$$e^{\ln x} = x$$

The change of base rule is: $\log_b x = \dfrac{\ln x}{\ln b}$

Geometry

The area of a triangle is $\dfrac{1}{2}$ (*base*)(*height*).

The area of a rectangle is (*base*)(*height*).

The area of a trapezoid is $\dfrac{1}{2}$ (*base*$_1$ + *base*$_2$)(*height*).

The area of a circle is πr^2.

The circumference of a circle is $2\pi r$.

The Pythagorean theorem states that the sum of the squares of the legs of a right triangle equals the square of the hypotenuse. This is more commonly stated as $a^2 + b^2 = c^2$, where c equals the length of the hypotenuse.

The volume of a right circular cylinder is $\pi r^2 h$.

The surface area of a right circular cylinder is $2\pi rh$.

The volume of a right circular cone is $\dfrac{1}{3} \pi r^2 h$.

The volume of a sphere is $\dfrac{4}{3} \pi r^3$.

The surface area of a sphere is $4\pi r^2$.

Trigonometry

Given a right triangle with sides x, y, and r and angle θ below:

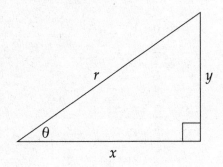

$$\sin\theta = \frac{y}{r}$$

$$\csc\theta = \frac{r}{y}$$

Thus, $\sin\theta = \dfrac{1}{\csc\theta}$

$$\cos\theta = \frac{x}{r}$$

$$\sec\theta = \frac{r}{x}$$

Thus, $\cos\theta = \dfrac{1}{\sec\theta}$

$$\tan\theta = \frac{y}{x}$$

$$\cot\theta = \frac{x}{y}$$

Thus, $\tan\theta = \dfrac{1}{\cot\theta}$

$\sin 2\theta = 2\sin\theta\,\cos\theta$

$\cos 2\theta = 1 - 2\sin^2\theta$

$\sin^2\theta + \cos^2\theta = 1$

$\cos 2\theta = \cos^2\theta - \sin^2\theta$

$\cos 2\theta = 2\cos^2\theta - 1$

$1 + \tan^2\theta = \sec^2\theta$

$\cos^2\theta = \dfrac{1 + \cos 2\theta}{2}$

$\sin^2\theta = \dfrac{1 - \cos 2\theta}{2}$

$1 + \cot^2\theta = \csc^2\theta$

$$\sin(A + B) = \sin A \cos B + \cos A \sin B$$

$$\sin(A - B) = \sin A \cos B - \cos A \sin B$$

$$\cos(A + B) = \cos A \cos B - \sin A \sin B$$

$$\cos(A - B) = \cos A \cos B + \sin A \sin B$$

You must be able to work in radians and know that $2\pi = 360°$.

You should know the following:

$\sin 0 = 0$	$\cos 0 = 1$	$\tan 0 = 0$
$\sin \dfrac{\pi}{6} = \dfrac{1}{2}$	$\cos \dfrac{\pi}{6} = \dfrac{\sqrt{3}}{2}$	$\tan \dfrac{\pi}{6} = \dfrac{1}{\sqrt{3}}$
$\sin \dfrac{\pi}{4} = \dfrac{1}{\sqrt{2}}$	$\cos \dfrac{\pi}{4} = \dfrac{1}{\sqrt{2}}$	$\tan \dfrac{\pi}{4} = 1$
$\sin \dfrac{\pi}{3} = \dfrac{\sqrt{3}}{2}$	$\cos \dfrac{\pi}{3} = \dfrac{1}{2}$	$\tan \dfrac{\pi}{3} = \sqrt{3}$
$\sin \dfrac{\pi}{2} = 1$	$\cos \dfrac{\pi}{2} = 0$	$\tan \dfrac{\pi}{2} = \infty$
$\sin \pi = 0$	$\cos \pi = -1$	$\tan \pi = 0$
$\sin \dfrac{3\pi}{2} = -1$	$\cos \dfrac{3\pi}{2} = 0$	$\tan \dfrac{3\pi}{2} = \infty$
$\sin 2\pi = 0$	$\cos 2\pi = 1$	$\tan 2\pi = 0$

Completely darken bubbles with a No. 2 pencil. If you make a mistake, be sure to erase mark completely. Erase all stray marks.

1.

YOUR NAME: _____
(Print)　　　　　Last　　　　　　　First　　　　　　M.I.

SIGNATURE: _____　　DATE: ___/___/___

HOME ADDRESS: _____
(Print)　　　　　　　　　Number and Street

_____
City　　　　　　State　　　　　Zip Code

PHONE NO.: _____

IMPORTANT: Please fill in these boxes exactly as shown on the back cover of your test book.

2. TEST FORM

_____

6. DATE OF BIRTH

Month	Day		Year	
○ JAN				
○ FEB	⓪	⓪	⓪	⓪
○ MAR	①	①	①	①
○ APR	②	②	②	②
○ MAY	③	③	③	③
○ JUN		④	④	④
○ JUL		⑤	⑤	⑤
○ AUG		⑥	⑥	⑥
○ SEP		⑦	⑦	⑦
○ OCT		⑧	⑧	⑧
○ NOV		⑨	⑨	⑨
○ DEC				

3. TEST CODE　　**4. REGISTRATION NUMBER**

⓪	Ⓐ	Ⓙ	⓪	⓪	⓪	⓪	⓪	⓪	⓪	⓪
①	Ⓑ	Ⓚ	①	①	①	①	①	①	①	①
②	Ⓒ	Ⓛ	②	②	②	②	②	②	②	②
③	Ⓓ	Ⓜ	③	③	③	③	③	③	③	③
④	Ⓔ	Ⓝ	④	④	④	④	④	④	④	④
⑤	Ⓕ	Ⓞ	⑤	⑤	⑤	⑤	⑤	⑤	⑤	⑤
⑥	Ⓖ	Ⓟ	⑥	⑥	⑥	⑥	⑥	⑥	⑥	⑥
⑦	Ⓗ	Ⓠ	⑦	⑦	⑦	⑦	⑦	⑦	⑦	⑦
⑧	Ⓘ	Ⓡ	⑧	⑧	⑧	⑧	⑧	⑧	⑧	⑧
⑨			⑨	⑨	⑨	⑨	⑨	⑨	⑨	⑨

7. GENDER
○ MALE
○ FEMALE

The Princeton Review

5. YOUR NAME

First 4 letters of last name				FIRST INIT	MID INIT
Ⓐ	Ⓐ	Ⓐ	Ⓐ	Ⓐ	Ⓐ
Ⓑ	Ⓑ	Ⓑ	Ⓑ	Ⓑ	Ⓑ
Ⓒ	Ⓒ	Ⓒ	Ⓒ	Ⓒ	Ⓒ
Ⓓ	Ⓓ	Ⓓ	Ⓓ	Ⓓ	Ⓓ
Ⓔ	Ⓔ	Ⓔ	Ⓔ	Ⓔ	Ⓔ
Ⓕ	Ⓕ	Ⓕ	Ⓕ	Ⓕ	Ⓕ
Ⓖ	Ⓖ	Ⓖ	Ⓖ	Ⓖ	Ⓖ
Ⓗ	Ⓗ	Ⓗ	Ⓗ	Ⓗ	Ⓗ
Ⓘ	Ⓘ	Ⓘ	Ⓘ	Ⓘ	Ⓘ
Ⓙ	Ⓙ	Ⓙ	Ⓙ	Ⓙ	Ⓙ
Ⓚ	Ⓚ	Ⓚ	Ⓚ	Ⓚ	Ⓚ
Ⓛ	Ⓛ	Ⓛ	Ⓛ	Ⓛ	Ⓛ
Ⓜ	Ⓜ	Ⓜ	Ⓜ	Ⓜ	Ⓜ
Ⓝ	Ⓝ	Ⓝ	Ⓝ	Ⓝ	Ⓝ
Ⓞ	Ⓞ	Ⓞ	Ⓞ	Ⓞ	Ⓞ
Ⓟ	Ⓟ	Ⓟ	Ⓟ	Ⓟ	Ⓟ
Ⓠ	Ⓠ	Ⓠ	Ⓠ	Ⓠ	Ⓠ
Ⓡ	Ⓡ	Ⓡ	Ⓡ	Ⓡ	Ⓡ
Ⓢ	Ⓢ	Ⓢ	Ⓢ	Ⓢ	Ⓢ
Ⓣ	Ⓣ	Ⓣ	Ⓣ	Ⓣ	Ⓣ
Ⓤ	Ⓤ	Ⓤ	Ⓤ	Ⓤ	Ⓤ
Ⓥ	Ⓥ	Ⓥ	Ⓥ	Ⓥ	Ⓥ
Ⓦ	Ⓦ	Ⓦ	Ⓦ	Ⓦ	Ⓦ
Ⓧ	Ⓧ	Ⓧ	Ⓧ	Ⓧ	Ⓧ
Ⓨ	Ⓨ	Ⓨ	Ⓨ	Ⓨ	Ⓨ
Ⓩ	Ⓩ	Ⓩ	Ⓩ	Ⓩ	Ⓩ

1. Ⓐ Ⓑ Ⓒ Ⓓ Ⓔ
2. Ⓐ Ⓑ Ⓒ Ⓓ Ⓔ
3. Ⓐ Ⓑ Ⓒ Ⓓ Ⓔ
4. Ⓐ Ⓑ Ⓒ Ⓓ Ⓔ
5. Ⓐ Ⓑ Ⓒ Ⓓ Ⓔ
6. Ⓐ Ⓑ Ⓒ Ⓓ Ⓔ
7. Ⓐ Ⓑ Ⓒ Ⓓ Ⓔ
8. Ⓐ Ⓑ Ⓒ Ⓓ Ⓔ
9. Ⓐ Ⓑ Ⓒ Ⓓ Ⓔ
10. Ⓐ Ⓑ Ⓒ Ⓓ Ⓔ
11. Ⓐ Ⓑ Ⓒ Ⓓ Ⓔ
12. Ⓐ Ⓑ Ⓒ Ⓓ Ⓔ
13. Ⓐ Ⓑ Ⓒ Ⓓ Ⓔ
14. Ⓐ Ⓑ Ⓒ Ⓓ Ⓔ
15. Ⓐ Ⓑ Ⓒ Ⓓ Ⓔ
16. Ⓐ Ⓑ Ⓒ Ⓓ Ⓔ
17. Ⓐ Ⓑ Ⓒ Ⓓ Ⓔ
18. Ⓐ Ⓑ Ⓒ Ⓓ Ⓔ
19. Ⓐ Ⓑ Ⓒ Ⓓ Ⓔ
20. Ⓐ Ⓑ Ⓒ Ⓓ Ⓔ
21. Ⓐ Ⓑ Ⓒ Ⓓ Ⓔ
22. Ⓐ Ⓑ Ⓒ Ⓓ Ⓔ
23. Ⓐ Ⓑ Ⓒ Ⓓ Ⓔ

24. Ⓐ Ⓑ Ⓒ Ⓓ Ⓔ
25. Ⓐ Ⓑ Ⓒ Ⓓ Ⓔ
26. Ⓐ Ⓑ Ⓒ Ⓓ Ⓔ
27. Ⓐ Ⓑ Ⓒ Ⓓ Ⓔ
28. Ⓐ Ⓑ Ⓒ Ⓓ Ⓔ
29. Ⓐ Ⓑ Ⓒ Ⓓ Ⓔ
30. Ⓐ Ⓑ Ⓒ Ⓓ Ⓔ
31. Ⓐ Ⓑ Ⓒ Ⓓ Ⓔ
32. Ⓐ Ⓑ Ⓒ Ⓓ Ⓔ
33. Ⓐ Ⓑ Ⓒ Ⓓ Ⓔ
34. Ⓐ Ⓑ Ⓒ Ⓓ Ⓔ
35. Ⓐ Ⓑ Ⓒ Ⓓ Ⓔ
36. Ⓐ Ⓑ Ⓒ Ⓓ Ⓔ
37. Ⓐ Ⓑ Ⓒ Ⓓ Ⓔ
38. Ⓐ Ⓑ Ⓒ Ⓓ Ⓔ
39. Ⓐ Ⓑ Ⓒ Ⓓ Ⓔ
40. Ⓐ Ⓑ Ⓒ Ⓓ Ⓔ
41. Ⓐ Ⓑ Ⓒ Ⓓ Ⓔ
42. Ⓐ Ⓑ Ⓒ Ⓓ Ⓔ
43. Ⓐ Ⓑ Ⓒ Ⓓ Ⓔ
44. Ⓐ Ⓑ Ⓒ Ⓓ Ⓔ
45. Ⓐ Ⓑ Ⓒ Ⓓ Ⓔ

Completely darken bubbles with a No. 2 pencil. If you make a mistake, be sure to erase mark completely. Erase all stray marks.

1.

YOUR NAME:
(Print)   Last   First   M.I.

SIGNATURE: _____ DATE: __ / __ / __

HOME ADDRESS:
(Print)   Number and Street

_____
City   State   Zip Code

PHONE NO.: _____

5. YOUR NAME

First 4 letters of last name				FIRST INIT	MID INIT
Ⓐ	Ⓐ	Ⓐ	Ⓐ	Ⓐ	Ⓐ
Ⓑ	Ⓑ	Ⓑ	Ⓑ	Ⓑ	Ⓑ
Ⓒ	Ⓒ	Ⓒ	Ⓒ	Ⓒ	Ⓒ
Ⓓ	Ⓓ	Ⓓ	Ⓓ	Ⓓ	Ⓓ
Ⓔ	Ⓔ	Ⓔ	Ⓔ	Ⓔ	Ⓔ
Ⓕ	Ⓕ	Ⓕ	Ⓕ	Ⓕ	Ⓕ
Ⓖ	Ⓖ	Ⓖ	Ⓖ	Ⓖ	Ⓖ
Ⓗ	Ⓗ	Ⓗ	Ⓗ	Ⓗ	Ⓗ
Ⓘ	Ⓘ	Ⓘ	Ⓘ	Ⓘ	Ⓘ
Ⓙ	Ⓙ	Ⓙ	Ⓙ	Ⓙ	Ⓙ
Ⓚ	Ⓚ	Ⓚ	Ⓚ	Ⓚ	Ⓚ
Ⓛ	Ⓛ	Ⓛ	Ⓛ	Ⓛ	Ⓛ
Ⓜ	Ⓜ	Ⓜ	Ⓜ	Ⓜ	Ⓜ
Ⓝ	Ⓝ	Ⓝ	Ⓝ	Ⓝ	Ⓝ
Ⓞ	Ⓞ	Ⓞ	Ⓞ	Ⓞ	Ⓞ
Ⓟ	Ⓟ	Ⓟ	Ⓟ	Ⓟ	Ⓟ
Ⓠ	Ⓠ	Ⓠ	Ⓠ	Ⓠ	Ⓠ
Ⓡ	Ⓡ	Ⓡ	Ⓡ	Ⓡ	Ⓡ
Ⓢ	Ⓢ	Ⓢ	Ⓢ	Ⓢ	Ⓢ
Ⓣ	Ⓣ	Ⓣ	Ⓣ	Ⓣ	Ⓣ
Ⓤ	Ⓤ	Ⓤ	Ⓤ	Ⓤ	Ⓤ
Ⓥ	Ⓥ	Ⓥ	Ⓥ	Ⓥ	Ⓥ
Ⓦ	Ⓦ	Ⓦ	Ⓦ	Ⓦ	Ⓦ
Ⓧ	Ⓧ	Ⓧ	Ⓧ	Ⓧ	Ⓧ
Ⓨ	Ⓨ	Ⓨ	Ⓨ	Ⓨ	Ⓨ
Ⓩ	Ⓩ	Ⓩ	Ⓩ	Ⓩ	Ⓩ

IMPORTANT: Please fill in these boxes exactly as shown on the back cover of your test book.

2. TEST FORM

_____

3. TEST CODE   **4. REGISTRATION NUMBER**

⓪	Ⓐ	Ⓙ	⓪	⓪	⓪	⓪	⓪	⓪	⓪	⓪
①	Ⓑ	Ⓚ	①	①	①	①	①	①	①	①
②	Ⓒ	Ⓛ	②	②	②	②	②	②	②	②
③	Ⓓ	Ⓜ	③	③	③	③	③	③	③	③
④	Ⓔ	Ⓝ	④	④	④	④	④	④	④	④
⑤	Ⓕ	Ⓞ	⑤	⑤	⑤	⑤	⑤	⑤	⑤	⑤
⑥	Ⓖ	Ⓟ	⑥	⑥	⑥	⑥	⑥	⑥	⑥	⑥
⑦	Ⓗ	Ⓠ	⑦	⑦	⑦	⑦	⑦	⑦	⑦	⑦
⑧	Ⓘ	Ⓡ	⑧	⑧	⑧	⑧	⑧	⑧	⑧	⑧
⑨			⑨	⑨	⑨	⑨	⑨	⑨	⑨	⑨

6. DATE OF BIRTH

Month	Day		Year	
◯ JAN				
◯ FEB	⓪	⓪	⓪	⓪
◯ MAR	①	①	①	①
◯ APR	②	②	②	②
◯ MAY	③	③	③	③
◯ JUN		④	④	④
◯ JUL		⑤	⑤	⑤
◯ AUG		⑥	⑥	⑥
◯ SEP		⑦	⑦	⑦
◯ OCT		⑧	⑧	⑧
◯ NOV		⑨	⑨	⑨
◯ DEC				

7. GENDER

◯ MALE
◯ FEMALE

The Princeton Review.

1. Ⓐ Ⓑ Ⓒ Ⓓ Ⓔ
2. Ⓐ Ⓑ Ⓒ Ⓓ Ⓔ
3. Ⓐ Ⓑ Ⓒ Ⓓ Ⓔ
4. Ⓐ Ⓑ Ⓒ Ⓓ Ⓔ
5. Ⓐ Ⓑ Ⓒ Ⓓ Ⓔ
6. Ⓐ Ⓑ Ⓒ Ⓓ Ⓔ
7. Ⓐ Ⓑ Ⓒ Ⓓ Ⓔ
8. Ⓐ Ⓑ Ⓒ Ⓓ Ⓔ
9. Ⓐ Ⓑ Ⓒ Ⓓ Ⓔ
10. Ⓐ Ⓑ Ⓒ Ⓓ Ⓔ
11. Ⓐ Ⓑ Ⓒ Ⓓ Ⓔ
12. Ⓐ Ⓑ Ⓒ Ⓓ Ⓔ
13. Ⓐ Ⓑ Ⓒ Ⓓ Ⓔ
14. Ⓐ Ⓑ Ⓒ Ⓓ Ⓔ
15. Ⓐ Ⓑ Ⓒ Ⓓ Ⓔ
16. Ⓐ Ⓑ Ⓒ Ⓓ Ⓔ
17. Ⓐ Ⓑ Ⓒ Ⓓ Ⓔ
18. Ⓐ Ⓑ Ⓒ Ⓓ Ⓔ
19. Ⓐ Ⓑ Ⓒ Ⓓ Ⓔ
20. Ⓐ Ⓑ Ⓒ Ⓓ Ⓔ
21. Ⓐ Ⓑ Ⓒ Ⓓ Ⓔ
22. Ⓐ Ⓑ Ⓒ Ⓓ Ⓔ
23. Ⓐ Ⓑ Ⓒ Ⓓ Ⓔ

24. Ⓐ Ⓑ Ⓒ Ⓓ Ⓔ
25. Ⓐ Ⓑ Ⓒ Ⓓ Ⓔ
26. Ⓐ Ⓑ Ⓒ Ⓓ Ⓔ
27. Ⓐ Ⓑ Ⓒ Ⓓ Ⓔ
28. Ⓐ Ⓑ Ⓒ Ⓓ Ⓔ
29. Ⓐ Ⓑ Ⓒ Ⓓ Ⓔ
30. Ⓐ Ⓑ Ⓒ Ⓓ Ⓔ
31. Ⓐ Ⓑ Ⓒ Ⓓ Ⓔ
32. Ⓐ Ⓑ Ⓒ Ⓓ Ⓔ
33. Ⓐ Ⓑ Ⓒ Ⓓ Ⓔ
34. Ⓐ Ⓑ Ⓒ Ⓓ Ⓔ
35. Ⓐ Ⓑ Ⓒ Ⓓ Ⓔ
36. Ⓐ Ⓑ Ⓒ Ⓓ Ⓔ
37. Ⓐ Ⓑ Ⓒ Ⓓ Ⓔ
38. Ⓐ Ⓑ Ⓒ Ⓓ Ⓔ
39. Ⓐ Ⓑ Ⓒ Ⓓ Ⓔ
40. Ⓐ Ⓑ Ⓒ Ⓓ Ⓔ
41. Ⓐ Ⓑ Ⓒ Ⓓ Ⓔ
42. Ⓐ Ⓑ Ⓒ Ⓓ Ⓔ
43. Ⓐ Ⓑ Ⓒ Ⓓ Ⓔ
44. Ⓐ Ⓑ Ⓒ Ⓓ Ⓔ
45. Ⓐ Ⓑ Ⓒ Ⓓ Ⓔ

Completely darken bubbles with a No. 2 pencil. If you make a mistake, be sure to erase mark completely. Erase all stray marks.

1.

YOUR NAME: _____
(Print)   Last   First   M.I.

SIGNATURE: _____   DATE: ___ / ___ / ___

HOME ADDRESS: _____
(Print)   Number and Street

_____
City   State   Zip Code

PHONE NO.: _____

IMPORTANT: Please fill in these boxes exactly as shown on the back cover of your test book.

2. TEST FORM

3. TEST CODE

4. REGISTRATION NUMBER

⓪	Ⓐ	Ⓙ	⓪	⓪	⓪	⓪	⓪	⓪	⓪	⓪	⓪
①	Ⓑ	Ⓚ	①	①	①	①	①	①	①	①	①
②	Ⓒ	Ⓛ	②	②	②	②	②	②	②	②	②
③	Ⓓ	Ⓜ	③	③	③	③	③	③	③	③	③
④	Ⓔ	Ⓝ	④	④	④	④	④	④	④	④	④
⑤	Ⓕ	Ⓞ	⑤	⑤	⑤	⑤	⑤	⑤	⑤	⑤	⑤
⑥	Ⓖ	Ⓟ	⑥	⑥	⑥	⑥	⑥	⑥	⑥	⑥	⑥
⑦	Ⓗ	Ⓠ	⑦	⑦	⑦	⑦	⑦	⑦	⑦	⑦	⑦
⑧	Ⓘ	Ⓡ	⑧	⑧	⑧	⑧	⑧	⑧	⑧	⑧	⑧
⑨			⑨	⑨	⑨	⑨	⑨	⑨	⑨	⑨	⑨

6. DATE OF BIRTH

Month	Day		Year	
⚬ JAN				
⚬ FEB	⓪	⓪	⓪	⓪
⚬ MAR	①	①	①	①
⚬ APR	②	②	②	②
⚬ MAY	③	③	③	③
⚬ JUN		④	④	④
⚬ JUL		⑤	⑤	⑤
⚬ AUG		⑥	⑥	⑥
⚬ SEP		⑦	⑦	⑦
⚬ OCT		⑧	⑧	⑧
⚬ NOV		⑨	⑨	⑨
⚬ DEC				

7. GENDER
⚬ MALE
⚬ FEMALE

5. YOUR NAME

First 4 letters of last name				FIRST INIT	MID INIT
Ⓐ	Ⓐ	Ⓐ	Ⓐ	Ⓐ	Ⓐ
Ⓑ	Ⓑ	Ⓑ	Ⓑ	Ⓑ	Ⓑ
Ⓒ	Ⓒ	Ⓒ	Ⓒ	Ⓒ	Ⓒ
Ⓓ	Ⓓ	Ⓓ	Ⓓ	Ⓓ	Ⓓ
Ⓔ	Ⓔ	Ⓔ	Ⓔ	Ⓔ	Ⓔ
Ⓕ	Ⓕ	Ⓕ	Ⓕ	Ⓕ	Ⓕ
Ⓖ	Ⓖ	Ⓖ	Ⓖ	Ⓖ	Ⓖ
Ⓗ	Ⓗ	Ⓗ	Ⓗ	Ⓗ	Ⓗ
Ⓘ	Ⓘ	Ⓘ	Ⓘ	Ⓘ	Ⓘ
Ⓙ	Ⓙ	Ⓙ	Ⓙ	Ⓙ	Ⓙ
Ⓚ	Ⓚ	Ⓚ	Ⓚ	Ⓚ	Ⓚ
Ⓛ	Ⓛ	Ⓛ	Ⓛ	Ⓛ	Ⓛ
Ⓜ	Ⓜ	Ⓜ	Ⓜ	Ⓜ	Ⓜ
Ⓝ	Ⓝ	Ⓝ	Ⓝ	Ⓝ	Ⓝ
Ⓞ	Ⓞ	Ⓞ	Ⓞ	Ⓞ	Ⓞ
Ⓟ	Ⓟ	Ⓟ	Ⓟ	Ⓟ	Ⓟ
Ⓠ	Ⓠ	Ⓠ	Ⓠ	Ⓠ	Ⓠ
Ⓡ	Ⓡ	Ⓡ	Ⓡ	Ⓡ	Ⓡ
Ⓢ	Ⓢ	Ⓢ	Ⓢ	Ⓢ	Ⓢ
Ⓣ	Ⓣ	Ⓣ	Ⓣ	Ⓣ	Ⓣ
Ⓤ	Ⓤ	Ⓤ	Ⓤ	Ⓤ	Ⓤ
Ⓥ	Ⓥ	Ⓥ	Ⓥ	Ⓥ	Ⓥ
Ⓦ	Ⓦ	Ⓦ	Ⓦ	Ⓦ	Ⓦ
Ⓧ	Ⓧ	Ⓧ	Ⓧ	Ⓧ	Ⓧ
Ⓨ	Ⓨ	Ⓨ	Ⓨ	Ⓨ	Ⓨ
Ⓩ	Ⓩ	Ⓩ	Ⓩ	Ⓩ	Ⓩ

1. Ⓐ Ⓑ Ⓒ Ⓓ Ⓔ
2. Ⓐ Ⓑ Ⓒ Ⓓ Ⓔ
3. Ⓐ Ⓑ Ⓒ Ⓓ Ⓔ
4. Ⓐ Ⓑ Ⓒ Ⓓ Ⓔ
5. Ⓐ Ⓑ Ⓒ Ⓓ Ⓔ
6. Ⓐ Ⓑ Ⓒ Ⓓ Ⓔ
7. Ⓐ Ⓑ Ⓒ Ⓓ Ⓔ
8. Ⓐ Ⓑ Ⓒ Ⓓ Ⓔ
9. Ⓐ Ⓑ Ⓒ Ⓓ Ⓔ
10. Ⓐ Ⓑ Ⓒ Ⓓ Ⓔ
11. Ⓐ Ⓑ Ⓒ Ⓓ Ⓔ
12. Ⓐ Ⓑ Ⓒ Ⓓ Ⓔ
13. Ⓐ Ⓑ Ⓒ Ⓓ Ⓔ
14. Ⓐ Ⓑ Ⓒ Ⓓ Ⓔ
15. Ⓐ Ⓑ Ⓒ Ⓓ Ⓔ
16. Ⓐ Ⓑ Ⓒ Ⓓ Ⓔ
17. Ⓐ Ⓑ Ⓒ Ⓓ Ⓔ
18. Ⓐ Ⓑ Ⓒ Ⓓ Ⓔ
19. Ⓐ Ⓑ Ⓒ Ⓓ Ⓔ
20. Ⓐ Ⓑ Ⓒ Ⓓ Ⓔ
21. Ⓐ Ⓑ Ⓒ Ⓓ Ⓔ
22. Ⓐ Ⓑ Ⓒ Ⓓ Ⓔ
23. Ⓐ Ⓑ Ⓒ Ⓓ Ⓔ

24. Ⓐ Ⓑ Ⓒ Ⓓ Ⓔ
25. Ⓐ Ⓑ Ⓒ Ⓓ Ⓔ
26. Ⓐ Ⓑ Ⓒ Ⓓ Ⓔ
27. Ⓐ Ⓑ Ⓒ Ⓓ Ⓔ
28. Ⓐ Ⓑ Ⓒ Ⓓ Ⓔ
29. Ⓐ Ⓑ Ⓒ Ⓓ Ⓔ
30. Ⓐ Ⓑ Ⓒ Ⓓ Ⓔ
31. Ⓐ Ⓑ Ⓒ Ⓓ Ⓔ
32. Ⓐ Ⓑ Ⓒ Ⓓ Ⓔ
33. Ⓐ Ⓑ Ⓒ Ⓓ Ⓔ
34. Ⓐ Ⓑ Ⓒ Ⓓ Ⓔ
35. Ⓐ Ⓑ Ⓒ Ⓓ Ⓔ
36. Ⓐ Ⓑ Ⓒ Ⓓ Ⓔ
37. Ⓐ Ⓑ Ⓒ Ⓓ Ⓔ
38. Ⓐ Ⓑ Ⓒ Ⓓ Ⓔ
39. Ⓐ Ⓑ Ⓒ Ⓓ Ⓔ
40. Ⓐ Ⓑ Ⓒ Ⓓ Ⓔ
41. Ⓐ Ⓑ Ⓒ Ⓓ Ⓔ
42. Ⓐ Ⓑ Ⓒ Ⓓ Ⓔ
43. Ⓐ Ⓑ Ⓒ Ⓓ Ⓔ
44. Ⓐ Ⓑ Ⓒ Ⓓ Ⓔ
45. Ⓐ Ⓑ Ⓒ Ⓓ Ⓔ

About the Author

David S. Kahn studied applied mathematics and physics at the University of Wisconsin and has taught courses in calculus, precalculus, algebra, trigonometry, and geometry at the college and high school levels. He has worked as an educational consultant for many years and tutored more students in mathematics than he can count! He has worked for The Princeton Review since 1989, and, in addition to AP calculus, he has taught math and verbal courses for the SAT, SAT II, LSAT, GMAT, and the GRE, trained other teachers, and written several other math books.

NOTES

NOTES

NOTES

NOTES

NOTES